RADIOLOGY OF THE NOSE, PARANASAL SINUSES AND NASOPHARYNX

VOLUMES OF
GOLDEN'S DIAGNOSTIC RADIOLOGY SERIES

RADIOLOGY OF THE NOSE, PARANASAL SINUSES AND NASOPHARYNX

GERALD D. DODD, M.D.,

Professor and Head,
Department of Diagnostic Radiology,
The University of Texas System Cancer Center M.D. Anderson Hospital and Tumor Institute,
Houston, Texas, and Professor of Radiology,
The University of Texas Medical School at Houston, Houston, Texas.

BAO-SHAN JING, M.D.,

Professor and Radiologist,
Department of Diagnostic Radiology,
The University of Texas System Cancer Center M.D. Anderson Hospital and Tumor Institute, Houston,
Texas, and Clinical Professor of Radiology,
The University of Texas Medical School at Houston, Houston, Texas

The Williams & Wilkins Company   Baltimore

SECTION 2
GOLDEN'S DIAGNOSTIC RADIOLOGY
Laurence L. Robbins, M.D., Editor

Copyright ©, 1977
The Williams & Wilkins Company
428 E. Preston Street
Baltimore, Md. 21202, U.S.A.

Made in the United States of America

Library of Congress Cataloging in Publication Data

Library of Congress Cataloging in Publication Data

Dodd, Gerald D
   Radiology of the nose, paranasal sinuses, and nasopharynx.
   (Golden's diagnostic radiology; section 2)
   Bibliography: p.
   Includes index.
   1. Nose—Radiography. 2. Nose, Accessory sinuses of—Radiography. 3. Nasopharynx—Radiography. I. Jing, Bao-Shan, joint author. II. Title.
RC78.G6 Sect. 2 [RF345] 616.07'572'08s
ISBN 0-683-02602-X   [616.2'12'07572]   76-44916

Composed and printed at the
Waverly Press. Inc.
Mt. Royal and Guilford Aves.
Baltimore, Md. 21202, U.S.A.

DEDICATION

To our wives

Foreword

It should be no wonder that a radiotherapist is writing the introduction to a book on diagnostic radiology of the nose, paranasal sinuses, and nasopharynx structures of the upper respiratory and digestive tracts, which are areas of great interest to radiotherapists.

The interest of radiotherapists in the radiologic findings in the head and neck structures dates as far back as Coutard, who first used lateral tissue films in 1922 to diagnose extension of tumors of the pharynx and larynx. Tomography, discovered in the 1920s was first used by Felix Leborgne, a radiotherapist from Uruguay, for the diagnosis and definition of extensions of tumors of the larynx. There were some papers on tomography in the United States in the 1930s, but there was no tomographic attachment at the New York Hospital in the first part of the 1940s when I was in training in general radiology. I do not believe there was any tomographic equipment in use at that time in the United States.

In 1947, when I took an extra year of training in radiotherapy in Europe, I became acquainted with tomography of the head and neck structures at the Curie Foundation and realized it was essential in the diagnosis and definition of extensions of tumors of the nose, paranasal sinuses, and nasopharynx. As a result, when I was appointed to the staff of M. D. Anderson Hospital in February 1948, I was instrumental in seeing that a tomographic attachment be installed on one of the diagnostic units. From that time on, there has been considerable interest at M. D. Anderson Hospital in the technique and interpretive skill of tomograms. For the last 20 years, sustained interest and demonstrated need have prompted the diagnostic radiologists, in particular Drs. Gerald Dodd and Bao-Shan Jing, to be greatly concerned with improvement of techniques and interpretive skill. Furthermore, Doctor Dodd has seen to it that the best polytomes have been acquired. The present day tomograms have a superb definition.

Writing has meaning and substance only if authors have knowledge and conviction of the usefulness of their writing. Not because we are colleagues in the same institution, but after reviewing most the English literature on head and neck, I can unreservedly state that there is no book that can remotely match this volume for technical quality and interpretive skill.

This volume contains information not only on tumors but also trauma and inflammatory conditions of the three closely associated structures. Diagnostic and therapuetic radiologists and trainees in both discipines should familiarize themselves thoroughly with the content of this book. It will be a working tool in the differential diagnosis and treatment planning of tumors of the nose, paranasal sinuses, and nasopharynx.

Gilbert H. Fletcher
Professor and Head,
Department of Radiotherapy
The University of Texas System
Cancer Center
M.D. Anderson Hospital and
Tumor Institute
Houston, Texas

Preface

The genesis of this volume may be traced to the policies instituted at The University of Texas M. D. Anderson Hospital and Tumor Institute by Dr. Gilbert H. Fletcher in 1948. Following his appointment as Head of the Department of Radiology, Dr. Fletcher introduced the radiologic techniques of Coutard, Baclesse and LeBorgne as a routine part of the clinical evaluation of patients with tumors of the head and neck. This institutional policy has remained in force, the examining techniques being augmented and diversified as experience increased and newer approaches became available. Although tumors of the nasal cavity and paranasal sinuses comprise less than 1% of all cancers, a substantial number have accumulated in the files of the M. D. Anderson Hospital over a period of 27 years. In reviewing this material one cannot help but be struck by the number in which a proper diagnosis was not reached until the tumor had spread beyond treatable bounds. The most common mistake was attribution of the radiologic findings to inflammatory disease. While complaints referable to the nasal passages and related organs are extremely common and for the most part of minor import, the number of local and systemic diseases which may involve these areas is numerous. Since the dominant radiologic findings are related to obliteration of the normal air spaces with or without concomitant bone involvement, appearances are often similar. It is therefore essential that a precise knowledge of the clinical situation be available to the radiologist as well as an awareness of the differential possibilities. While the literature on the subject is ample, much of it emphasizes non-radiologic aspects and is found in publications which are not normally consulted by the radiologist. Although there is increasing interest in radiology of the head and neck, no recent text devoted to the subject is available and training programs do not normally categorize a particular segment of the curriculum for this purpose. Hopefully, this volume will succeed in bridging the gap between non-radiologic literature and the requirements of the practicing radiologist. While techniques are changing rapidly and computerized tomography will undoubtedly have a profound effect upon both diagnosis and staging, the finding and techniques discussed in this volume are fundamental and basic to the proper identification of those situations likely to be encountered in clinical practice.

Gerald D. Dodd, M.D.
Bao-Shan Jing, M.D.

Acknowledgments

The authors and the publisher wish to acknowledge the contribution of important illustrative materials to this volume by the following physicians:

Godfrey E. Arnold
Melvin H. Becker
Michael Bleshman
Betsy Brown
George Campbell
Leonard D. Doubleday
Warren E. Emley
G. H. Fletcher
Jess C. Galbreath
Thomas Harle
Yoshitaka Kawabe
Donald J. Lawrence

Marton Majoros
Antonio J. Maniglia
Warren McFarland
Robert C. Newell
Arthur S. Patchefsky
Bernard S. Pogorel
Stanley Siegleman
E. B. Singleton
Emanuel M. Skolnik
Stanford B. Trachtenberg
Milton L. Wagner
Joseph W. Walike

Contents

PART I

The
Nose

Chapter 1

Anatomy of the Nose

GROSS ANATOMY

The external nose is pyramidal in shape, with its upper angle, or base, in continuity with the forehead; the inferior or free angle is termed the apex. The supporting framework of the external nose consists of the two nasal bones, the frontal processes of the maxillary bones, the upper lateral cartilages, the two paired lower lateral cartilages, and the anterior edge of the cartilaginous nasal septum (Fig. 1.1). Extending inferiorly and anteriorly from the nasion is the rounded dorsum of the nose, the upper part of which, known as the bridge of the nose, is formed by the two nasal bones.

The nasal cavity lies behind the external nose within the bony framework of the face. It is divided by the nasal septum into two symmetrical and approximately equal chambers, the nasal fossae, which extend from the nostrils (anterior nares) to the choanae (posterior nares) (Fig. 1.2). The anterior nares represent the bony apertures leading from the external nose to the nasal cavity. Anatomically the anterior nares are bounded superiorly by the nasal bones, laterally by the nasal process of the maxilla, and inferiorly by the junction of the horizontal processes of the maxillae.

The choanae are two oval openings, each measuring approximately 2.5 cm vertically and 1.5 cm transversely, which communicate with the nasopharynx. On the lateral wall are the superior, middle, and inferior turbinates; below and lateral to each turbinate is the corresponding meatus. In the anterior part of the superior meatus are the openings of the posterior ethmoid cells. Above each superior turbinate is the sphenoethmodial recess into which the sphenoid sinus drains. Contained within the middle meatus is a rounded elevation, the bulla ethmoidalis, into which or just above which the middle ethmoid cells open. Below and in front of the bulla ethmoidalis is a curved cleft, the hiatus semilunaris. The hiatus semilunaris is bounded inferiorly by a sharp concave margin of the uncinate process of the ethmoid bone and leads into a curved channel, the infundibulum, which is bounded above by the bulla ethmoidalis, and below by the lateral surface of the uncinate process of the ethmoid bone. The anterior ethmoid cells open into the anterior part of the infundibulum. In slightly over 50% of subjects, the infundibulum is directly continuous with the frontonasal duct. The ostium of the maxillary sinus opens into the middle meatus below the bulla ethmoidialis. Occasionally, an accessory ostium of the maxillary sinus is present below the posterior end of the middle turbinate. In the inferior meatus lies the ostium of the nasolacrimal duct.

The roof of the nasal cavity is formed, from front to back, by the nasal cartilage, the nasal bone, the nasal process of the frontal bone, the cribriform plate of the ethmoid bone, and the body of the sphenoid bone. The cribriform plate of the ethmoid forms the major part of the roof of the nasal cavity, and transmits the filaments of the olfactory nerve as they descend from the undersurface of the olfactory bulb to their distribution in the mucous membrane of the olfactory region of the nasal cavity. The floor of the nasal cavity is formed by the palatal process of the maxilla and the horizontal part of the palatine bone; near the anterior end of the floor are the openings of the incisive foramina.

In the anterior portion of each nasal cavity is a slight dilation, the vestibule, which is lined with epidermis containing hairs and sebaceous glands. Beyond the vestibule lies the nasal cavity proper, which is lined with mucous membrane. Each nasal fossa above and behind the vestibule is divided into a superior, or olfactory

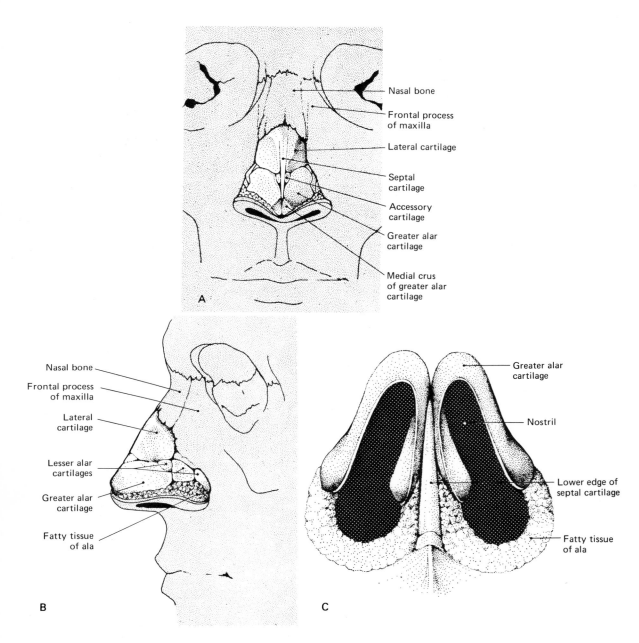

Figure 1.1. Bones and cartilages of the external nose. *A*, front view; *B*, side view; *C*, seen from below.

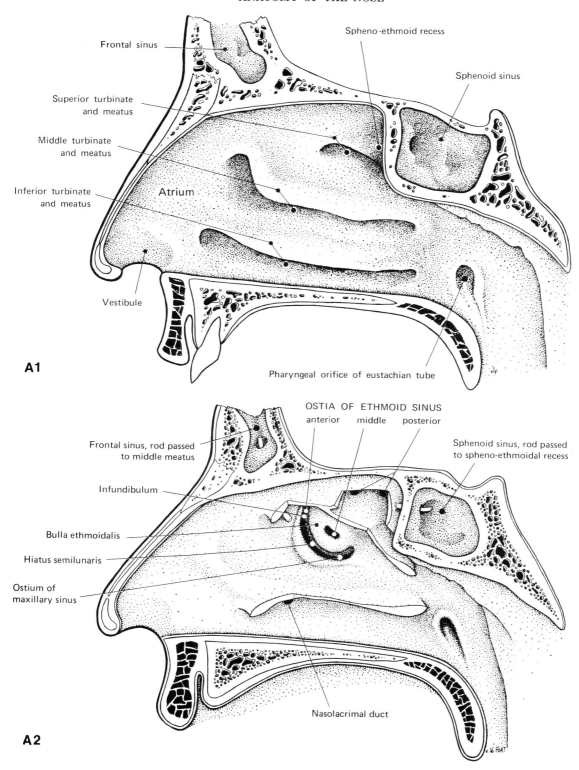

Figure 1.2. Nasal cavity. *A*, lateral wall of the nasal cavity: *1*, lateral wall of the nasal cavity with turbinates intact; *2*, lateral walls of the nasal cavity with turbinates removed.

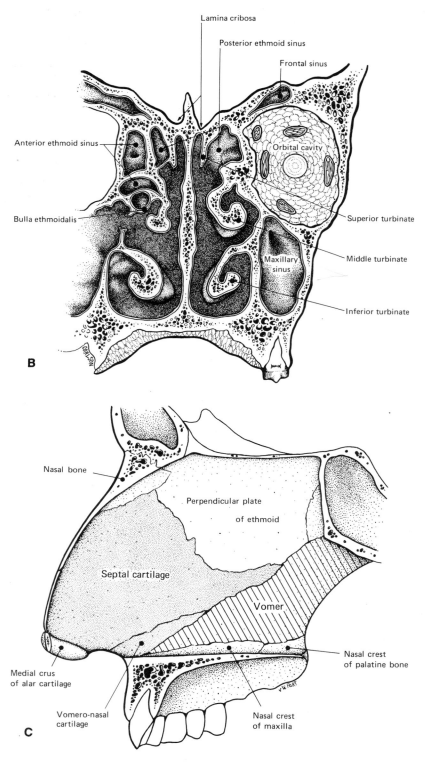

Figure 1.2 (continued). *B*, frontal view of nasal cavity. *C*, lateral view of nasal septum.

portion, and an inferior, or respiratory portion. The olfactory region is a narrow slit-like space bounded laterally by the superior turbinate and medially by the corresponding portion of the nasal septum; it is situated above an imaginary horizontal plane tangential to the inferior border of the superior turbinate. The respiratory region includes the remainder of the nasal fossa below the imaginary plane.

The nasal mucous membrane is continuous with that of the paranasal sinuses, the nasolacrimal ducts, and the nasopharynx. In the respiratory region, it is composed of columnar ciliated epithelium interspersed with goblet or mucin cells.

LYMPHATIC DRAINAGE OF THE NASAL CAVITY

The lymphatics of the vestibule of the nose communicate anteriorly with those of the skin and subcutaneous tissue of the face and posteriorly with lymphatics of the respiratory region. In the nasal cavity proper, there are two separate lymphatic regions: the olfactory and the respiratory (Fig. 1.3). Channels from the olfactory region pass upward across the cribriform plate of the ethmoid bone and communicate with the lymphatics of the subarachnoid space.

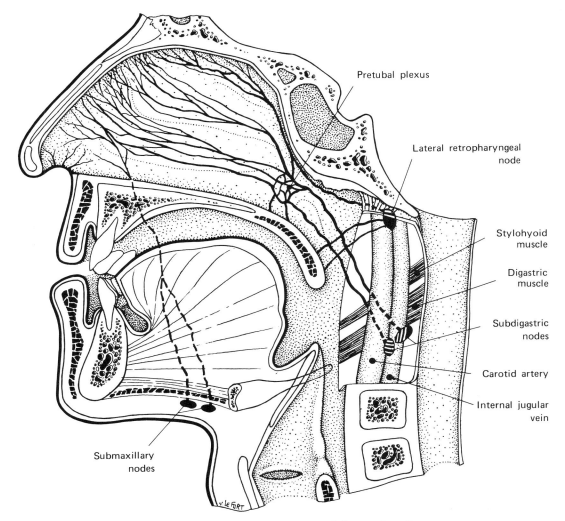

Figure 1.3. Lymphatic drainage of the nasal cavity.

Other collecting trunks from the olfactory region run beneath the mucosa of the nasopharynx and end in the lateral retropharyngeal nodes. Additional trunks are directed posteriorly to the pretubal lymphatic plexus just anterior to the pharyngeal orifice of the eustachian tube on the lateral wall of the nasopharynx.

The lymphatics of the respiratory region drain both anteriorly and posteriorly. Those from the anterior portion pass inferiorly and anteriorly to reach sequentially the anterior facial vein, the submaxillary nodes, and, eventually the deep cervical nodes.

The lymphatics from the posterior portion of the respiratory region run posteriorly and laterally beneath the superior surface of the soft palate toward the pharyngeal orifice of the eustachian tube where they join the pretubal lymphatic plexus. From this plexus, the collecting trunks divide into lateral and posterior groups. The lateral collecting trunks pass laterally, inferiorly, and posteriorly across the posterior surfaces of the stylohyoid styloglossus and digastric muscles to terminate in the subdiagastric nodes of the jugular chain. The posterior collecting trunks diverge from the deep surface of the levator veli palatini and run beneath the mucosa of the nasopharynx to empty to the lateral retropharyngeal nodes (Fig. 1.3).

REFERENCES

Goss, C. M.: *Gray's Anatomy of Human Body*. 26th ed., p. 184–186, 1196–1203, Lea and Febiger, Philadelphia, 1954.

Rouviere, H.: *Anatomy of the Human Lymphatic System*. Translated by M. J. Tobias. Edwards Brothers, Inc., Ann Arbor, Michigan, 1938.

Chapter 2

Roentgen Technique

The cartilaginous structure of the external nose is readily examined clinically, and radiological investigation is seldom required. However, roentgen examination is a valuable adjunct in the diagnosis and management of pathological changes in the nasal bones and nasal cavity.

NASAL BONES

Lateral Projection

When possible the examination should be performed in a sitting position with the patient's head in the true lateral projection. The nasal bones are positioned at the center of the film. The horizontal central ray is directed through the base of the nose at a right angle to the film. Right and left lateral views should be taken with respiration suspended during exposures (Fig. 2.1). The two projections provide a slightly different view of each nasal bone and give useful complementary information.

Occlusal Projection

The patient is placed in a sitting position with the head so adjusted that the plane of the forehead is in a precise vertical position. An anterior or posterior tilt of the head usually hides all or part of the nasal bones behind the teeth or the frontal bone. A dental film is placed between the patient's anterior teeth. The vertical central ray is directed downward in the midline through the base of the nose at a right angle to the film (Fig. 2.2). This projection gives the best view of the nasal bones and the anterior part of the nasal septum.

Occipitomental (Waters) Projection

The technique is described in Part II, Chapter 9. The details of the nasal bones are usually not well shown on this projection, but occasionally, deviation of the nasal bones and/or septum may be apparent.

NASAL CAVITY

Routine Projections

Posteroanterior (Caldwell) View

The technique is described in Part II, Chapter 9. In this view, the turbinates, the nasal septum, and the nasal fossae are clearly seen.

Submentovertical, or Base, View

The technique is described in Part II, Chapter 9. The nasal septum and the relative position of the inferior turbinates can be seen in this view.

Lateral View

The technique is described in Part II, Chapter 9. This view is of value in delineating the nasal bones, and in the location of foreign bodies in the nasal cavity.

Occipitomental (Waters) View

This view is of little value in the demonstration of the finer details of the nasal cavity.

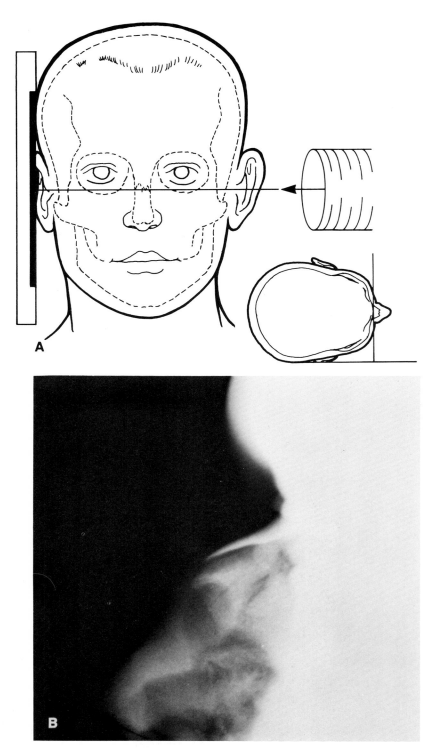

Figure 2.1. Lateral projection of the nasal bones. *A*, line drawing to show the radiographic position for the lateral projection; *B*, radiograph of the nasal bones in the lateral projection.

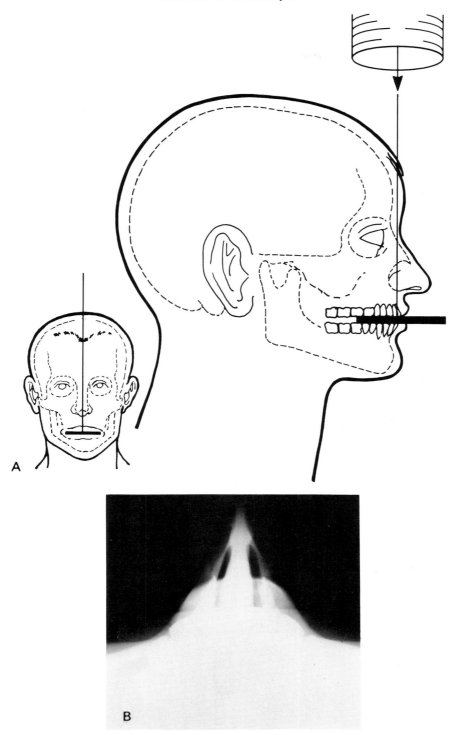

Figure 2.2. Occlusal projection of the nasal bones. *A*, line drawing to show the radiographic position for the occlusal projection; *B*, radiograph of the nasal bones as seen in the occlusal projection.

Anteroposterior Tomograms

Anteroposterior tomograms are of great value in the demonstration of lesions of the nasal cavity. The purposes of this particular technique are 3-fold: (1) to delineate the extent and degree of the fracture involving the nasal bone and the surrounding structures; (2) to determine the location of the lesion and the extent of bony destruction; and (3) to provide a definite diagnosis when a lesion is clinically suspected, but cannot be substantiated by more conventional studies. The technique is discussed in detail in Part II, Chapter 9.

Nasopharyngogram

Because of the complexity of the structures of the nasal cavity and the surrounding paranasal sinuses, routine projections and anteroposterior tomograms sometimes fail to delineate the pathological changes with certainty. Under such conditions, a contrast nasopharyngogram may be of value. The technique of this procedure is discussed in Part III, Chapter 21.

REFERENCES

Dingman, R. O., and Natvig, P.: *Surgery of Facial Fractures*. W. B. Saunders, Philadelphia, 1964.

Freimanis, A. K.: Fractures of the facial bones. Radiol. Clin. North Am. *4:* 341, 1966.

Jing, B. S., and McGraw, J. P.: Contrast nasopharyngography in diagnosis of tumors. Arch. Otolaryngol. *81:* 365, 1965.

Merill, V.: *Atlas of Roentgenographic Positions*. 2nd ed. C. V. Mosby Company, St. Louis, Missouri, 1959.

Chapter 3

Roentgen Anatomy

NASAL BONES

Figure 3.1 shows the normal roentgen anatomy of the nasal bones as seen in different projections. Normally each nasal bone develops from a single ossification center, but additional centers may occur. These are usually indicated by vertical or transverse suture lines. Occasionally, a medially placed, small bone is seen between the frontal and the nasal bones and a supernumerary flat bone may occur along the margin of the aperture of the nostril (Fig. 3.2). Care should be taken not to confuse these developmental variations with fractures.

On the lateral projection two thin radiolucent lines are usually seen running parallel to the dorsum of the nose. These represent the right and left nasomaxillary sutures. Less obvious is a longitudinal radiolucent line representing the bony groove for passage of the nasal branch of the nasocilliary nerve (Fig. 3.3A). These lines are normal and should not be confused with fractures. A small foramen for passage of a vein is often seen in the center of the nasal bones in the lateral projection (Fig. 3.3B).

On the occlusal view the incisive foramen is often seen behind the incisor teeth. This foramen contains Stenson's and Scarpa's canals for the passage of the terminal branches of the descending palatine arteries and the nasopalatine nerves.

NASAL CAVITY

Figure 3.4 shows the normal roentgen anatomy of the nasal cavity in different projections.

The inferior turbinates are rarely symmetrical and vary considerably in size. Their scroll-like appearance is characteristic and they are easily recognized.

The middle turbinates are small in size and are less easily visualized on routine projections because of overlapping of the shadows of the ethmoid sinuses. Occasionally congenital absence of the middle turbinates may occur. Beneath the middle turbinate, in the middle meatus, there is an area which is devoid of bone. This is best shown by tomography and should not be mistaken for an area of destruction secondary to a new growth.

The superior turbinates are the smallest of the turbinates and can be discerned only in the posterior cuts of the posteroanterior tomograms.

The ethmoid labyrinth can be recognized by the honeycombed appearance of the ethmoid cells.

The floor of the nasal cavity is slightly higher than that of the maxillary sinuses and can be clearly demonstrated by posteroanterior tomograms. The lateral wall of the nasal cavity is almost wholly formed by bone but it may be difficult to see in plain films because of its thinness and obliquity.

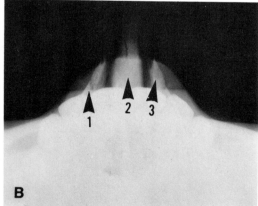

Figure 3.1. Normal nasal bones. *A*, lateral projection: *1*, nasal bones; *2*, groove for nerve; *3*, Nasomaxillary suture. *B*, occlusal projection: *1*, frontal process of the maxilla; *2*, nasal septum superimposed with anterior nasal spine of the maxilla; *3*, nasal bone.

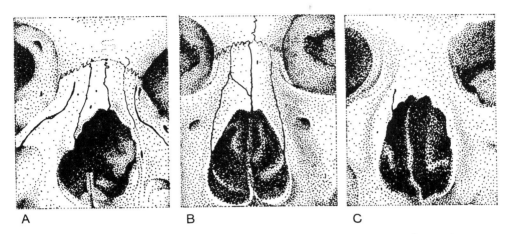

Figure 3.2. Developmental variations of the nasal bone. *A*, frontal process of the maxilla having a supernumerary suture on each side; *B*, right nasal bone having an oblique suture; *C*, fusion of the two nasal bones with absence of the midline suture. (Modified from Pendergrass, Schaeffer and Hodes.)

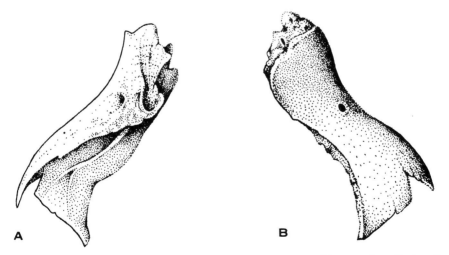

Figure 3.3. Normal markings of the nasal bones. *A*, inner surface of the nasal bone to show the longitudinal groove for the nerve; *B*, outer surface of the nasal bone to show the foramen for vein.

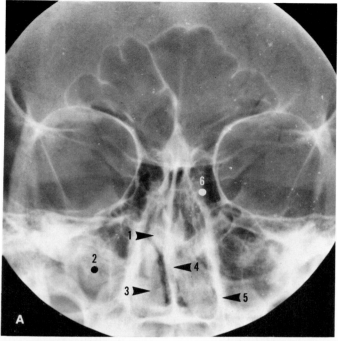

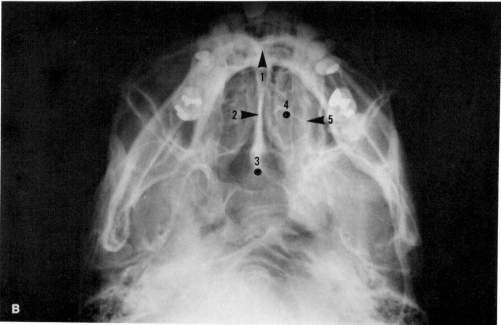

Figure 3.4. Normal nasal cavity. *A*, posteroanterior view: *1*, middle turbinate; *2*, maxillary sinus; *3*, inferior turbinate; *4*, nasal septum; *5*, lateral wall of the nasal cavity; *6*, ethmoid sinus. *B*, submentovertical view: *1*, mandible; *2*, nasal septum; *3*, sphenoid sinus; *4*, inferior turbinate superimposed with ethmoid cells; *5*, medial wall of maxillary sinus.

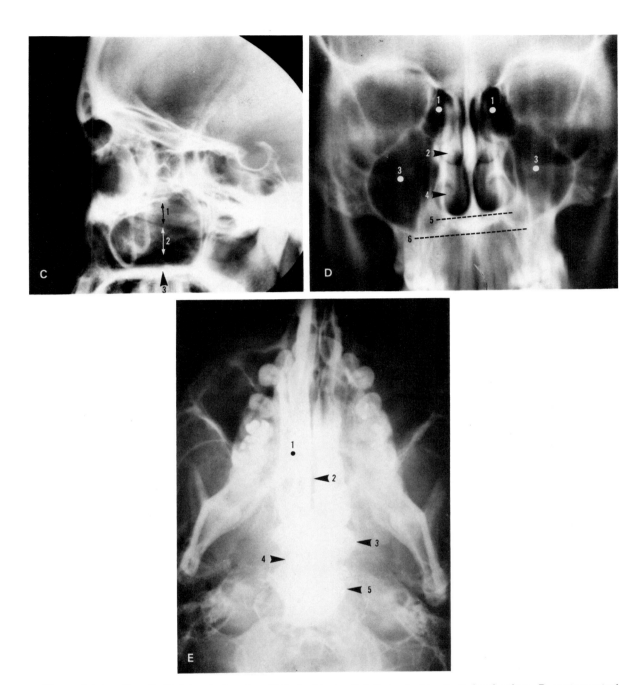

Figure 3.4 (continued). *C*, lateral view: *1*, middle turbinate; *2*, inferior turbinate; *3*, hard palate. *D*, posteroanterior tomograms: *1*, ethmoid sinus; *2*, middle turbinate; *3*, maxillary sinus; *4*, inferior turbinate; *5*, floor of the nasal cavity; *6*, floor of the maxillary sinus. *E*, submentovertical projection of the nasopharyngogram: *1*, right nasal fossa; *2*, nasal septum; *3*, pharyngeal orifice of the eustachian tube; *4*, torus tubarius; *s*, fossa of Rosenmüller.

REFERENCES

Goss, C. M.: *Gray's Anatomy of Human Body*. 26th ed. p. 1196, Lea and Febiger, Philadelphia, 1954.

Jing, B. S., and McGraw, J. P.: Contrast nasopharyngography in diagnosis of tumors. Arch. Otolaryngol. *81:* 365, 1965.

Pendergrass, E. O., Schaeffer, J. P., and Hodges, P. J.: *The Head and Neck in Roentgen Diagnosis*. 2nd ed., p. 193–199, 545–561, Charles C Thomas, Springfield, Illinois, 1956.

Schaeffer, J. P.: *The Nose, Paranasal Sinuses, Nasolacrimal Passageways, and Olfactory Organ in Man*. Balkiston, Philadelphia, 1920.

Chapter 4

Inflammatory and Nonneoplastic Diseases of the Nasal Bone and Nasal Cavity

CHOANAL ATRESIA

General Considerations

The absence of one or both choanae of the nasal cavity is known as choanal atresia. This condition is usually congenital and is seen in newborn infants or young children. Pathogenetically, choanal atresia is probably caused by the persistence of the buccopharyngeal membrane. The occlusion may be unilateral or bilateral, partial or complete, membranous, membranosseous, or osseous (Fig. 4.1).

Clinically the striking findings are difficult breathing through the mouth and a nasal cavity filled with inspissated mucus on the affected side. The condition should be suspected in all newborns having difficulty in breathing or exhibiting asphyxial attacks.

Roentgen Findings

Routine Projections

1. On the posteroanterior view there is a decrease in the size and radiolucency of the affected side of the nasal cavity.

2. On the lateral view, if the atresia is osseous in nature, the site of obstruction may be visualized.

Anteroposterior Tomograms

The characteristic appearance of the affected side of the nasal cavity and the site of obstruction can be shown to good advantage.

Catheter Test

A small catheter is placed in the nasal cavity and an attempt is made to pass it into the nasopharynx. If atresia is present, it will prevent the passage of the catheter. Roentgenograms in the lateral projection will reveal the point of obstruction (Fig. 4.2A).

Contrast Nasopharyngogram

This is the preferred method of demonstrating choanal atresia. Roentgenograms should be taken in both the lateral and submentovertical projections. With complete atresia pooling of the contrast medium in the posterior part of the nasal cavity occurs; their is no communication with the nasopharynx (Fig. 4.2B). When the atresia is partial, the extent and degree of stenosis can be clearly demonstrated.

Differential Diagnosis

Stenosis of the Nasopharynx

This is usually the result of chronic granulomatous disease or severe, uncontrolled infection, such as lues or diphtheria, with a resultant cicatritial stenosis. Inexpert surgery in the region of the nasopharynx and the oropharynx may also be an etiological factor. The stenosis is usually located in the region of the pharyngeal isthmus. The soft palate and tonsillar pillars are adherent, in varying degree, to the posterior pharyngeal wall, causing a partial or complete obstruction between the nasopharynx and oropharynx.

The symptomatology depends upon the degree of obstruction. The prominent symptom is mouth breathing. When a patient develops breathing difficulty after surgical procedures involving the nasopharynx or oropharynx, the possibility of a postsurgical stenosis should be considered.

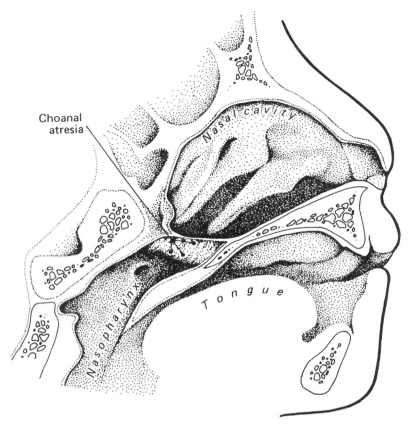

Figure 4.1. Line drawing of choanal atresia.

Tumors of the Nasopharynx

Adenoids and tumors of considerable size in the nasopharynx may cause obstruction of the choanae.

Tumors of the Posterior Nasal Cavity

Polyps, cysts, or other tumors of the posterior aspect of the nasal cavity may, if sufficiently large, produce nasal obstruction and mouth breathing.

INFLAMMATORY DISEASES OF THE NASAL BONES

Radiological studies are infrequently required in the evaluation of inflammatory diseases involving the soft tissues and cartilages of the nose. When there is bony involvement, however, roentgen examination is useful in determining the location and extent of the lesions. Diseases which attack the nasal bones include lupus vulgaris, syphilis, and leprosy.

Lupus Vulgaris

Lupus vulgaris is an indolent form of tuberculosis which may spread via hematogenous or lymphatic pathways or by direct extension from adjacent structures. It commonly involves the face, but may also affect the nasal, oral, or pharyngeal mucosa. In the nose, the early le-sion is a reddish, firm nodule at the mucocutaneous junction of the nasal septum. In advanced cases, there may be extensive involvement of the floor and the turbinates of the nasal cavity as well as spread to the cartilaginous portion of the septum. The involved septum may be perforated. Eventually, the nose becomes deformed due to sinking of the nasal bridge.

Syphilis

In syphilitic infections of either the congenital or tertiary variety, the nose is commonly affected. The pathological lesion is a syphilitic gumma, which invades the mucous membrane, periosteum, or bone. The syphilitic ulceration and destruction ultimately result in atrophic

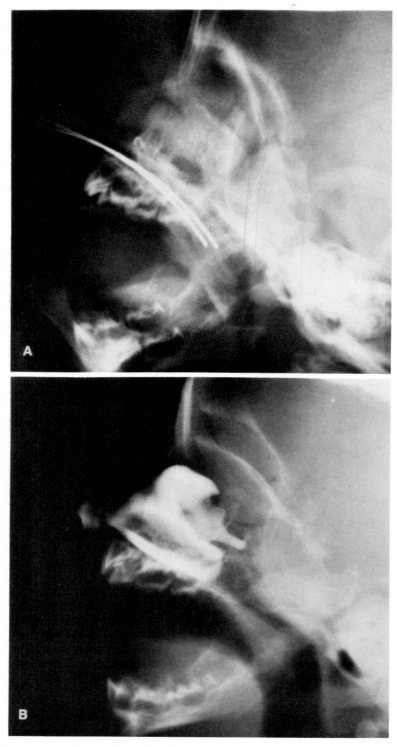

Figure 4.2. Choanal atresia. *A*, lateral view of the face showing catheter in each nasal fossa. There is atresia of both choanae which prevents the passage of catheters into the nasopharynx. *B*, lateral view of contrast nasopharyngogram showing complete atresia on the right side with pooling of contrast medium in the nasal cavity. The left choana is normal. (Courtesy of Dr. Milton L. Wagner and Dr. E. B. Singleton.)

changes in the mucous membrane, secondary ozena, perforation of the nasal septum, and destruction of the nasal bones. Eventually there is collapse of the nasal bridge, producing the typical saddle-nose deformity (Fig. 4.3).

Leprosy

Leprosy is a chronic communicable disease caused by *Mycobacterium leprae* (Fig. 4.4C). There is a variety of forms of the disease, which can be broadly classified into three groups: tuberculoid, intermediate, and lepromatous. Characteristic of all forms is the preferential involvement of the skin, mucous membranes, and peripheral nerves. The lesion consists of granulomatous nodules which may undergo necrosis and ulceration, followed by fibrosis and contraction. In the tuberculoid and intermediate forms, because of host defense mechanisms, the lesion is usually localized and mild. Infiltration of the skin and mucous membranes occurs, with subsequent granulation tissue formation, but there is no obvious destruction of cartilage or bone.

When the host resistance is less effective, the lepromatous type develops. This form of the disease tends to be more generalized, involving cartilage, bone, and the reticuloendothelial systems in addition to the skin, peripheral nerves, and mucous membranes. The initial lesions are often seen in the face, ear lobes, and perioral and nasal regions. In advanced cases the granulomatous reaction produces nodules and elevations in the skin of the face, resulting in the characteristic leonine facies. The loss of the lateral portion of the eyebrows is typical. In the nasal region, the granulomatous nodules spread over the nasal septum and turbinates, coalesce, and ulcerate. The cartilaginous portion of the nasal septum may be perforated and destroyed (Fig. 4.4, *A* and *B*). The framework of the nose may become flat and deformed, assuming a saddle shape which is the main facial deformity in lepromatous leprosy. Other bony changes commonly seen include atrophy of the anterior nasal spine and absorption of the anterior alveolar process of the maxilla.

NASAL POLYPS

General Considerations

Polyps are frequently found in the nasal cavity. Their precise etiology is obscure, but they probably result from the combined effects of allergy and chronic infection. Occasionally either alone may account for their development.

Histologically, the polyps usually consist of edematous stroma covered by hyperplastic, ciliated epithelium of the pseudostratified variety. Occasionally squamous metaplasia may occur, but the basic pathological picture is one of edema and inflammatory infiltration.

Nasal polyps may occur singly or in large numbers and may be pedunculated or sessile. The nasal mucosa alone may be affected, but more often there is an associated sinusitis, usually affecting the ethmoid and maxillary chambers. However, all sinus chambers may be involved.

Clinically, the predominant symptoms are nasal obstruction and anosmia. In allergic cases, paroxysmal sneezing and watery nasal discharge are often present. When there is an underlying chronic infectious sinusitis, headache, malaise, and nasal discharge, mucopurulent or purulent, may be quite pronounced.

Roentgen Findings

The roentgen manifestations are variable, depending upon the extent of the disease. There may be: (1) a small well defined soft tissue shadow in the nasal fossa; (2) a large soft tissue mass filling the entire nasal cavity with expansion and thinning of the lateral walls of the nasal cavity (Fig. 4.5); (3) an associated thickening of the lining mucosa of the paranasal sinuses, particularly the maxillary antra. (see also Fig. 13.9).

Differential Diagnosis

Lesions in the nasal cavity which should be differentiated from nasal polyps are papillomas, fibromas, angiofibromas, hemangiomas, and fibromyxomas,

BENIGN CYSTS OF THE NASAL BONE

Benign cysts of the nasal bones are rare and generally of congenital origin. More common are meningoceles, encephaloceles and dermoid cysts.

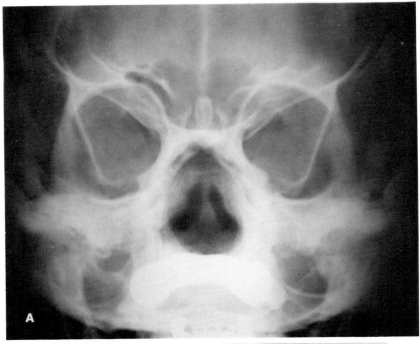

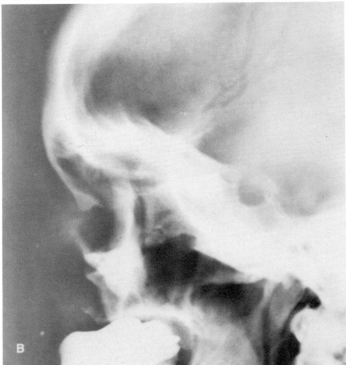

Figure 4.3. Saddle nose resulting from congenital syphilis. *A*, posterioanterior view of the face showing extensive destruction of the nasal bones and nasal septum. There is sclerosis of the remaining nasal bones and septum. *B*, lateral view of the face showing empty nasal cavity with loss of the normal structures.

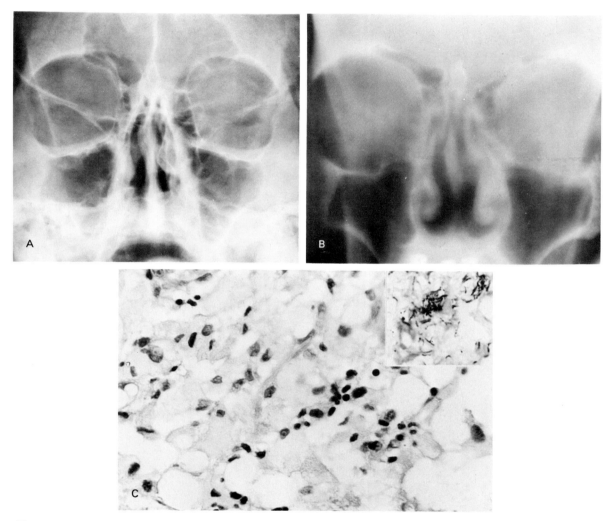

Figure 4.4. Leprosy of nose. *A*, Waters view: thickening of the lining mucosa of the nasal cavity. *B*, anteroposterior tomogram (mid cut): perforation and destruction of the nasal septum. *C*, histologic section of biopsy of the nasal septum showing diffuse infiltrate of large foamy cells (H&E, × 400). *Inset*: *Mycobacterium leprae* characteristically grouped in packets (Fite-Faraco, × 1500).

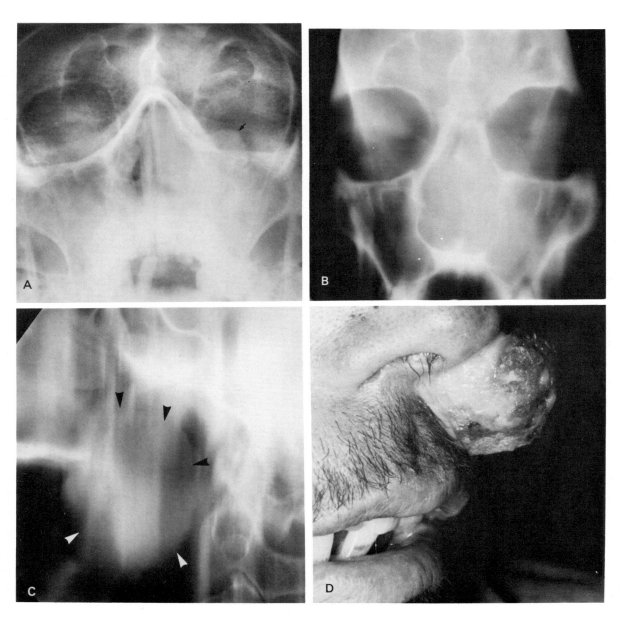

Figure 4.5. Huge nasal polyp. *A*, Waters view: anterior extremity of the polyp outside of the left nostril. *B*, anteroposterior tomogram: complete opacity of the nasal cavity with destruction of the nasal septum and thinning of the lateral walls of the nasal cavity. There is cloudiness with thickening of the lining mucosa of the maxillary sinuses, especially on the left. *C*, lateral tomogram: posterior extremity of the polyp in the nasopharynx with a clear zone between the roof of the nasopharynx and the nasopharyngeal surface of the polyp. *D*, photograph: anterior extremity of the polyp protruding through the left nostril.

MENINGOCELES AND ENCEPHALOCELES

General Considerations

Meningoceles and encephaloceles involving the nasal bones and nasal cavity are rare. The majority are of congenital origin but, on occasion, may be traumatic in nature. In the congenital variety there is herniation of the meninges and/or brain substance through bony defects in the facial bones or base of the skull.

Meningoceles and encephaloceles may occur in the occipital, sincipital, and basal regions. The sincipital type is situated about the dorsum of the nose, the orbits, and the forehead. The herniations may pass through defects between the nasal, frontal, ethmoid, and lacrimal bones.

In the basal type, the herniations may follow several routes. They may pass through a defect in the cribriform plate to lie within the nasal cavity medial to the middle turbinate, where they simulate a nasal polyp. They may also pass through defects between the posterior ethmoid cells and sphenoid to protrude into the nasopharynx or pass through a patent craniopharyngeal canal into the nasopharynx, where they resemble adenoidal tissue. Lastly, they may pass through the superior and the inferior orbital fissures into the sphenopalatine fossae.

Roentgen Findings

The roentgen findings depend upon the type and location of the meningocele or encephalocele.

Sincipital Type (Fig. 4.6)

This type is characterized by: (1) a circumscribed bony defect, with or without sclerotic margins, more or less in the midline of the facial bones; (2) associated soft tissue mass protruding through the bony defect; (3) hypertelorism—a widened nasal root with increased interorbital distance; and (4) absence of the frontal sinus with those occurring in the nasofrontal region.

Basal Type

1. The bony defect may not be evident.
2. Soft tissue mass may be seen at the base of the skull, in the nasopharynx, or in the upper nasal cavity (Fig. 4.7).

In meningoceles and encephaloceles of the ethmoid region of the nasal fossa the following may be observed:

1. Low position of the cribriform plate may indicate the site of bony defect.
2. Defect in the cribriform plate may be evident.
3. Soft tissue mass may be seen in the ethmoid region of the nasal fossa medial to the middle turbinate.
4. Hypertelorism may be present.
5. Unilateral opacity of the sinuses may be present.

Differential Diagnosis

Meningoceles and encephaloceles of the nasofrontal area and ethmoid region of the nasal fossa must be differentiated from dermoid cysts, nasal gliomas, and nasal polyps.

DERMOID CYSTS

General Considerations

Dermoid cysts are of developmental origin, are lined by epidermis, and contain epithelial debris, sebaceous material, and hairs. Several theories have been advanced regarding the origin of dermoid cysts of the nose, but no single theory offers a satisfactory explanation. In general, the most widely accepted theory is that of Grunwald. Embryologically, the anterior wall of the external nose is formed by a cartilaginous capsule. The frontal and nasal bones are developed later by intramembranous ossification of the fibrous tissue anterior to the cartilaginous capsule. A small hiatus exists between the frontal and nasal bones, which is filled by a membrane, the nasofrontal fontanel. Between the cartilaginous capsule and the nasal bones, a prenasal space is formed extending from the brain to the tip of the nose. A small projection of the dura protrudes into the prenasal space and, in the early embryo, is in contact with the skin. As the development progresses, the nasofrontal fontanel is replaced by the nasal process of the frontal bone which separates the dura from the skin. The projection of the dura becomes encir-

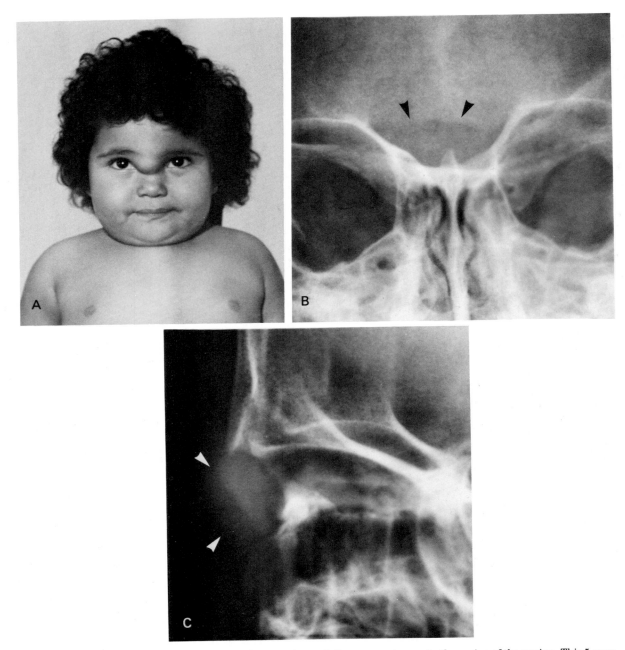

Figure 4.6. Nasofrontal encephalocele. *A*, photograph: a soft tissue mass is seen in the region of the nasion. This 5-year-old girl had had a soft tissue tumor mass over the bridge of the nose since birth. The tumor mass was soft and depressible into the skull, but came out immediately when pressure was removed. Cranioplasty was employed, with an acrylic prosthesis to close the bony defect in the region of the nasion. *B*, posteroanterior view of the skull: there is a circumscribed bony defect, measuring 2 cm in transverse diameter, in the region of the nasion. There is absence of the frontal sinuses. A mild hypertelorism is present. *C*, lateral view of the skull: a bony defect is again noted in the region of the nasion, with faintly visible soft tissue mass.

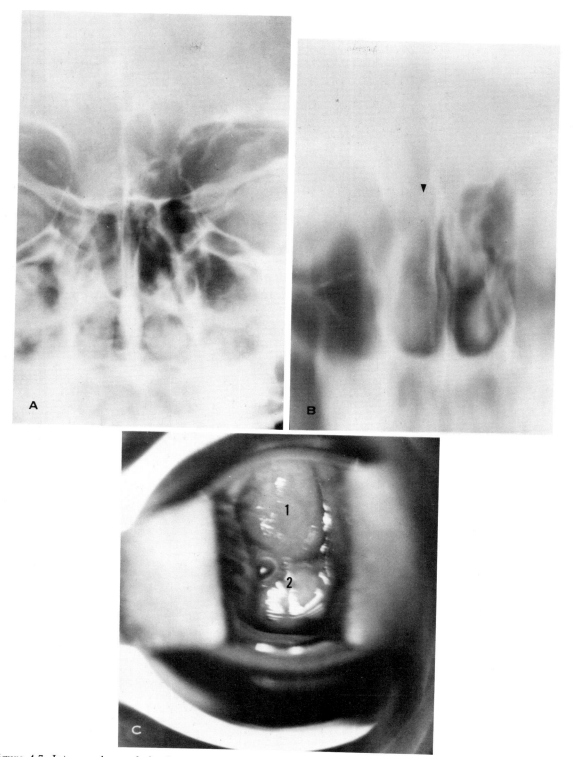

Figure 4.7. Intranasal encephalocele (encephalocele of ethmoid region). *A*, posteroanterior view: opacity of the right ethmoid and maxillary sinuses. There is a soft tissue density in the right nasal fossa. *B*, posteroanterior tomogram: a soft tissue mass filling the right nasal fossa with loss of the definition of the turbinates. There is a suggestive defect in the right cribriform plate (*arrow*). Opacity of the right ethmoid is again noted. *C*, photograph (intranasal view): a smooth mass (*1*) appears to originate superolaterally above the middle meatus (*2*) and fills the right nasal fossa.

A 30-year-old female was seen with the chief complaints of epistaxis, right nasal obstruction, frontal headache, and persistent nasal discharge. At operation, a dehiscence 2 × 2.5 cm was found in the right cribriform plate with brain tissue protruding through the defect into the right nasal fossa. The final diagnosis was encephalocele. (Reprinted with permission from: D. Blumenfeld and E. Skolnik: Archives of Otolaryngology, 82: 527, 1965).

cled by the foramen cecum of the frontal bone and ultimately it seals and obliterates the foramen. If obliteration fails, the foramen can serve as a pathway along which the neural tissues may extend into the nose, resulting in a nasal glioma, meningocele, or encephalocele. If any portion of skin fails to separate from the dura, it may be carried into the prenasal space to form a dermoid cyst. When the connection of the cyst with the skin persists, a sinus tract often forms.

Littlewood, on the contrary, advocates that dermoid cysts arise from persistence of the epithelial elements included in the nasal septum during the fetal development. Other theories include derivation of the dermoid cyst from aberration of the skin appendages or from a retention cyst which is formed by rudiments of the skin entrapped in one of the embryonic clefts or fissures.

Dermoid cysts may be little larger than the stalk itself or may expand to a relatively large mass. They may be found at any point from the nasofrontal suture to the base of the columella along the midline of the nose or the nasal septum. The most common location is at the lower margin of the nasal bone. They may be present at birth, but may become manifest only later during adolescence. Clinically, the typical dermoid cyst presents as a dimple or a fistula containing a hair over the dorsum of the nose. However, in some cases growth occurs inwardly and external manifestations may be lacking. Whether the cyst presents as an external soft tissue mass depends upon its location and size. If large enough, the cyst may extend into the frontal or ethmoid sinuses.

Roentgen Findings

The chief roentgen abnormalities may be found in the following two areas.

1. The nasal bones (Fig. 4.8): (1) broadening of the bridge of the nose, with ocular hypertelorism; (2) sharply demarcated bony defect with well defined or slightly sclerotic margin in the region of the nasofrontal suture or the nasal bone; (3) thinning and erosion of the adjacent nasal bones and nasal process of the frontal bone.

2. The nasal septum (Fig. 4.9): (1) bifid septum; (2) fusiform soft tissue mass within the nasal septum; (3) broadening or disruption of the bridge of the nose.

In either type infrafrontal or interethmoid cystic areas may indicate extension of the cyst to the frontal or ethmoid region.

Differential Diagnosis

If a dimple or fistula is present, the diagnosis of a dermoid cyst is often obvious. However, when they are absent, the diagnosis may be difficult. In the differential diagnosis conditions to be considered are meningocele, encephalocele, nasal glioma, neurofibroma, and epidermoid.

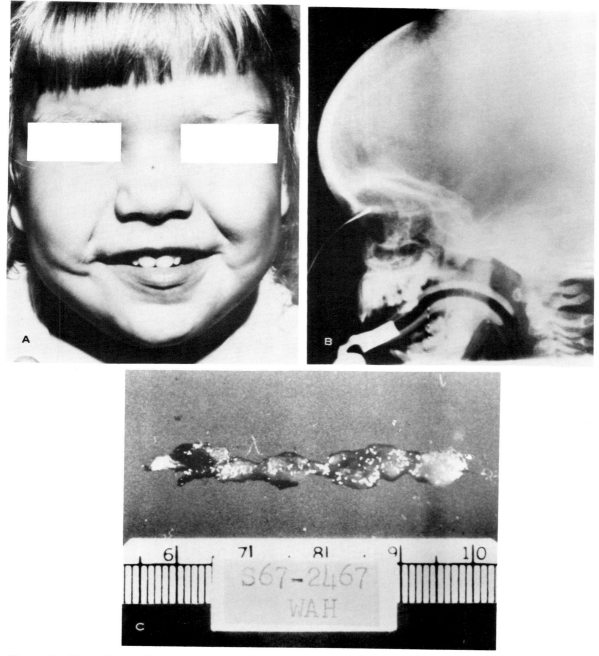

Figure 4.8. Dermoid cyst of the nasal bone. *A*, photograph of the face: widening of the bridge of the nose with ocular hypertelorism. There is a dimple in the mid-dorsum of the nose. *B*, lateral view of the skull: a metallic probe in the bony defect with the tip reaching the cribriform plate. *C*, photograph of the excised specimen: a long and narrow cystic mass. (Reprinted with permission from: J. W. Walike: Archives of Otolaryngology, 93: 487, 1971.)

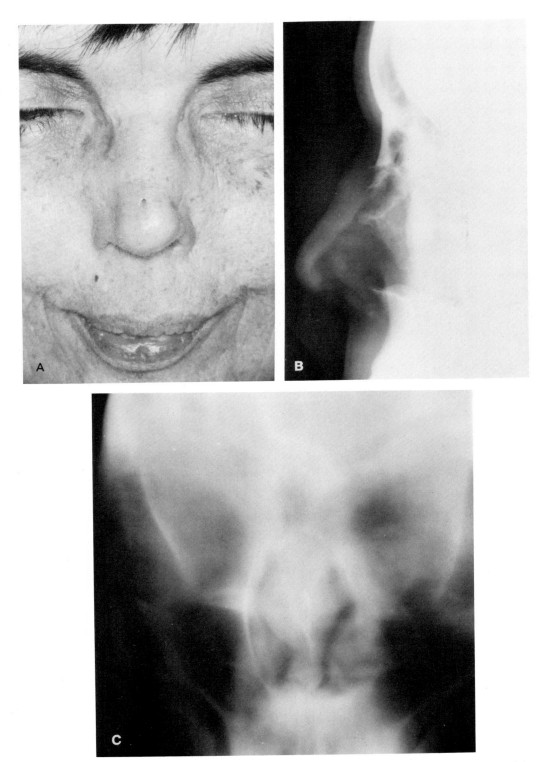

Figure 4.9. Dermoid cyst of the nasal septum. *A*, photograph of the face: a dimple in the lower dorsum of the nose. *B*, lateral view of the nose: an oval-shaped radiolucency with sclerotic margins in the region of the nasal bone. A soft tissue swelling is noted in the mid-dorsum of the nose. *C*, anteroposterior tomogram of the nose: bifid septum with a fusiform soft tissue mass within and around the nasal septum. There is broadening and disruption of the bridge of the nose.

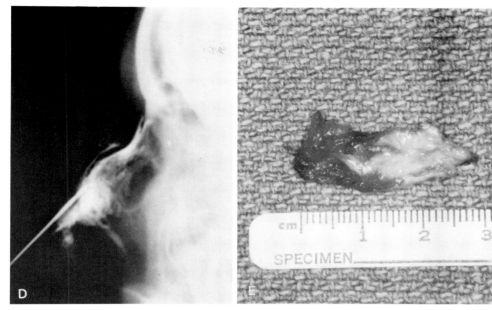

Figure 4.9 (continued). *D*, fistulous tract injection: partial opacification of the cyst. *E*, photograph of the excised specimen: it was about 2.8 cm in length with a terminal cystic sac.

REFERENCES

Ballenger, J. J.: *Disease of the Nose, Throat and Ear*. 11th ed., Lea and Febiger, Philadelphia, 1969.

Beinfield, H. H.: Surgery of bilateral atresia of the posterior nares in the newborn, Arch. Otolaryngol., *70:* 1, 1959.

Beinfield, H. H.: Surgical management of complete and incomplete bony atresias of the posterior nares (choanae). Am. J. Surg., *89:* 957, 1955.

Blumenfeld, D. and Skolnik, E. M.: Intranasal encephaloceles. Arch. Otolaryngol., *82:* 527–531, 1975.

Chodosh, P. L. and Wills, W.: Tuberculosis of the upper respiratory tract. Laryngoscope, *80:* 679, 1970.

Colp, R.: Dermoid cysts of the floor of the mouth. *Surg. Gynecol. Obstet., 40:* 183, 1925.

Eggston, A. and Wolff, D.: *Histopathology of the Ear, Nose and Throat,* Williams & Wilkins, Baltimore, 1947.

Enna, C. D., Jacobson, R. R., and Rausch, R. O.: Bone changes in leprosy: a correlation of clinical radiographic features. Radiology, *100:* 293, 1971.

Evans, J. N. G. and Maclachlan, R. F.: Choanal atresia. J. Laryngol. Otol., *85:* 903, 1971.

Graham, H. B.: Frequent but neglected evidence of syphilis from the side of the nose, accessory sinuses and ear. Am. J. Syphilis, *3:* 26, 1919.

Grunwald, L.: Beitrage zur Kenntnis kongenitalar Greschwerlote und Missbildungen an Ohr und Nase. Z. Ohrenhjk, *60:* 270, 1910.

Hornstein, O. P. and Gorlin, H. M.: Infectious oral disease. In *Thomas Oral Pathology*, 6th ed., edited by R. J. Gorlin and H. M. Goldman, p. 708–774, C. V. Mosby Co., St. Louis, 1970.

Hoshaw, T, C. and Walike, J. W.: Dermoid cysts of the nose. Arch. Otolaryngol., *93:* 487, 1971.

James, M. A.: Acute and chronic inflammations of the nose. In *Diseases of the Ear, Nose and Throat,* 2nd ed., edited by J. Ballantyne, W. G. Scott-Brown, and J.

Groves, p. 173–230, Butterworth Co., London, 1965.

Job, C. K., Karat, S., and Karat, A. B. A.: Pathological study of nasal deformity in lepromatous leprosy. Leprosy India, *40:* 42, 1968.

Job, C. K., Karat, A. B. A., and Karat, S.: The histopathological appearance of leprous rhinitis and pathogenesis of septal perforation in leprosy. J. Laryngol. Otol., *80:* 718, 1966.

Johnson, G. F., and Weisman, P. A.: Radiological features of dermoid cysts of the nose. Radiology, *82:* 1016, 1964.

Katz, A. and Lewis, J. S.: Nasal gliomas. Arch. Otolaryngol, *94:* 351, 1971.

Khan, M. A., and Gibb, A. G.: Median dermoid cysts of the nose—familial occurrence. J. Laryngol., *84:* 709, 1970.

Littlewood, A. H. M.: Congenital nasal dermoid cysts and fistulas. Plas. Reconstr. Surg., *27:* 471, 1961.

Luongo, R. A.: Dermoid cyst of the nasal dorsum. Arch. Otolaryngol., *17:* 755, 1933.

Moller-Christensen, V., Bakke, S. N., Melsom, R. S. and Waaler, E. Changes in the anterior nasal spine and the alveolar process of the maxillary bone in leprosy. Int. J. Leprosy, *20:* 335, 1952.

Nakamura, T., Grant, J., Hubbard, R.: Nasoethmoidal meningoencephalocele. Arch. Otolaryngol, *100:* 62, 1974.

New, E. B. and Erich, J. B.: Dermoid cysts of the head and neck. Surg. Gynecol. Obstet., *65:* 48, 1937.

Pendergrass, E. P., Schaeffer, J. P. and Hodes, P. J.: *The Head and Neck in Roentgen Diagnosis*. 2nd ed., p. 545–561, Charles C Thomas, Springfield, Illinois, 1956.

Schaeffer, J. P.: The various types of congenital atresias of the nose and their genetic interpretation. Am. Laryngol. Assoc. Trans. (Ser A), *56:* 126, 1934.

Schmidt, P., and Luyendijk, W.: Intranasal meningoencephalocele. Arch. Otolaryngol., *99:* 402, 1974.

Snyder, G., McCarthy, R., Toomey, J., and Rothfield, N.:

Nasal septal perforation in septemic lupus erythematous. Arch. Otolaryngol., *99:* 456, 1974.

Spencer, F. R.: Chronic granulomas of the nose and nasal accessory sinuses. In *Otolaryngology,* edited by G. M. Coates, H. P. Schenk and V. Miller, Chap. 19, p. 1–24, W. F. Prior Company, Baltimore, 1960.

Wilson, G. E. and Stern, W. K.: Tuberculosis of the nose and paranasal sinuses. In *Otolaryngology,* edited by G.

M. Coates, H. P. Schenk and V. Miller, Chap. 20, p. 1–7, W. F. Prior Co. Baltimore 1960.

Wilson, C. P.: Observations on the surgery of the nasopharynx. Ann. Otol. Rhinol. Laryngol., *66:* 5, 1957.

Yassin, A. EL., Shennawg, M., Enany, G. E., Wassef, N. F. and Shoeb, S.: Leprosy of the upper respiratory tract. J. Laryngol. and Otol. *89:* 505, 1975.

Chapter 5

Traumatic Lesions Involving the Nasal Bone and Nasal Cavity

FOREIGN BODIES IN THE NASAL CAVITY

The gamut of foreign bodies which may be found in the nasal cavity includes virtually any object which can be introduced through the nostrils. The patient is usually referred for roentgen examination of the nasal cavity and paranasal sinuses because of a discharge from the nose. In infants and children, a persistent unilateral sanguineous or purulent nasal discharge should always lead to a suspicion of a foreign body in the nasal cavity.

Opaque foreign bodies are readily detected by roentgen examination. Nonopaque objects may cause obliteration of the nasal cavity, and yet may not be revealed by radiological investigation.

RHINOLITHS

Rhinoliths are the results of an incrustation of a foreign body nucleus by the lime salts of the nasal cavity. This occurs when a foreign body remains in a nasal cavity for considerable length of time. The nuclei may be either exogenous or endogenous. Those of endogenous origin include displaced teeth, fragments of bone, blood clots, desquamated epithelium, clumps of bacteria, or inspissated mucus. The commonest exogenous foreign bodies include fruit seeds, buttons, and fragments of paper or other substances. The rhinoliths consist chiefly of calcium and magnesium salts, principally phosphate and carbonate. They are usually unilateral and situated in the lower portion of the nasal cavity. The most common locations are the inferior meatus and the space between the inferior turbinate and the septum. Rhinoliths vary from small granules to bodies the size of a hen's egg.

Rhinoliths are more common in adults than in children and are found more frequently in the female than the male.

Clinically there may be no symptoms for years, but sooner or later a unilateral nasal obstruction with foul smelling discharge develops and epistaxis may occur. Occasionally, swelling of the nose may develop.

Roentgen Findings

1. Calcified mass in a nasal fossa. Such masses may vary in size and shape but usually conform to the available space. A coral like appearance may be seen occasionally. (Fig. 5.1).
2. Displacement and perforation of the nasal septum.
3. Thinning and expansion of the lateral wall of the nasal cavity with large masses.
4. Destruction of the nasal bony wall. This change is uncommon and is usually due to superimposed osteomyelitis.

FRACTURES OF THE NASAL BONES

General Considerations

Fractures of the nasal bones are common. The variety and extent of the fractures vary with the site of impact and with the direction and intensity of the force. Fractures of the nasal bones

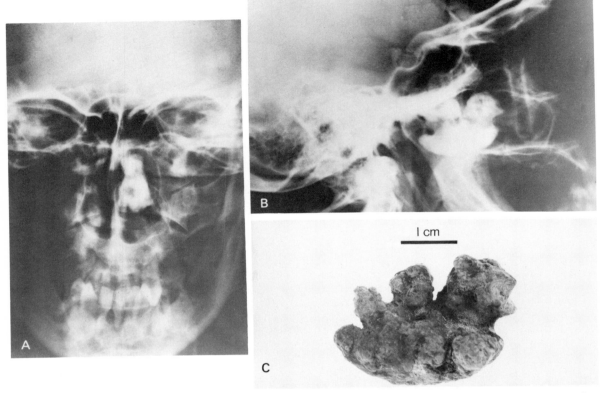

Figure 5.1. Rhinoliths. *A*, Caldwell view: there is a large irregular calcified mass in the left nasal fossa without associated bony destruction. *B*, lateral view: the calcified mass is located in the posterior aspect of the nasal cavity, situated in the inferior meatus, between the inferior turbinate and the nasal septum and in the middle meatus. It is protruding through the choana into the nasopharynx. *C*, specimen: the calcified mass has a polyhedral appearance with intact margins. (Reprinted with permission from: M. Bleshman: Radiology, 113: 615, 1974.)

range from a simple linear fracture to smash fractures with involvement of the regional facial bones (Fig. 5.2).

Isolated Fractures of the Nasal Bones

Isolated fractures of the nasal bones are caused by direct force from the front, or from the side. The fractures occur most frequently in the lower half of the nasal bones.

Simple Linear Fractures. This type of fracture is fairly common and is usually caused by a slight blow or fall. There is no displacement of the fragments, but edema and swelling of the soft tissue is often present (Fig. 5.2B).

Lateral Fractures. This is a more common type of fracture of the nasal bones. It occurs after a heavy blow from the side. There is displacement of the nasal bone fragment beneath the frontal process of the maxilla on the side of blow, and lateral displacement with impaction or overriding on the maxilla on the contralateral side. In severe cases, comminution may occur and the septum may be dislocated (Fig. 5.2, *C* and *D*).

Frontal Fractures. This type of fracture occurs after a heavy blow from the front which impinges squarely on the nasal bridge, splays the nasal bones laterally, and depresses them posteriorly. In the severe type, the septum is crushed so that the mucosa and cartilages are telescoped (Fig. 5.2, *E* and *F*).

Frontolateral Fractures. This is a combination of the lateral and frontal fractures and is best described as a displacement of the nasal cap. It is often seen after a glancing blow. There is fracture of both nasal bones and the nasal septum, with lateral shifting of the entire bony framework. The nasal bone on the side of the injury is displaced medially and the opposite side is displaced laterally (Fig. 5.2, *G* and *H*).

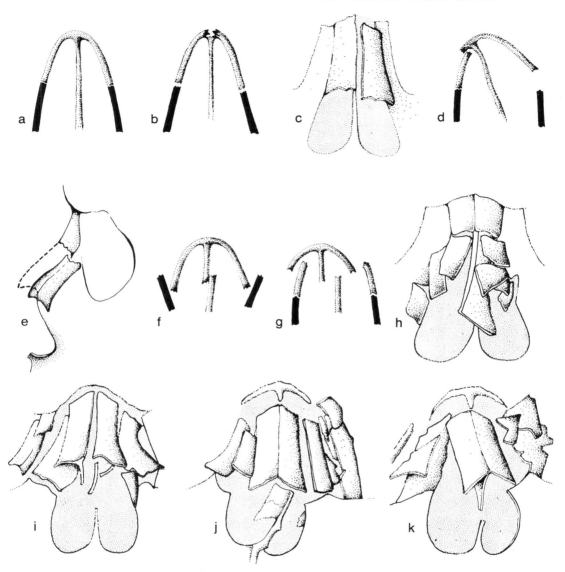

Figure 5.2. Line drawing of various types of fractures of the nasal bones (modified from Dingman and Natvig). *A*, normal nasal bones and septum. *B*, simple linear fracture—separation of the nasal bones in the midline and from the frontal processes of the maxillae. The nasal septum is intact. *C*, lateral fracture—separation of one nasal bone with inferolateral displacement and overriding of the maxilla. *D*, lateral fracture—lateral displacement with impaction. *E*, frontal fracture—fracture of two nasal bones with posteroinferior displacement. *F*, frontal fracture—depressed fracture of nasal bones and nasal septum with lateral splays of frontal processes of the maxillae. *G*, frontolateral fracture—displacement of the nasal cap with overriding. *H*, frontolateral fracture—comminuted fractures of the nasal bones, anterior parts of the frontal processes, and the nasal septum, with depression. *I–K*, smash fracture.

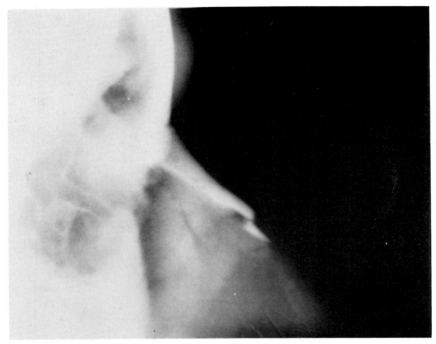

Figure 5.3. Simple fracture of the tip of the nasal bone. Lateral view: a transverse fracture of the tip of the nasal bone with posteroinferior displacement.

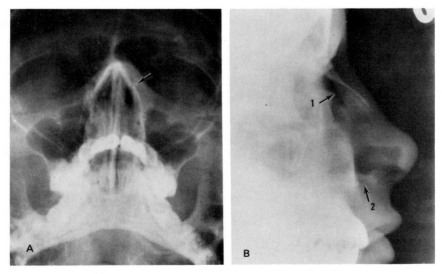

Figure 5.4. Simple linear fracture of the left nasal bone. *A*, Waters view: a fracture line in the left nasal bone associated with regional soft tissue swelling. The bridge of the nasal bones is intact. *B*, lateral view: a longitudinal fracture separating the left nasal bone from the frontal process of the left maxilla (*1*). In addition, there is a comminuted fracture of the anterior nasal spine of the maxilla (*2*). (Courtesy of Dr. Leonard D. Doubleday.)

Smash Fractures of the Nasal Bones

Motor vehicle accidents or a direct, violent frontal force may result in smash fractures of the nasal bones, the frontal process of the maxilla, the lacrimal bones, and the nasal septum. Lateral splaying of fragments may result in severe soft tissue damage, with lateral displacement of the lacrimal bones and damage to the ethmoid sinuses, cribriform plate, and the orbital processes of the frontal bone. (Fig. 5.2, *I* to *K*).

Roentgen Diagnosis

In the interpretation of the roentgenograms of the nose, it is essential not only to have a thorough knowledge of the roentgen anatomy of the nasal bones, but also of the facial bones.

In fractures of the nasal bones, a precise knowledge of the type of injury is often helpful in establishing an accurate diagnosis. Helpful points to determine this include the direction, intensity, and type of injuring force, and the history of preexistent nasal disease, previous operations and prior deforming injuries.

The evaluation of the roentgen findings should include: (1) presence of fracture line or lines; (2) depression of the fragments of one or both nasal bones; (3) deviation of the fragments of the nasal bones or of the nasal septum; and (4) condition of the adjacent facial bones.

Demonstration of the Fracture Line

Fracture of the nasal bones is usually transverse with ventral or dorsal displacement (Fig. 5.3). The lateral view is usually sufficient to determine whether a fracture is present and to reveal the presence or absence of displacement (Figs. 5.4 to 5.8).

If the fracture line is thin, longitudinal, and more or less parallel to the dorsum of the nose, it must be differentiated from the longitudinal groove for the nasal branch of the nasocillary nerve, and from the nasomaxillary sutures. These normal anatomical features are relatively straight, less radiolucent than a fracture line, and parallel to the dorsum.

Determination of the Displacement of the Fragments of the Fracture

Lateral or medial deviation of the fragments of the fracture is sometimes difficult to determine. Usually the displacement of the nasal bones and the septum can be seen on Waters' projection, but occasionally the occlusal view may be of value in determining the presence or absence of minor degrees of deviation.

Involvement of the Adjacent Facial Bones

In a smash fracture of the nasal bones, involvement of the regional facial bones is often present. Posteroanterior and lateral views of the skull and Waters' view are necessary to determine the presence and extent of the fractures. Frontal tomograms of the paranasal sinuses are helpful to establish the full import of the facial injuries (see also figs. 16.2 to 16.5).

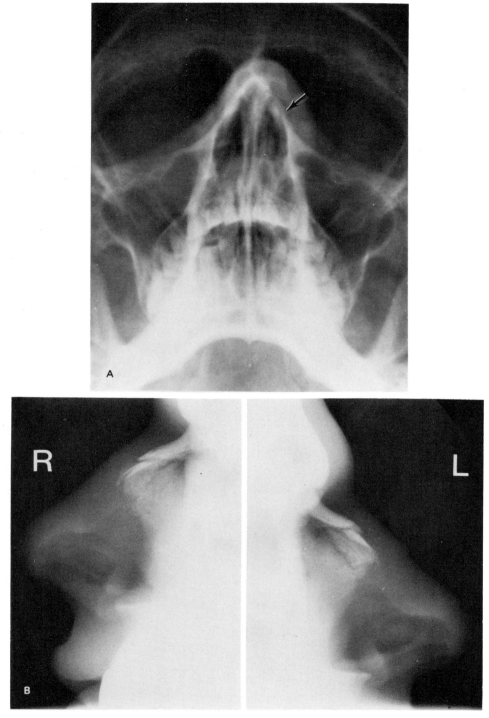

Figure 5.5. Frontal fracture of the nasal bones. *A*, Waters view: there is a faintly visible fracture of the left nasal bone associated with regional soft tissue swelling. *B*, lateral view: the fracture line crosses the nasal bone obliquely, with slight posteroinferior displacement on each side. (Courtesy of Dr. Thomas Harle.)

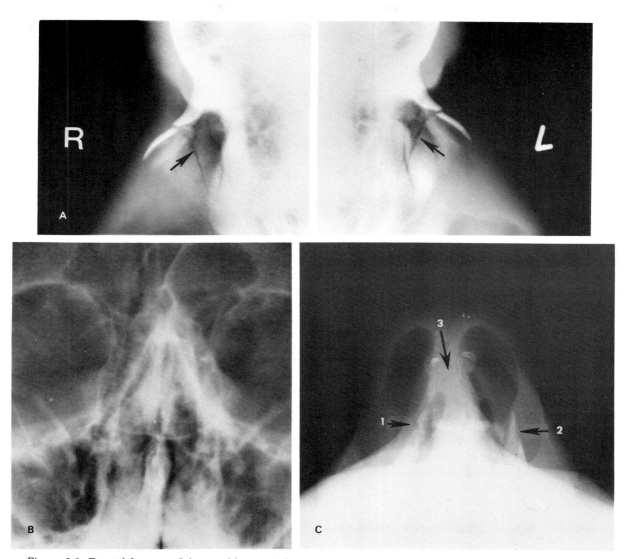

Figure 5.6. Frontal fracture of the nasal bones. *A*, lateral view of the nasal bones: a transverse fracture of the nasal bone with extension to the frontal process of the maxillary bone on each side (*arrows*). The fracture line crosses over the groove for the nasociliary nerve and the nasomaxillary suture. *B*, Waters view of the nasal bones: fracture of the nasal bones with flattening of the bridge of the nasal bones. *C*, occlusal view: there is a fracture of the right nasal bone with medial displacement of the fragments of the fracture (*1*). A fracture of the left nasal bone is noted, with slight lateral displacement of the fragments of the fracture (*2*). There is also impaction with slight left lateral deviation of the septum (*3*).

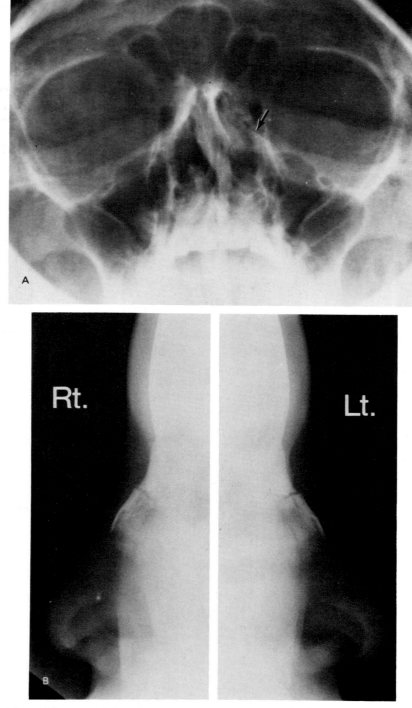

Figure 5.7. Frontolateral fracture of the nasal bones. *A*, Waters view: fracture of the nasal bones and nasal septum with slight displacement of the nasal cap to the right and overriding; *B*, lateral view: transverse fractures of the nasal bone on each side.

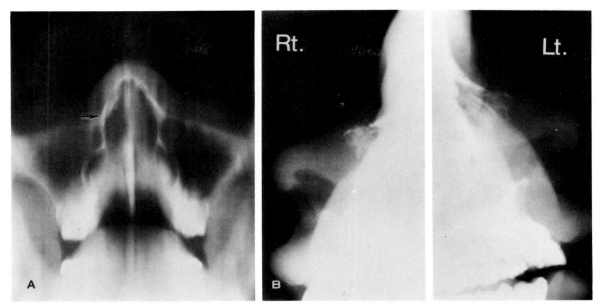

Figure 5.8. Depressed frontolateral fracture of the nasal bones. *A*, tomogram in Waters' position: depressed fracture of the nasal bones and nasal septum, with slight splaying of the frontal process of the right maxilla. There is flattening of the bridge of the nose, associated with regional soft tissue swelling. *B*, lateral view: comminuted fractures of the nasal bones are seen on each side. (Courtesy of Dr. Thomas Harle.)

REFERENCES

Ballenger, J. J.: *Disease of the Nose, Throat, and Ear.* 11th ed., p. 80–84, Lea and Febiger, Philadelphia, 1969.

Becker, O. J.: Nasal fractures. Arch. Otolaryngol., *48:* 344, 1948.

Bleshman, M., Bonakdarpour, A., and Ronis, M.: Rhinoliths. Radiology *113:* 615, 1974.

Dingman, R. O. and Natvig, P: *Surgery of Facial Fractures.* W. B. Saunders, Philadelphia, 1964.

Freimanis, A. K.: Fractures of the facial bones. Radiol. Clin. North Am., *4:* 341, 1966.

Keogh, C. A.: Affection of the external nose and nasal cavity. In *Disease of the Ear, Nose and Throat,* 2nd ed. edited by W. G. Scott-Brown, J. Ballantyne, and J. Groves, p. 75–102, Butterworths, London, 1965.

Pendergrass, E. P., Schaeffer, J. P., and Hodes, P. J.: *The Head and Neck in Roentgen Diagnosis,* 2nd ed., Charles C Thomas, Springfield, Illinois, 1956.

Samuel, E.: Rhinoliths. Br. J. Radiol, *16:* 186, 1943.

Zatzkin, H. R.: *The Roentgen Diagnosis of Trauma.* Year Book Medical Publishers, Inc., Chicago, Illinois, 1965.

Chapter 6

Benign Tumors of the Nasal Bone and Nasal Cavity

The classification of benign tumors of the nasal bone and nasal cavity is the same as that of the paranasal sinuses and will be discussed in detail in that section. Exceptions are tumors of neurogenic and vascular origin, which are more common in the nose.

NASAL GLIOMAS

General Considerations

Nasal gliomas (encephalochoristoma nasofrontalis) are quite rare; less than 70 cases have been reported in the world literature. There is a marked divergence of opinion regarding the exact origin of these tumors; however, most observers believe they are of embryonal origin and not neoplastic. The most widely accepted explanation considers the tumor an encephalocele which has been isolated from the brain during embryonic development by closure of the sutures of the skull. A persistent connection with subarachnoid space or the ventricular system often indicates an origin in an encephalocele.

Histologically, nasal gliomas are composed of nests of glial cells in a fine fibrillar stroma. The cells are usually astrocytes with their processes and may be multinuclear. Intranasal gliomas are covered with respiratory epithelium and extranasal gliomas are covered by skin.

Depending upon the location, nasal gliomas have been divided into three groups: intranasal, extranasal, and combined forms. The extranasal glioma is the commonest type. The tumor is usually located to the right or left of the midline of the base of the nose. Averaging 2 to 3 cm in diameter, it is usually moderately firm, elastic, circumscribed, and relatively smooth. It is often fixed to the skin and attached to the nasal bone beneath. There is no pulsation of the mass and no change in size on straining. The overlying skin may show discoloration, and a defect may occur in the underlying bone. The base of the nose is frequently broadened and the eyes are far apart.

In the intranasal variety the tumors, averaging 2 to 4 cm in diameter, tend to occur high in the nasal fossa lateral to the septum; occasionally they may arise from the middle turbinate. They may fill the nasal fossa and displace the septum to the opposite side; occasionally the tumor may bulge from the nostril. Frequently the tumors become congested, causing nasal obstruction and bleeding. The tumors may be confused with polyps, but tend to be firmer and less radiolucent.

When both extranasal and intranasal components are present there is a communication between the two, usually through a defect in the nasal bone or at the lateral margin of the nasal bone. When an intracranial communication exists, the connection usually passes through a defect in the cribriform plate or the region of the nasofrontal suture.

The condition is usually present at birth, but because recognition is dependent upon sufficient growth, nasal gliomas may be overlooked in infancy and discovered only in later years. There is no sex predilection.

Roentgen Findings

There is no characteristic roentgen finding in nasal gliomas.

1. In the extranasal variety there is often a soft tissue mass in the region of the base of the nose. This is best seen on soft tissue roentgeno-

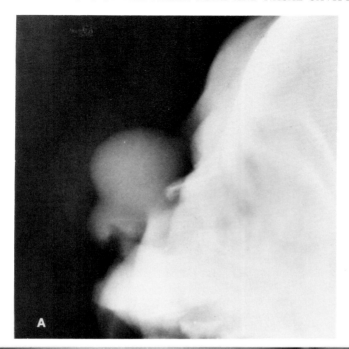

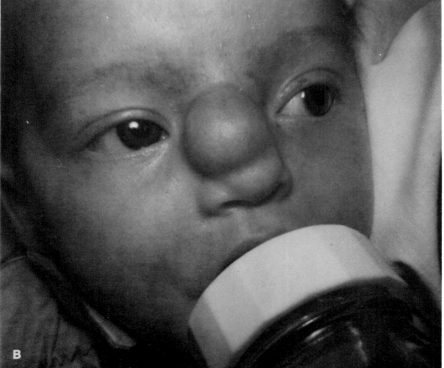

Figure 6.1. Extranasal glioma. *A*, lateral view of the face showing a well defined rounded soft tissue mass in the region of the dorsum of the nose; *B*, photograph showing a smooth rounded mass in the dorsum of the nose, more on the right side.

Six-week-old female with a soft tissue mass over the bridge of the nose, more on the right side, since birth. The tumor mass was excised and the pathological report was nasal glioma.

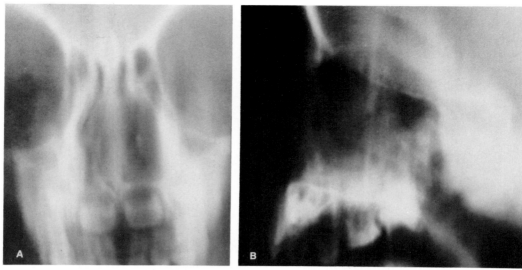

Figure 6.2. Intranasal glioma. *A*, anteroposterior tomogram demonstrating a rounded mass measuring 2 cm in transverse diameter filling the entire left nasal fossa and displacing the nasal septum to the right. *B*, lateral tomogram demonstrating again a soft tissue mass in the nasal fossa with possible attachment to the cribriform plate. There is no evidence of bony destruction.

Two-year-old baby girl with a tumor mass in the left nasal fossa since birth. Operative findings: a tumor, approximately 2 cm in diameter, filled the left nasal fossa. It was attached to the nasal septum and the middle turbinate. Superiorly, a very narrow stalk passed into the cribriform plate. Pathological report: nasal glioma.

grams. A bony defect with a well defined margin may be present in the region of the base of the nose (Fig. 6.1).

2. In the intranasal type, a soft tissue mass is seen in the nasal fossa. When large enough, the tumor mass will fill the entire nasal fossa, with displacement of the nasal septum and expansion of the lateral wall. Occasionally the site of the lesion can be delineated (Fig. 6.2).

3. When an intracranial communication exists, conventional films and tomography may be of great value in demonstrating the bony defect through which communication occurs.

Differential Diagnosis

Tumors of the nasal region which should be considered in the differential diagnosis of nasal glioma are tumors of neurogenic origin such as encephaloceles, meningoceles, and neurofibromas; tumors of ectodermic origin such as sebaceous cysts, epidermoid cysts, dermoid cysts, abscesses, nasal polyps, and cysts of the lacrimal duct apparatus; and tumors of mesodermic origin, i.e., hemangiomas, lipomas, fibromas, etc.

HEMANGIOMAS

General Considerations

Hemangiomas are considered developmental vascular malformations rather than neoplasms. The lesions are thought to be congenital and growth occurs mostly during early life; however, an increase in size may not be manifest clinically for years. The age incidence in the reported cases ranges from a few months to 60 years of age. The malformation is more common in females.

Hemangiomas of the nasal cavity usually arise from the soft tissues of the lateral wall of the ethmoid region. Growth is usually slow, with infiltration and destruction of the surrounding tissues. There is a tendency to invade the paranasal sinuses, usually the maxillary or

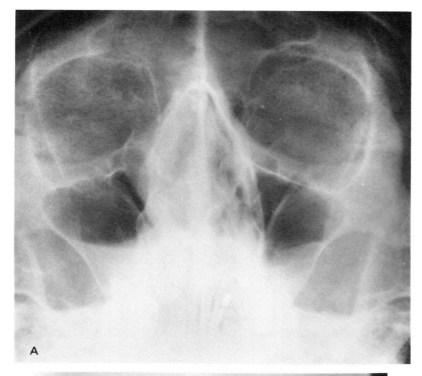

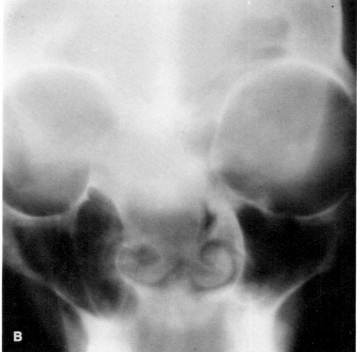

Figure 6.3. Sclerosing cavernous hemangioma of the nasal cavity. *A*, Waters view: marked opacity of the nasal cavity and cloudiness of the ethmoid sinuses. There is thickening of lining mucosa of the left maxillary sinus and probable fluid level in the right maxillary sinus. *B*, anteroposterior tomogram: a well defined homogenous expanding soft tissue mass in the right ethmoid sinus and the right nasal fossa above the inferior turbinate. There is thinning of the superior aspect of the right lateral wall of the nasal cavity and bowing of the nasal septum to the left. The right lamina papyracea is slightly displaced laterally.

ethmoid chambers, with eventual filling of the nose, sinuses, and nasopharynx.

Hemangioma of the nasal bone is an unusual entity. A history of previous trauma to the nose has been recorded in six of 11 reported cases.

The symptomatology is variable, depending upon the location and size of the tumor. In some cases, no symptoms occur until a relatively large size is attained. Hemangiomas of the nasal cavity are often associated with nasal obstruction, discharge, and epistaxis. In advanced cases, exophthalmos and facial disfiguration may be present.

In hemangioma of the nasal bone there may be a lump in the dorsum of the nose. The skin over the swelling is, as a rule, freely movable and may show telangiectases.

Roentgen Findings

1. In hemangioma of the nasal cavity, the changes are (Fig. 6.3): (1) opacification of the nasal fossa. Clinically a pulsating soft tissue mass may be apparent; (2) displacement and thinning of the nasal septum; (3) expansion and erosion of the lateral wall of the nasal cavity.

2. In hemangioma of the nasal bone (Fig. 6.4) the changes are: (1) a well defined expansile lesion with a sunburst appearance in the lateral projection; (2) possible regional soft tissue swelling of the dorsum of the nose.

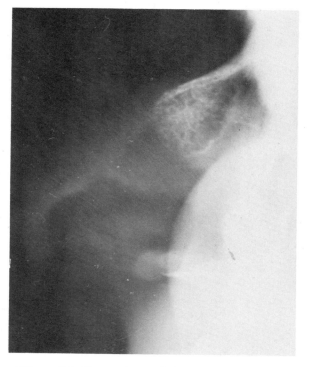

Figure 6.4. Hemangioma of the nasal bone. Lateral view: a well-circumscribed expansile lesion with sunburst appearance in the nasal bone. (Reprinted with permission from Archives of Otolaryngology, 88: 67, 1968.)

REFERENCES

Altany, F. E., and Pickrell, K. L.: Nasal gliomas. Arch. Surg. 71: 275, 1962.

Borsanyi, S.: Nasal glioma. Arch. Otolaryngol., 72: 376, 1955.

Davis, E., Morgan, L. R.: Hemangioma of bone. Arch. Otolaryngol. 99: 443, 1974.

Hymans, V. J.: Papillomas of the nasal cavity and paranasal sinuses. Ann. Otol., Rhinol, Laryngol. 80: 192, 1971.

Lampertico, P. and Ibanez, M. L.: Nasal glioma (encephal-ochoristoma nasofrontalis). Arch. Otolaryngol., 79: 628, 1964.

Schmidt, M. B.: Üeber seltene Spaltbildungen im Bereiche des mittleren Stirnfrostazen. Arch Pathol. Anat., 162: 340, 1900.

Siegleman, S.: Hemangioma of the nasal bones. Arch. Otolaryngol., 88: 67, 1968.

Walker, E. A., and Resler, D. R.: Nasal glioma. Laryngoscope, 73: 93, 1962.

Chapter 7

Malignant Tumors of the Nasal Cavity

Malignant tumors of the nasal cavity may be divided into superior and inferior groups. Those of the superior group arise from the nasal cavity above the lower border of the superior turbinate; the tumors comprising the inferior group originate from the lateral nasal wall, the middle and inferior turbinates, or the nasal septum. The sites of origin and routes of spread are shown in Figure 7.1.

MALIGNANT TUMORS OF THE SUPERIOR GROUP

The pathological classification of malignant tumors of the nasal cavity is similar to that of the paranasal sinuses and is discussed in that section. However, the olfactory neuroblastoma (olfactery esthesioneuroblastoma) is peculiar to the superior group of nasal fossa neoplasms and requires independent description.

General Considerations

Olfactory neuroblastoma is a relatively rare tumor of the nasal cavity which arises from the epithelial and neural tissues of the olfactory mucosa. First described by Berger in 1924, approximately 110 cases have now been reported in world literature. The tumors are polypoid in nature, arising high in the nasal fossa from the olfactory region, an area which includes the roof of the nasal cavity, the anterior and middle ethmoid regions, the superior turbinate, and the superior portion of the nasal septum. The growths may vary in size from a small, polypoid mass to one which completely obstructs the nasal fossa and nasopharynx. Histologically the tumors vary in pattern but have certain basic characteristics. All are composed of nerve cells, either neurocytes or neuroblasts. In addition, neurofibrils and pseudorosettes are present in the majority and a few include true neuroepithelial rosettes. These patterns may not be very distinctive and may be difficult to recognize. For this reason olfactory neuroblastomas have been confused with undifferentiated carcinoma, lymphoma, and other neoplasms. Depending upon the dominant histological pattern, attempts have been made to subdivide these tumors into three types: the olfactory esthesioneuroepithelioma, the olfactory esthesioneurocytoma, and the olfactory esthesioneuroblastoma. However, the rarity of the neoplasms makes the correlation between histological subtype and biological behavior somewhat tenuous. Clinically, the tumor is a slowly growing malignancy. The early symptoms are nasal stuffiness and sneezing. As the lesion progresses there may be unilateral nasal obstruction of the ostia of the paranasal sinuses. Pain over the face and facial deformity may result. Excessive lacrimation and rhinorrhea may occur. The tumor is locally invasive and frequently involves the base of the skull and the paranasal sinuses. Regional lymph node metastases occur, and metastasis to the lungs has been reported.

The disease is very rare in the Negro and occurs more frequently in males. The age incidence has a wide range, from 9 to 79 years, with peaks occurring between 11 to 20 years and 31 to 40 years of age (Skolnik).

Roentgen Findings

1. In the early stage, a tumor mass may be seen in the superior portion of the nasal fossa, with opacity of the ethmoid sinus (Fig. 7.2). It is often difficult to differentiate from a primary lesion of the ethmoid sinus.

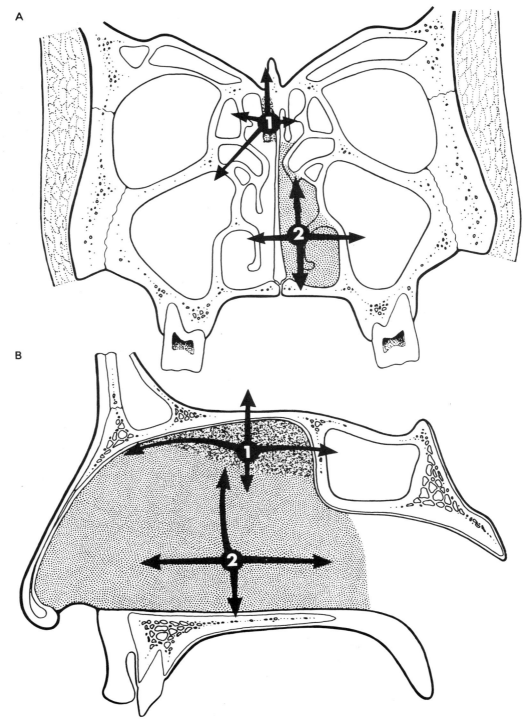

Figure 7.1. Sites of origin and routes of spread of tumors of the nasal cavity. *A1,B1*: Superior group—tumors arising from the nasal cavity above lower border of the superior turbinate. *A2,B2*: Inferior group—tumors originating in the lateral wall, including the middle and inferior turbinates, or the septum.

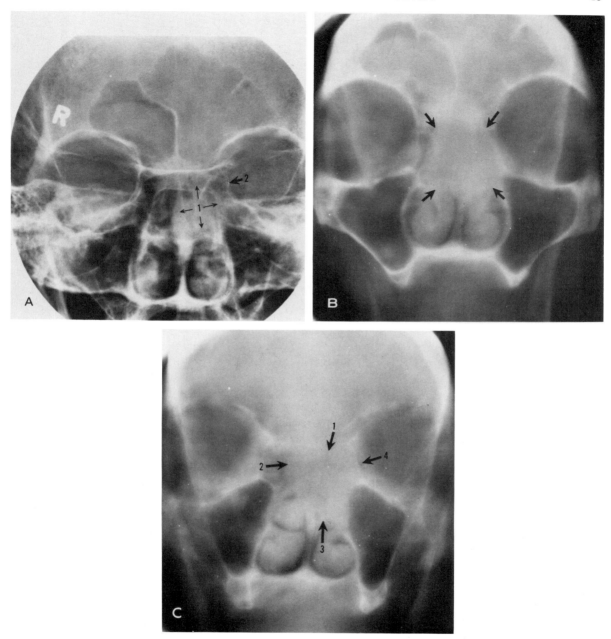

Figure 7.2. Early olfactory neuroblastoma of the nasal cavity. *A*, Caldwell view: opacity of superior aspect of left nasal fossa and left ethmoid (*1*). The left lamina papyracea is slightly displaced laterally (*2*). There is clouding of frontal sinuses, with a retention cyst on the right. *B*, anteroposterior tomogram (*anterior cut*): a large soft tissue mass in the superior part of the left nasal fossa and ethmoid extending to both middle turbinates (*arrows*). Both ethmoidomaxillary plates are intact. *C*, anteroposterior tomogram (*middle cut*): tumor mass in the superior part of the left nasal fossa with extension laterally to the left ethmoid (*1*), medially to right nasal fossa (*2*), and inferiorly to the left middle turbinate (*3*). The left lamina papyracea is displaced laterally and partly destroyed (*4*). The cribriform plate is intact.

2. In the late stage, there is opacity of the nasal fossa and of the ipsilateral paranasal sinuses, with destruction of the ethmoid and middle and inferior turbinates. Extension to the sphenoid sinus, maxillary sinus and orbital cavity often occurs. Involvement of the contralateral nasal cavity and ethmoid sinus is not uncommon. Destruction of the cribriform plate may occur (see also fig. 18.29).

3. The roentgen findings are nonspecific. The final diagnosis is usually made by tissue biopsy.

Differential Diagnosis

Tumors which should be differentiated from olfactory neuroblastoma are: (1) neurogenic tumors—neurofibroma, ganglineuroma, meningocele, and encephalocele; (2) carcinoma and sarcomas—undifferentiated carcinoma, transitional cell carcinoma, melanoma, rhabdomyosarcoma, and malignant lymphoma; (3) benign cysts and tumors—inverting papilloma, hemangioma, and nasal polyps.

MALIGNANT TUMORS OF THE INFERIOR GROUP

The inferior group of malignant tumors of the nasal cavity consists mostly of carcinomas and sarcomas. The behavior of these tumors is analogous to that of like tumors of the paranasal sinuses. In the early stages of development the tumors may be confined to the nasal cavity. With advanced neoplasms, invasion of the surrounding structures is very frequent. The tumors may extend superiorly to the ethmoid sinus, laterally to the maxillary sinus, and inferiorly to the hard palate. If the lesion arises from the nasal septum, it may involve the opposite nasal fossa.

Roentgen Findings

1. A soft tissue mass is seen in the affected nasal fossa which may extend beyond the bony confines. The latter is highly suggestive of a malignant process.

2. Destruction of the septum or of the lateral wall of the nasal fossa usually indicates the invasive nature of the lesion and is reliable evidence of a malignant process (Figs. 7.3 to 7.5).

3. There is invasion of the surrounding structures in advanced lesions (Fig. 7.6).

Differential Diagnosis

Lesions of the nasal cavity which should be considered in the differential diagnosis are nasal polyps, inverting papillomas, and chronic granulomas. When a tumor arises from the lateral wall of the nasal cavity, it may sometimes be difficult to differentiate from a primary lesion of the maxillary sinus.

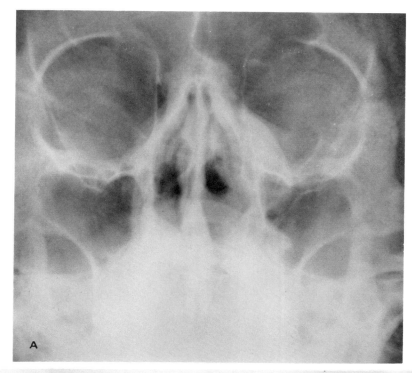

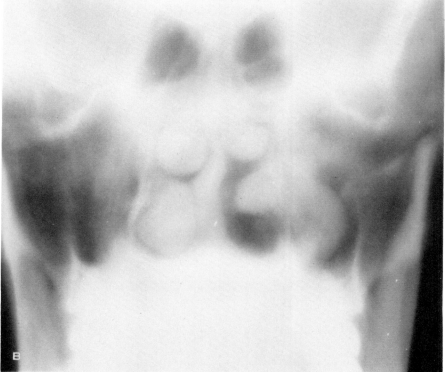

Figure 7.3. Undifferentiated carcinoma of the left inferior turbinate. *A*, Waters view: a soft tissue density in the region of the left ala nasi. *B*, anteroposterior tomogram: a large tumor mass involving left inferior turbinate and extending to the medial aspect of the maxillary sinus. There is destruction of the inferior portion of the left lateral wall of the nasal cavity. A defect in the medial aspect of the tumor mass results from tissue biopsy. (Reprinted with permission from Annals of Otology, Rhinology, and Laryngology, 79: 584, 1970.)

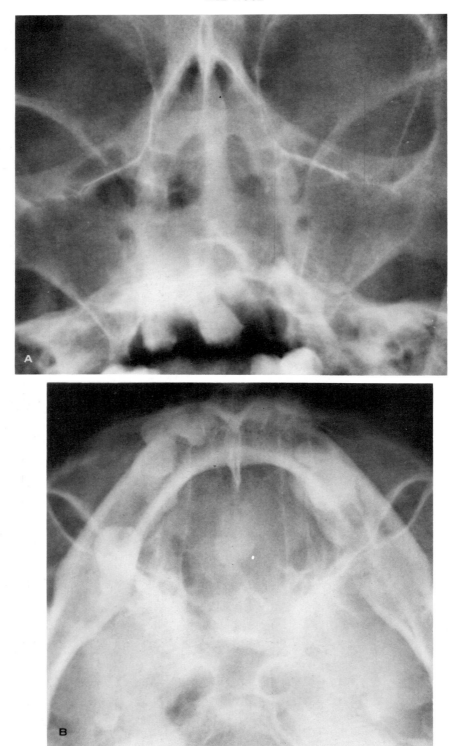

Figure 7.4. Squamous cell carcinoma of the nasal septum. *A*, Waters view: opacity of both maxillary sinuses. There is marked thickening of the nasal mucosa with irregular densities in each nasal fossa. *B*, base view: destruction of the bony nasal septum. There is a large soft tissue mass filling most of the posterior nasal cavity.

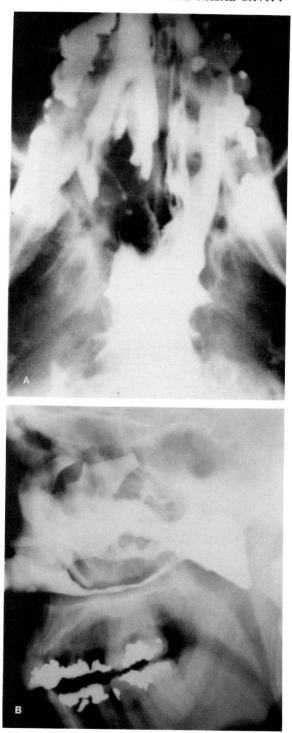

Figure 7.5. Unclassified malignant tumor of the right middle turbinate. *A*, submentovertical view of contrast naso-pharyngogram: a large filling defect in the posterior portion of the nasal fossa and the anterior aspect of the nasopharynx on the right; *B*, lateral view of contrast nasopharyngogram: tumor mass involving the right middle turbinate with extension to the anterior part of the nasopharynx.

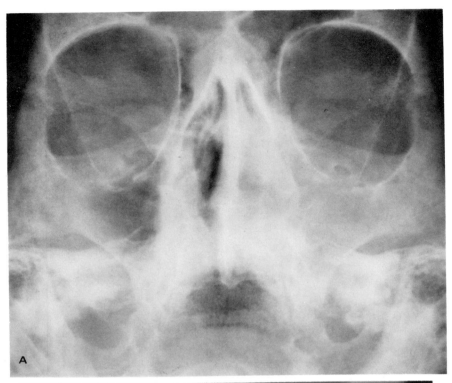

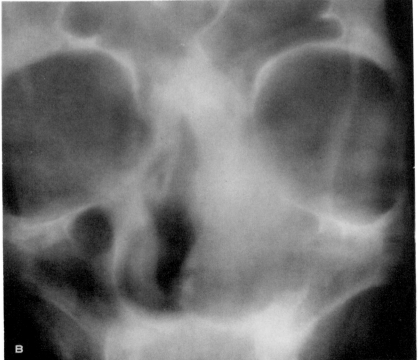

Figure 7.6. Malignant melanoma of the left nasal fossa. *A*, Waters view: opacity of left nasal fossa and maxillary and ethmoid sinuses. *B*, anteroposterior tomogram: a large soft tissue mass filling the entire left nasal fossa with extension to left maxillary and ethmoid sinuses. The nasal septum is partially destroyed.

REFERENCES

Alava, A., and Gallager, H. S.: Olfactory esthesioneuro-epithelioma. Arch. Pathol., *67:* 43, 1959.

Bidstrup, R. J., and Wilkins, S. A., Jr.: Olfactory esthesioneuroma. South. Med. J., *63:* 1426, 1970.

Boone, M. L. M., Harle, T. S., Higholt, H. W., and Fletcher, G.: Malignant disease of the paranasal sinuses and nasal cavity. Am. J. Roentgenol. Radium Ther. Nucl. Med., *102:* 627, 1968.

Bortnick, E.: Neoplasms of the nasal cavity. Otolaryngol. Clin. North Am., *6:* 801, 1973.

Frazell, E. L., and Lewis, J. S.: Cancer of the nasal cavity and accessory sinuses. Cancer, *16:* 1293, 1963.

Holdcraft, J. and Gallagher, J. C.: Malignant melanomas of the nasal and paranasal mucosa. Ann. Otol., *78:* 1, 1969.

Jesse, R. H., Butler, J. J., Healey, J. E., Jr., Fletcher, G. H., and Chau, P. M.: Paranasal sinuses and nasal cavity. In *Cancer of the Head and Neck,* 1st ed., Chap. 10, p. 329–356, edited by William S. MacComb and Gilbert Fletcher. Williams & Wilkins Co., Baltimore, 1967.

Jing, B. S.: Roentgen diagnosis of malignant disease of paranasal sinuses and nasal cavity. Ann. Otol. Rhinol Laryngol., *79:* 584, 1970.

Kholi, G. S. and Sachdeva, O. P.: Malignant melanoma of the nose and paranasal sinuses. J. Laryngol. Otol. *88:* 693, 1974.

Larsson, L.-G. and Martensoon, G.: Carcinoma of the paranasal sinuses and the nasal cavities. Acta Radiol., *42:* 149, 1953.

Lewis, J. S.: Sarcomas of the paranasal sinuses and the nasal cavity. Ann. Otol. *78:* 776, 1969.

Lewis, J. S., Hutter, R. V. P., Tollefsen, H. R., and Foote, F. W., Jr.: Nasal tumors of olfactory origin. Arch. Otololaryngol., *81:* 169, 1965.

MacComb, W. S., and Martin, H. E.: Cancer of the nasal cavity. Am. J. Roentgenol., *47:* 11, 1942.

Mendelhoff, J.: The olfactory neuroepithelial tumors. Cancer, *10:* 944, 1957.

Ogura, J. H., and Schenck, N. L.: Unusual nasal tumors: problems in diagnosis and treatment. Otol. Clin. North Am., *6:* 813, 1973.

Skolnik, E. M., Massari, F. S, and Tenta, L. T.: Olfactory neuroepithelioma. Arch. Otolaryngol., *84:* 644, 1966.

PART II

The Paranasal Sinuses

Chapter 8

Anatomy of the Paranasal Sinuses

GROSS ANATOMY

The paranasal sinuses are air-filled cavities produced by evagination of the mucous membrane of the nasal cavity into the substance of adjacent facial and skull bones. They communicate with the nasal cavity through the ostia, and their mucosal linings are continuous with nasal mucosa.

The walls of the sinuses are composed of compact bone, the cancellous bone being displaced by the sinuses. At birth, the sinuses are small or absent. They grow slowly until puberty, after which they grow rapidly to their adult size. In old age, resorption of the diploe of the bone leads to further enlargement.

The Maxillary Sinuses

The paired maxillary sinuses, or antra of Highmore, are the largest of the paranasal sinuses, and are situated within the bodies of the maxillary bones. In each the lateral wall (infratemporal wall) is formed by the infratemporal bony wall of the maxilla, and the medial wall is the lower part of the lateral wall of the nasal cavity. The roof is the floor of the orbit, which is frequently ridged by the infraorbital canal. The narrow floor lies over the alveolar process of the maxilla and over the roots of the molar and premolar teeth, with its most dependent portion overlying the second premolar and first molar teeth. The lower level of the floor is usually 0.5 to 1.0 cm below the level of the floor of the nasal cavity. The maxillary sinuses may be small, overlying only the molar area, or may be sufficiently large to cover the molar, premolar, and canine regions. The roots of the teeth, particularly the first two molars, may produce conical elevations in the floor and may even penetrate the bone and protrude into the chambers. The anterior wall (facial wall) is the anterior bony wall of the maxilla, whereas the posterior wall is the posterior wall of the body of the maxilla,

separating the sinus from the pterygomaxillary fissure and the pterygopalatine fossa (Figs. 8.1 and 8.2).

The sinus chamber consists of a central cavity with four extensions: (1) the zygomatic extension, (2) the palatine extension into the floor of the nasal cavity, (3) the tuberosity extension above the third molar, and (4) the alveolar extension into the alveolar process of the maxilla. The maxillary ostium passes through the upper part of the medial wall of the sinus and opens into the hiatus semilunaris in the middle meatus of the nasal cavity. Occasionally, an accessory ostium is seen above the posterior part of the inferior turbinate (Fig. 8.3). The measurements of an average sinus chamber in the adult are: width, 2.5 cm; height (opposite the first molar tooth), 3.5 cm; and depth (anteroposterior), 3 cm.

At birth, the maxillary sinus is a small slit-like space adjacent to the lateral wall of the nasal cavity, with the greatest diameter in the anteroposterior plane. At 6 to 9 months, the sinus is large enough to be visualized in Waters and lateral views. At the end of the 1st year, the chamber assumes an adult rhomboidal configuration and is sufficiently well pneumatized that opacification is indicative of disease. From the 1st to the 3rd year, the growth of the face continues at a rather rapid rate, with the maxillary sinus showing an appreciable increase in size during this period. After the 8th year, the expansion of the sinus to adult size and shape, especially in the anteroposterior and vertical diameters, is relatively fast as the permanent teeth erupt and the maxilla grows in length (Fig. 8.4).

The Frontal Sinuses

The frontal sinuses are two irregular cavities which lie between the inner and outer tables of

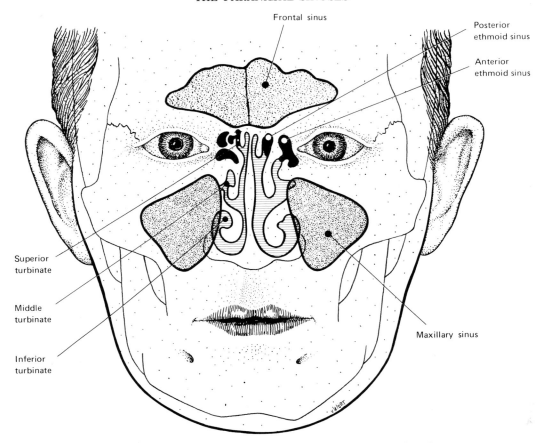

Figure 8.1. Anatomy of the paranasal sinuses—anteroposterior view.

the frontal bone above the supraorbital margins and the roof of the nose. Each of the frontal sinuses is composed of a central cavity which forms the main body of the sinus, plus vertical and horizontal extensions. The vertical extension spreads superiorly into the vertical plate of the frontal bone, whereas the horizontal extension spreads laterally and posteriorly into the orbital plate of the frontal bone. Along the roof and the wall of the sinus are numerous bony ridges from which arise membranous septae. The latter extend into the sinus cavity and may partially or completely subdivide the chamber. The two frontal sinuses are separated by a bony septum which usually deviates to one side of the middle line. They are seldom symmetrical and may vary considerably in size (Figs. 8.1 and 8.2).

The frontal sinus normally drains through the frontonasal duct and infundibulum into the hiatus semilunaris of the middle meatus. Occasionally, however, the duct may open directly at the side of the nasal septum (Fig. 8.3). The average measurements in the adult sinus are as follows: height, 3 cm; width, 2.5 cm; and depth, 2.5 cm.

The frontal sinuses are absent at birth. At or shortly after the 2nd year of life the frontal bone shows beginning pneumatization. This may be unilateral or bilateral. The ordinary site for the beginning of pneumatization is the area just above the superior ethmoid level. This progresses rather slowly through the 6th year. However, by the 7th and 8th years the chambers are fairly well developed, reaching the level of the superior orbital margins. Between the 8th and 10th years the frontal sinuses show a considerable increase in size, rising above the level of the superior orbital margins. At 12 years the sinuses have an adult configuration and gradually increase in size during the teenage period. Adult proportions are reached only after puberty (Fig. 8.4).

Congenital absence of one of the frontal si-

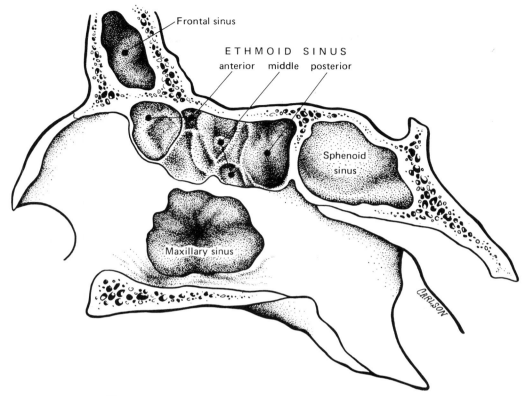

Figure 8.2. Anatomy of the paranasal sinuses — lateral view.

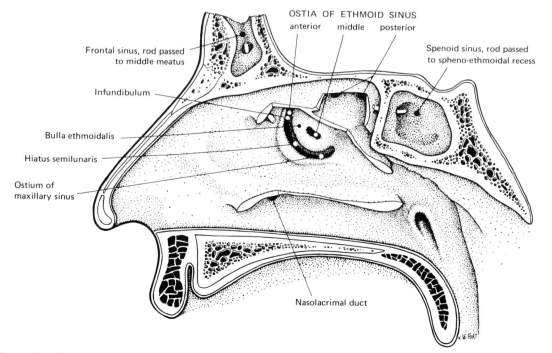

Figure 8.3. Lateral wall of the nasal cavity with removal of turbinates to show ostia of the paranasal sinuses.

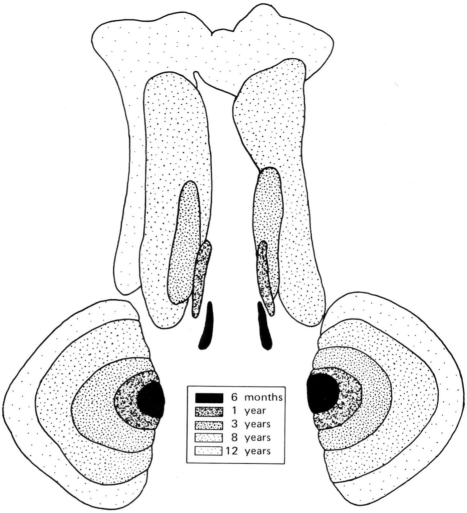

Figure 8.4. Line drawing showing development of the maxillary and frontal sinuses at various ages (after Caffey).

nuses is fairly common. Bilateral absence is rare, and occasionally is associated with persistence of the metopic suture.

The Ethmoid Sinuses

The ethmoid sinuses consist of two groups of three to 18 thin walled air cells situated in the ethmoidal labyrinth. The labyrinth lies between the upper part of the nasal cavity and the orbit and is separated from the anterior cranial fossa by the orbital plate or roof of the ethmoid bone. The medial wall of the ethmoid sinus is formed by a thin lamella to which are attached the superior and middle turbinates; the lateral wall is composed of the lamina papyracea of the ethmoid bone which separates the air cells from the orbit. Inferolaterally, the ethmoid articulates with the maxilla, forming the ethmoidomaxillary plate.

The ethmoid cells are frequently not confined to the ethmoidal labyrinth. They may expand into neighboring bones, forming frontoethmoidal, maxilloethmoidal, and sphenoethmoidal cells. They are also found anterior to the frontonasal duct (agger nasi cells).

Each sinus is divided into three cell groups, the anterior, middle, and posterior. The anterior group is located immediately above the hiatus semilunaris and drains into the infundibulum or the hiatus semilunaris of the middle meatus of the nasal cavity. The middle group is posterior to the hiatus semilunaris and opens into the middle meatus in the region of the bulla ethmoidalis. The posterior group, which lies just anterior to the sphernoid sinuses, drains into the superior meatus under cover of the superior turbinate (Figs. 8.1 to 8.3).

At birth, only one or two tiny ethmoid cells can be seen. During the first 2 months considerable growth of the cells occurs, but the sinuses are still too small to allow satisfactory demonstration of disease. By the end of the 10th month the ethmoid cells may be visualized in both the Waters and lateral views. At 1 year of age, a small orbital ethmoid cell and anterior ethmoid cells are visualized in their usual location. By the end of the 2nd year the cells are numerous and sufficiently large to permit the detection of gross changes. At 12 years, the ethmoid sinuses have an adult configuration.

The Sphenoid Sinuses

The sphenoid sinuses lie side by side within the body of the sphenoid bone separated by a bony septum. They vary in size and shape and are rarely symmetrical (Fig. 8.2). When exceptionally large, they may extend into the pterygoid plates or greater sphenoid wings, and may also spread into the dorsum sellae and clinoid processes. The sphenoid ostium opens into the sphenoethmoidal recess through the upper part of the anterior wall of the sinus (Fig. 8.3). In the adult the combined measurements of both sinuses are: height, 2.2 cm; transverse width, 2 cm; anteroposterior depth, 2.2 cm.

At birth, the sphenoid sinuses are usually absent. By the end of the 2nd year, beginning pneumatization of the sphenoid bone can be noted in the occasional case. From the 2nd to 6th year of age the pneumatization of the sphenoid sinus progresses rapidly. At age 12, the sinuses assume an adult configuration (Fig. 8.5). Complete absence of the sphenoid sinuses is rare, but the degree of pneumatization varies considerably.

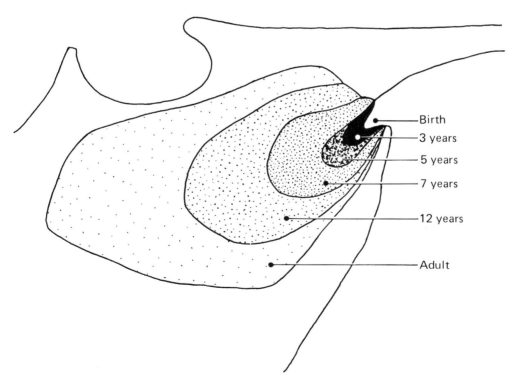

Birth
3 years
5 years
7 years
12 years
Adult

Figure 8.5. Line drawing showing development of the sphenoid sinus at various ages (after Caffey).

NERVE SUPPLY

Of the 12 cranial nerves only the fifth, or trigeminal, is closely related to the paranasal sinuses. A thorough knowledge of its anatomy is essential to an understanding of the clinical features of the diseases involving the sinuses (Fig. 8.6).

Ophthalmic Division Of The Trigeminal Nerve

The ophthalmic, or first, division of the tri-

geminal nerve leaves the anterosuperior part of the trigeminal (Gasserian) ganglion and enters the orbit through the superior orbital fissure. The ophthalmic nerve is sensory to the eyeball and cornea via the cicilary nerves, to the frontal, ethmoid, and sphenoid sinuses via the supraorbital and ethmoid nerves, and to the skin and conjunctival surface of the upper eyelid and the skin and mucous membrane of the external nose.

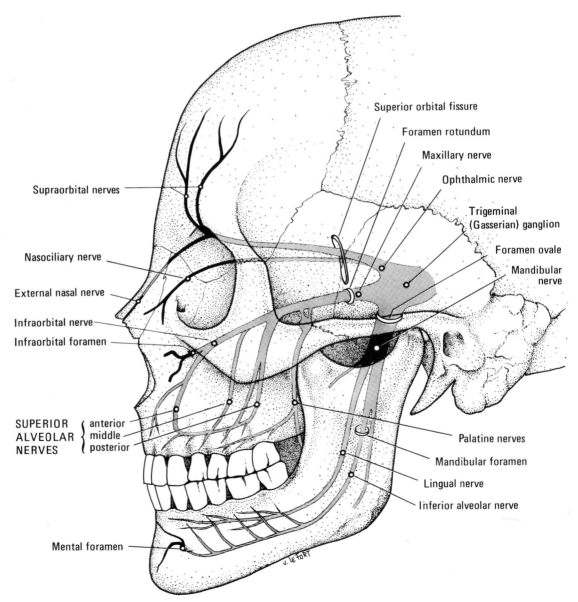

Figure 8.6. Course and branches of the trigeminal nerve.

Maxillary Division Of The Trigeminal Nerve

The maxillary, or second, division of the trigeminal nerve arises from the midportion of the trigeminal ganglion and leaves the cranial cavity through the foramen rotundum; it then crosses the pterygopalatine fossa and enters the orbit via the inferior orbital fissure. In the posterior part of the orbit the maxillary division becomes the infraorbital nerve, lying within the infraorbital groove and, continuing anteriorly, passing into the infraorbital canal. It emerges through the infraorbital foramen to supply sensory branches to the upper teeth and gum, the face, the lower eyelid, the skin of the side and vestibule of the nose, and the upper lip. Before entering the orbit branches are given off to the sphenopalatine ganglion, which in turn supplies sensory fibers to the mucoperiosteum of the nasal cavity, the palate, and roof of the pharynx and, via the sphenopalatine and posterior superior alveolar nerves, to the maxillary, ethmoid, and sphenoid sinuses.

Of the branches of the maxillary division, the infraorbital nerve is probably the most important in relation to malignant disease of the nasal cavity and paranasal sinuses. This branch traverses the infraorbital groove and canal in the roof of the maxillary sinus. Tumors of the antrum may directly invade the nerve, with subsequent extension of the disease along the perineural spaces into the middle cranial fossa through the foramen rotundum. Other branches of the maxillary division closely related to the maxillary sinus are the anterior, middle, and posterior divisions of the superior alveolar nerve. By posterior extension along the perineural spaces of any one of the branches of the superior alveolar nerve, malignant disease may gain access to the middle cranial fossa through the foramen rotundum.

Mandibular Division Of The Trigeminal Nerve

The mandibular, or third, division of the trigeminal nerve has sensory and motor roots.

Both leave the middle cranial fossa and enter the infratemporal fossa through the foramen ovale. After giving rise to a recurrent meningeal nerve and the nerve to the internal pterygoid muscle, the mandibular division divides into anterior and posterior branches. The anterior branch is primarily motor, supplying the external pterygoid, masseter, and temporal muscles; the one sensory component, the long buccal nerve, supplies a part of the mucous membrane of the oral cavity. The posterior branch is primarily sensory, giving rise to the lingual, inferior alveolar, and auriculotemporal nerves. From the inferior alveolar nerve a single motor component, the myohyoid nerve, arises.

Involvement of the mandibular division is a late sequela of the lateral extension of a malignant lesion of the paranasal sinuses. Extension into the lateral pharyngeal space may involve branches of this nerve, particularly the lingual nerve. Posterior extension along the perineural spaces of the branches of the mandibular division may result in metastasis to the middle cranial fossa through the foramen ovale.

LYMPHATIC DRAINAGE

Knowledge of the lymphatic drainage of the sphenoid and frontal sinuses is lacking.

The lymphatic drainage of the maxillary and ethmoid sinuses is partially understood. In general, the lymphatic channels pass through the natural ostia of the sinuses into the lymphatic trunks of the nasal cavity, and thence into the lateral retropharyngeal and internal jugular nodes. An exception is the lymphatic channels of the lateral aspect of the maxillary sinus, which drain into the more clinically accessible submaxillary nodes.

REFERENCES

Caffey, J.: *Pediatric X-Ray Diagnosis.* 3rd ed., Year Book Publishers, Chicago, Illinois, 1956.

Etter, L. E.: *Atlas of Roentgen Anatomy of the Skull.* Charles C Thomas, Springfield, Illinois, 1955.

Goss, C. M.: *Gray's Anatomy of Human Body.* 26th ed., p. 213–236, 977–999, 1198–1204. Lea and Febiger, Philadelphia, Pennsylvania, 1954.

MacComb, W. S. and Fletcher G. H.: *Cancer of the Head and Neck.* 1st ed., p. 329–356. The Williams & Wilkins Company, Baltimore, Maryland, 1967.

Reuviere, H.: *Anatomy of the Human Lymphatic System.* Translated by M. J. Tobias. Edward Brothers, Inc., Ann Arbor, Michigan, 1938.

Shapiro, H. H.: *Applied Anatomy of the Head and Neck.* 2nd ed., revised, J. B. Lippincott Company, Philadelphia, Pennsylvania, 1947.

Young, B. R.: *The Skull, Sinuses and Mastoids. A Handbook of Roentgen Diagnosis.* Year Book Publishers, Chicago, Illinois, 1951.

Chapter 9

Roentgen Technique

GENERAL CONSIDERATIONS

There is no universal agreement as to how many views constitute an adequate examination of the paranasal sinuses. Ordinarily all of the conventional projections should be made with the patient in a sitting position, with the incident beam in the horizontal plane. If illness necessitates placement of the patient in a prone position, a vertical beam is used. The advantages of examining the patient in the sitting position are 3-fold: (1) air-fluid levels within the sinus can be clearly demonstrated; (2) the patient feels more comfortable during the exami-

nation; and (3) the patient's head does not move or change in position with respiration. Immobility of the head is imperative in all projections. Dental plates, hairpins, and other radiopaque foreign bodies should be removed from the head. The x-ray tube should have a fine focal spot, and preferably a rotating anode. The film-target distance should be not less than 36 to 40 inches and a Potter-Bucky diaphragm and a small cone should be used to obtain maximal detail and contrast (Fig. 9.1*A*).

RADIOGRAPHIC PROJECTIONS

Conventional Projections

Although a number of projections have been described, it is not necessary to make use of all of them. At the University of Texas M. D. Anderson Hospital and Tumor Institute at Houston, the following projections are routinely employed.

Posteroanterior (Caldwell) View

The patient is placed in a sitting position with the nose and forehead against the cassette. The canthomeatal line is perpendicular to and the nasion at the center of the cassette. The central beam is projected posteroanteriorly at an angle of 15° caudally to the nasion (Fig. 9.1*B*).

The posteroanterior view is superior to other projections for demonstration of the frontal sinuses and of the superior ethmoid cells. The posterior aspect of the roof of the maxillary sinus and the floor of the apex of the orbit project above the inferior orbital rim (Fig. 9.1*C*).

Occipitomental (Waters) View

This view is obtained with the patient in a sitting position. The chin is placed against the

cassette with the head tilted backward. The tip of the nose should be approximately 1 to 1.5 cm away from the cassette. The junction of the nose and the upper lip is centered at the middle of the film and the canthomeatal line is at a 37° angle to the cassette. The central beam is directed perpendicularly to the cassette through the junction of the nose and the upper lip. When the patient is properly positioned, the petrous portion of the temporal bone is projected just below the floor of the maxillary sinus (Fig. 9.2*A*).

The occipitomental view is essential in the study of the paranasal sinuses. Developmental and pathological changes in the maxillary sinuses are best demonstrated in this projection (Fig. 9.2*B*), and valuable information may also be obtained concerning the facial bones, especially in the case of injuries. The frontal sinuses are well visualized and, with the mouth of the patient open, the posterior-inferior portion of the sphenoid sinus can be demonstrated.

Lateral View

The head of the patient is placed in a *true* lateral position. The canthomeatal line is parallel to the superior and inferior margins of the

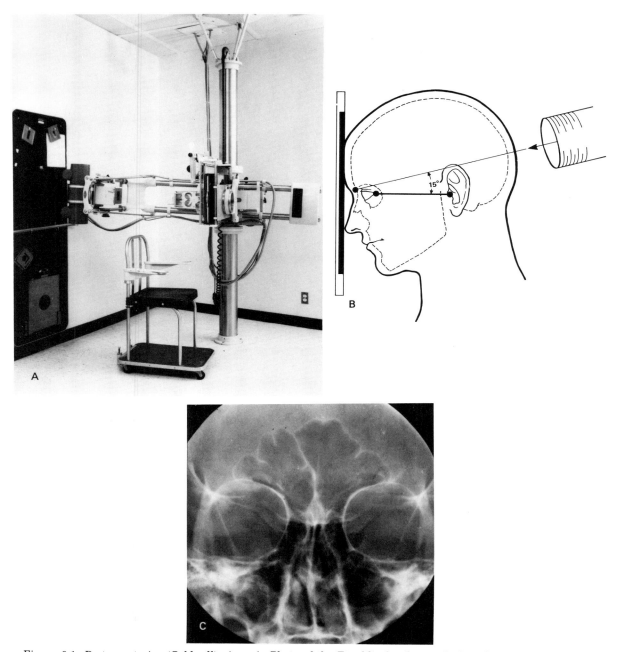

Figure 9.1. Posteroanterior (Caldwell) view. *A*, Photo of the Franklin head unit; *B*, line drawing to illustrate the radiographic position for the posteroanterior view; *C*, radiograph of the paranasal sinuses in the posteroanterior view.

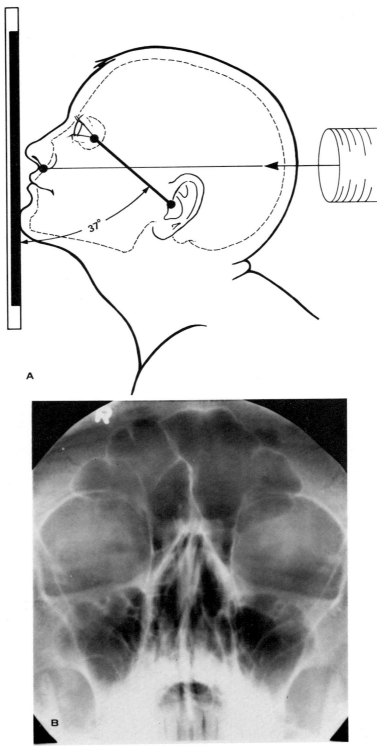

Figure 9.2. Occipitomental (Waters) view. *A*, Line drawing to show the radiographic position for the occiptomental view; *B*, radiograph of the paranasal sinuses in the occiptomental view.

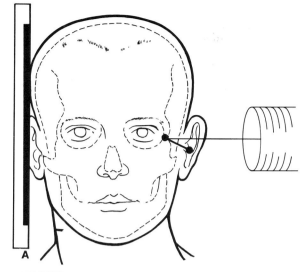

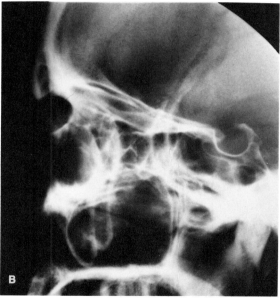

Figure 9.3. Lateral view. *A*, Line drawing to illustrate the radiographic position for the lateral view; *B*, radiograph of the paranasal sinuses in the lateral view.

cassette and the outer canthus is placed at the center. The central beam is directed to the center of the film at a 90° angle (Fig. 9.3*A*). The lateral view of the paranasal sinuses differs from that of the skull in that the central beam is directed more anteriorly in the former. In addition, the exposure of the paranasal sinuses requires approximately 8 to 10 kvp less than the lateral skull film.

A good lateral view provides considerable in-formation about all of the paranasal sinuses. It demonstrates the thickness of the anterior wall and the anteroposterior depth of the frontal sinuses, the anteroposterior depth of the sphenoid sinuses, and the structures of the ethmoidal cells. It also shows the thickness of the walls of the maxillary sinus and the relationship of the antral floor to the teeth. In addition, the nasopharynx and palate are well visualized (Fig. 9.3*B*).

Modified Submentovertical, or Base, View

This examination is carried out with the patient in a sitting position facing the x-ray tube. The head is extended until the vertex is resting against the cassette and the infraorbitomeatal line is parallel to it. The central beam is directed midway between the ascending rami of the mandible at a right angle to the center of the cassette (Fig. 9.4*A*).

Because of varying degrees of flexibility of the neck, the ideal position may not be obtainable. For practical purpose, at M. D. Anderson Hospital, the cassette holder is angled 30 to 35° toward the patient, and the head of the patient is extended until the infraorbitomeatal line is parallel to the cassette. The central beam is projected 1 1/2 inches anterior to the external auditory meatus and at a right angle to the cassette (Fig. 9.4*B*). Stereoscopic views are usually obtained with either projection.

The submentovertical view is the best projection for demonstration of the sphenoid sinuses and ethmoid cells, particularly the posterior group. It furnishes an excellent general view of the base of the skull and of many of the foramina which pass through it (Fig. 9.4*C*). The most important finding in this view is the demonstration of the triple lines of Etter, i.e., the posterolateral wall of the maxillary sinus, the orbital surface of the greater wing of the sphenoid, and the anterior border of the middle cranial fossa (Fig. 9.4*D*). It is important to note that in the detection of tumors of the maxillary sinus, reliance upon the conventional base (Hirtz) view can be a source of error because of superimposition of structures. The modified base view with stereoscopic technique gives a clear delineation of the posterolateral and the medial walls of the maxillary sinus.

Semiaxial View

The patient is placed in a sitting position with the nose and forehead against the cassette. The

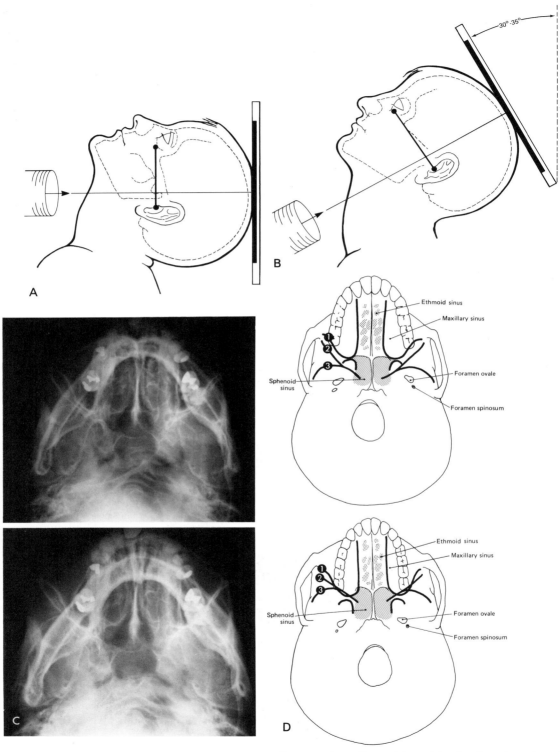

Figure 9.4. Submentovertical view. *A*, Line drawing to illustrate the radiographic position for the submentovertical view. *B*, Line drawing to illustrate the radiographic position for the modified submentovertical view. *C*, Radiographs of the base of the skull in the modified submentovertical view (stereoscopic views); *D*, Line drawings of triple lines (stereoscopic views): *1*, posterolateral wall of the maxillary sinus; *2*, orbital surface of the greater wing of the sphenoid bone; *3*, anterior border of the middle cranial fossa.

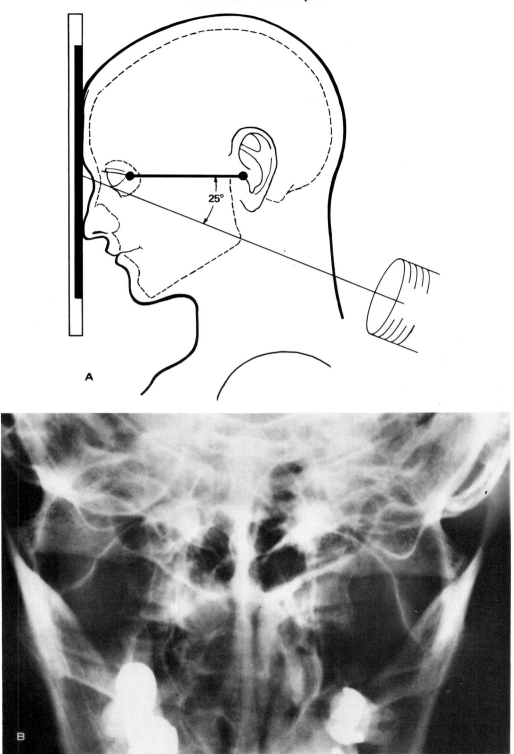

Figure 9.5. Semiaxial view. *A*, Line drawing to illustrate the radiographic position for the semiaxial view; *B*, radiograph of the posterosuperior walls of the maxillary sinuses in the semiaxial view.

canthomeatal line is perpendicular to the film. The x-ray beam is centered at the nasion and the tube is angulated approximately 25° cephalad (Fig. 9.5A). This view clearly demonstrates the posterosuperior wall of the maxillary sinus (Fig. 9.5B) and is useful in detecting bone destruction or facial fractures with extension to the maxillary sinus.

Other Views

Oblique (Rhese) View. The patient is placed in a sitting position facing the cassette. The head is rotated 53° from the frontal position toward the side under study. The zygoma, the superior orbital ridge, and the lateral surface of the nose of the side under study are all in contact with the cassette, and the canthomeatal line is perpendicular to it. The central beam is angled caudally 16°, passing through the center of the orbit in the posteroanterior plane (Fig. 9.6A).

This view is occasionally used for further study of the ethmoid cells. Unfortunately, superimposition of the cells often prevents clear visualization of the individual chambers. This view also reveals the optic foramen in profile (Fig. 9.6B).

Overangulated Submentovertical View. The position of the patient is the same as in the modified submentovertical view except that the head is extended sufficiently to project the mandible anteriorly to the frontal sinus and the cassette holder is angled 30 to 35° away from the patient. The central beam runs tangentially to the posterior wall of the frontal sinuses and 7° posteriorly to a point midway between the angles of the mandibles (Fig. 9.7A).

This view shows the anterior and posterior walls of the frontal sinuses to good advantage. The thickness of the walls and the depth of the sinuses are clearly visualized (Fig. 9.7B).

Tomography

Tomographic Projections

Anteroposterior Tomography. The patient is placed in the supine position and the head is immobilized. The canthomeatal line is perpendicular to the radiographic table. A measurement is made from the anterior skin surface of the maxillary sinus to the table top and the x-ray beam is centered at the nasion (Fig. 9.8, A and B). Anteroposterior sections are obtained at 0.5-cm intervals to a depth of 8 cm below the skin surface. Additional sections are taken if indicated.

Basal Tomography. The patient is placed in the supine position with the shoulder and the trunk elevated by a wooden platform or a Styrofoam positioning block. The head is extended until the vertex is resting upon the radiographic table and the canthomeatal line is parallel to the table top. The central beam is directed perpendicularly midway between the ascending rami of the mandible (Fig. 9.8C).

The base line tomographic section is made at the level of the external auditory meatus. If the patient is correctly positioned, the section through the ear will demonstrate the foramen ovale on each side. Multiple sections are then taken at 0.5-cm intervals from the alveolar recess of the maxillary sinus to the frontal sinus.

Lateral Tomography. The patient is placed in the lateral recumbent position and the head is supported in a true lateral projection by clamps and/or Styrofoam block. The canthomeatal line is parallel to the radiographic table. The distance from the midline of the face to the table top is measured and the central beam is centered at the nasion. The first tomographic section is made through the midline of the face. Additional sections are obtained at 1-cm intervals on each side of the midline to a depth of 4 cm.

Indications for Tomography

Tomography should be considered a supplemental technique which does not supplant the standard projections. In general it is not indicated in the diagnosis of inflammatory disease, but when pathological changes are obscured by adjacent dense shadows, tomography affords better visualization than conventional methods. Tomograms are also of value in the delineation of facial fractures which involve the paranasal sinuses and in the evaluation of postoperative changes in the paranasal sinuses. The presence of foreign bodies, especially in the maxillary sinus, may be easily detected by tomograms. Of particular importance is the establishment of a three-dimensional concept in the mapping of tumors of the paranasal sinuses; the site of origin, route of spread, and the extension of lesions can be clearly delineated. Bone destruction is more readily appreciated in the absence of superimposed densities.

Opaque Contrast Displacement Method

As advocated by Proetz, instillation of an

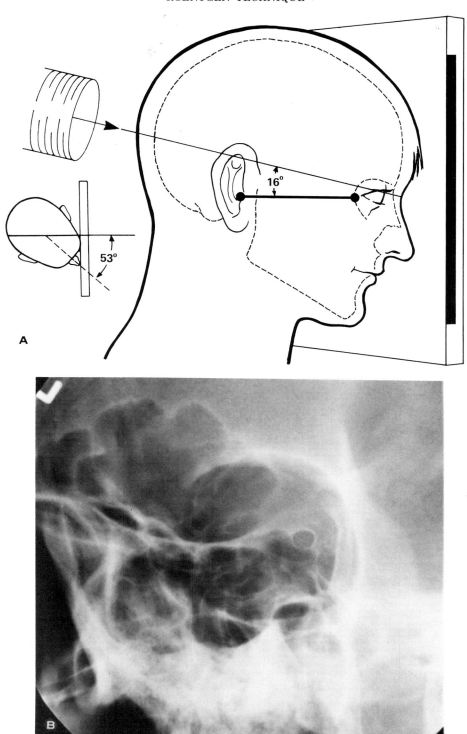

Figure 9.6. Oblique view. *A*, Line drawing to illustrate the radiographic position for the oblique view; *B*, radiograph of the ethmoid cells in the oblique view.

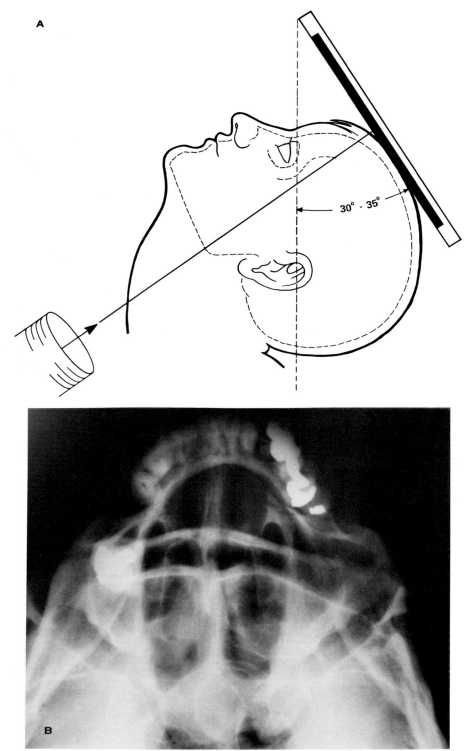

Figure 9.7. Overangulated submentovertical (Welin) view. *A*, Line drawing to illustrate the radiographic position in the overangulated submentovertical view; *B*, radiograph of the walls of the frontal sinuses in the overangulated submentovertical view.

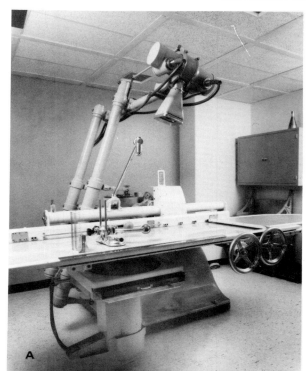

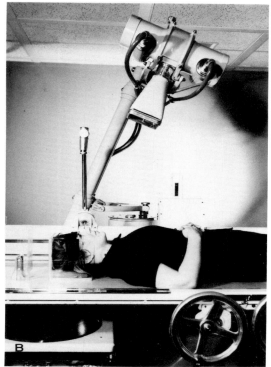

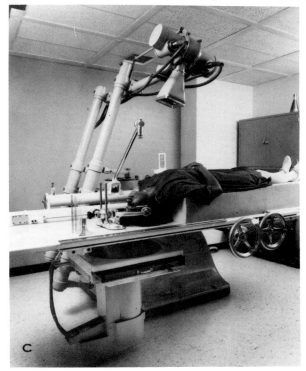

Figure 9.8. Tomographic machine (Massiot-Phillip's Polytome) and position of the patient. *A*, Photo of the tomographic machine; *B*, position of the patient for anterior-posterior tomogram of the paranasal sinuses; *C*, position for basal tomogram of the paranasal sinuses.

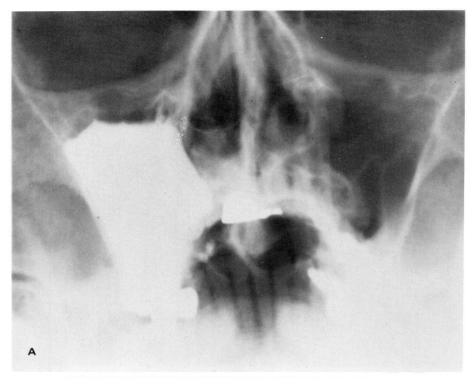

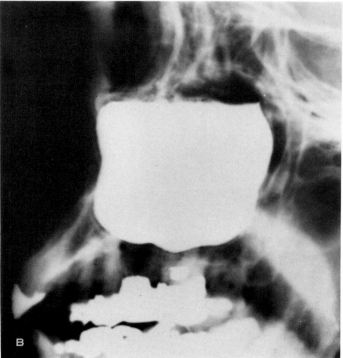

Figure 9.9. Normal opaque contrast examination of the maxillary sinus. *A*, Waters view: there is trapping of air in the superior aspect of the sinus; *B*, lateral view: trapping of air is again noted.

opaque contrast medium can evaluate the patency of the ostia and the condition of the mucosa of the paranasal sinuses (Fig. 9.9). However, in our opinion, the additional information obtained by this method is insufficient to warrant its routine use. Furthermore, such normal findings as a root of a tooth which bulges into the floor of the maxillary sinus, an aberrant ethmoid cell encroaching upon the opacified sinus, or the trapping of a small amount of air within the antrum may all simulate filling defects. Additionally, occlusion of the ostia of the paranasal sinuses is usually associated with pathological changes which are demonstrable by conventional films or tomography. These changes may be concealed by the superimposed density of the contrast medium. Nonfilling of the paranasal sinuses as a result of obstruction of the ostia by swollen mucosa or hypertrophy of a turbinate may also be misinterpreted when detected by the displacement method.

Although the clinical value of the displacement method has been disputed, the technique may be of value in determining the size and shape of a sinus cavity which cannot be appreciated by conventional examination because of excessive opacity. It is also possible to evaluate the emptying time of the paranasal sinuses by follow-up studies 48 to 72 hr after instillation of the contrast medium. Normally the sinuses empty themselves within 72 hr. It is claimed that retention of contrast media in the sinuses after this length of time is proof of pathology. In addition, the displacement method may be helpful in revealing the origin and size of aberrant sinus cells.

REFERENCES

Dodd, G. D., Collins, L. C., Egan, R. L., and Herrera, J. R.: The systematic use of tomography in the diagnosis of carcinoma of the paranasal sinuses. Radiology, *72:* 379, 1959.

El Gammal, T., and Keates, T. E.: The lamina dura (L.D.) of the maxillary antrum: its value in roentgen diagnosis. Am. J. Roentgenol., *105:* 830, 1961.

Etter, L. E.: Opacification studies of normal and abnormal paranasal sinuses. Am. J. Roentgenol., *89:* 1137, 1963.

Lloyd, G. A.: Axial tomography of the orbit and paranasal sinuses. Br. J. Radiol., *44:* 373, 1971.

MacMilliam, A. S.: Technique of paranasal sinus radiography. Semin. Roentgenol., *3:* 115, 1968.

Merill, V.: *Atlas of Roentgenographic Positions.* 2nd ed., C. V. Mosby, St. Louis, Missouri, 1959.

Potter, G. D.: *Sectional Anatomy and Tomography of the Head.* Grune & Stratton, New York, 1971.

Potter, G. D., and Seaman, W. B.: Polytomography of tumors of the base of the skull and paranasal sinuses. In *Proceedings of the International Workshop on Cancer of the Head and Neck,* edited by J. Conley, p. 83. Butterworth, Washington, 1967.

Proetz, A. W.: *The Displacement Method of Sinus Diagnosis and Treatment: A Practical Guide to the Use of Radiopaques in the Nasal Sinuses.* Annals Publishing Co., St. Louis, Missouri, 1931.

Samuel, E.: The radiology of the paranasal sinuses. In *Textbook of X-Ray Diagnosis by British Authors,* 4th ed., K. K. Lewis and Company, Ltd., London, 1961.

Shapire, R., and Janzen, A. H.: *The Normal Skull.* Paul B. Hoeber, Inc., Medical Division of Harper & Brothers, New York, New York, 1960.

Welin, C. S. H.: Paranasal sinuses. In *Roentgen Diagnosis.* 2nd American ed., edited by L. G. Rigler, p. 517–537. Grune and Stratton, New York, 1969.

Young, B. R.: *The Skull, Sinuses and Mastoids. A Handbook of Roentgen Diagnosis.* Year Book Publishers, Chicago, Illinois, 1951.

Chapter 10

Roentgen Anatomy

Because of the complexity and superposition of the paranasal sinuses and surrounding structures, meticulous roentgen examinations are often required. A thorough knowledge of roentgen anatomy is essential for interpretive purposes.

MAXILLARY SINUSES

The normal roentgen anatomy of the maxillary sinuses as seen in the standard projections is shown in Figures 10.1, *B* to *E*, 10.2, *A*, to *D*, and 10.3 *C*.

In general these sinuses are subject to fewer variations than the other paranasal chambers. The maxillary sinuses tend to be equal and symmetrical, although occasionally one sinus is much smaller than the other. Not infrequently, the maxillary sinus may be partially or completely compartmentalized by bony septa which either wall off a section of the sinus or form an incomplete partition (Fig. 10.4). Rarely, a large agger ethmoid cell encroaches upon the posterior aspect of the maxillary sinus, decreasing its size and producing a so-called double sinus. Very rarely, a preantral cell or a separate compartment is seen.

SPHENOID SINUSES

The normal roentgen anatomy of the sphenoid sinuses is shown in Figures 10.1, *C* and *D*, 10.2, *E* and *F*, and 10.3*B*.

Variations in the size and shape of the sphenoid sinuses are frequent. Rarely, the two sinuses are equal in size; usually one is much larger than the other, particularly in the anteroposterior dimension. Occasionally the size of the sphenoid sinuses is reduced by expansion of the posterior ethmoid cells into the sphenoid bone. Not infrequently, sphenoid pneumatization is so extensive that the dorsum sellae, the posterior clinoids, the clivus, the greater or lesser wings of the sphenoid, and the pterygoid plates are aerated. Compartmentation, either complete or incomplete, occurs when bony septa are formed in the sphenoid cavities.

Inequalities in the anteroposterior dimensions of sphenoid sinuses are readily demonstrated in the base and lateral views, whereas the posteroanterior view is useful to show variations in the height and width of the cavities. The conventional views may reveal lateral extension of the sphenoid sinuses into the pterygoid plates and the greater wings of the sphenoid; however, tomography is of value for confirmation of the origin of the defect and delineation of its extent (Fig. 10.5).

FRONTAL SINUSES

The normal roentgen anatomy of the frontal sinuses is shown in Figures (10.1, A, *B*, *C* and *G*, and 10.2*B*.

Each frontal sinus consists of a central cavity with varying extensions into the vertical and orbital plates. Either extension may develop at the expense of the other. Because the resorption of the diploë in the vertical portion is seldom regular enough to result in perfect symmetry, there is considerable variation in the size,

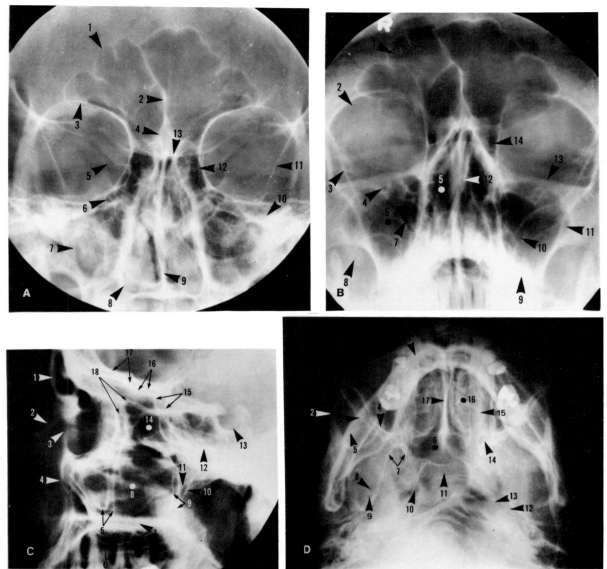

Figure 10.1. Normal conventional projections. *A*, Posteroanterior view: *1*, frontal sinus; *2*, bony septum; *3*, superior orbital margin; *4*, crista galli; *5*, superior orbital fissure; *6*, foramen rotundum; *7*, maxillary sinus; *8*, nasal fossa; *9*, nasal septum; *10*, petrous ridge of temporal bone; *11*, oblique orbital line; *12*, ethmoid sinus; *13*, limbus sphenoidalis. *B*, Waters view: *1*, frontal sinus; *2*, orbit; *3*, oblique orbital line; *4*, infraorbital foramen; *5*, nasal fossa; *6*, maxillary sinus; *7*, superior orbital fissure; *8*, intratemporal fossa; *9*, petrous ridge of temporal bone; *10*, foramen rotundum; *11*, zygomatic recess of the maxillary sinus; *12*, nasal septum; *13*, inferior orbital rim; *14*, ethmoid sinus. *C*, Lateral view: *1*, frontal sinus; *2*, nasal bone; *3*, orbital cavity; *4*, anterior wall of maxillary sinus; *5*, zygomatic process of maxilla; *6*, floor of maxillary sinus; *7*, hard palate; *8*, maxillary sinus; *9*, posterior wall of maxillary sinus; *10*, pterygoid plate; *11*, pterygomaxillary fissure; *12*, sphenoid sinus; *13*, sella turcica; *14*, ethmoid cells; *15*, anterior border of middle cranial fossa; *16*, orbital surface of orbital plate; *17*, cerebral surface of orbital plate; *18*, cribriform plate. *D*, Modified submentovertical view: *1*, mandible; *2*, zygomatic arch; *3*, orbital surface of the greater wing of the orbit; *4*, posterolateral wall of the maxillary sinus; *5*, anterior border of the middle cranial fossa; *6*, sphenoid sinus; *7*, pterygoid plates; *8*, foramen ovale; *9*, foramen spinosum; *10*, foramen lacerum; *11*, clivus; *12*, eustachian tube; *13*, carotid canal; *14*, palatine foramina; *15*, medial wall of maxillary sinus; *16*, ethmoid sinus; *17*, nasal septum. *E*, Semiaxial view: *1*, floor of the middle cranial fossa; *2*, maxillary sinus; *3*, posterosuperior wall of the maxillary sinus; *4*, inferior orbital fissure; *5*, infratemporal ridge of the greater wing of the sphenoid. *F*, Oblique view: *1*, frontal sinus; *2*, floor of the anterior cranial fossa; *3*, ethmoid cells; *4*, superior orbital fissure; *5*, optic foramen; *6*, orbital plate. *G*, Overangulated base view: *1*, maxillary teeth; *2*, anterior wall of the frontal sinus; *3*, posterior wall of the frontal sinus.

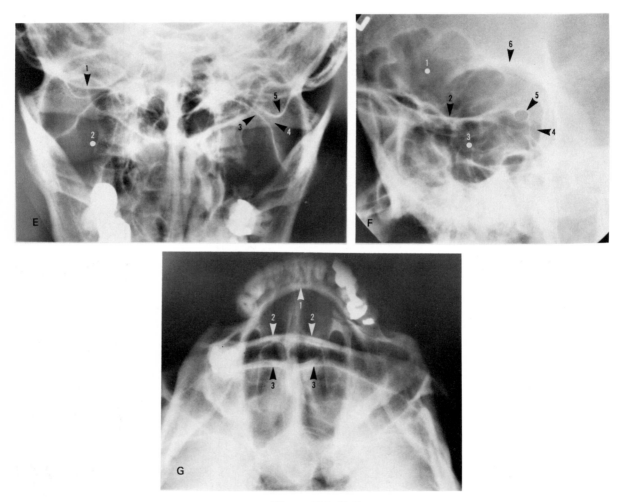

Figure 10.1. E–G.

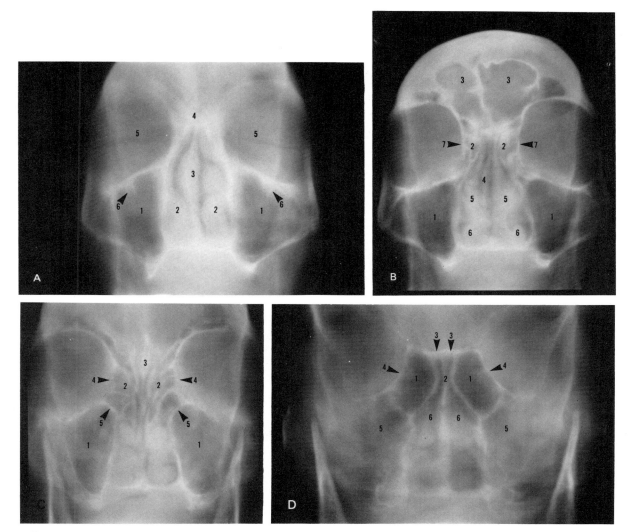

Figure 10.2. Anteroposterior tomograms of the paranasal sinuses (Polytome). *A*, Anterior section (19.5-cm cut): *1*, maxillary sinus; *2*, inferior turbinate; *3*, nasal septum; *4*, nasal bones; *5*, orbit; *6*, infraorbital canal. *B*, Anterior section (18.5-cm cut): *1*, maxillary sinus; *2*, ethmoid sinus (anterior ethmoid cells); *3*, frontal sinus; *4*, nasal septum; *5*, middle turbinate; *6*, inferior turbinate; *7*, lacrimal bone. *C*, Middle section (17.5-cm cut): *1*, maxillary sinus; *2*, ethmoid sinus (middle ethmoid cells); *3*, crista galli; *4*, lamina papyracea; *5*, ethmoid-maxillary plate. *D*, Middle section (16.5-cm cut): *1*, ethmoid sinus (posterior ethmoid cells); *2*, perpendicular plate of ethmoid; *3*, cribriform plate; *4*, lamina papyracea; *5*, maxillary sinus, posterior wall; *6*, nasal fossa. *E*, Posterior section (15.0-cm cut): *1*, sphenoid sinus; *2*, foramen rotundum; *3*, pterygopalatine canal; *4*, sphenoid rostrum and ala of vomer; *5*, nasopharynx; *6*, pterygoid plates; *7*, lesser wing of the sphenoid; *8*, floor of the sella turcica. *F*, Posterior section (14.5-cm cut): *1*, sphenoid sinus; *2*, clinoid process; *3*, middle cranial fossa; *4*, carotid sulcus; *5*, greater wing of the sphenoid; *6*, sphenoid rostrum and ala of vomer; *7*, nasopharynx; *8*, pterygoid fossa.

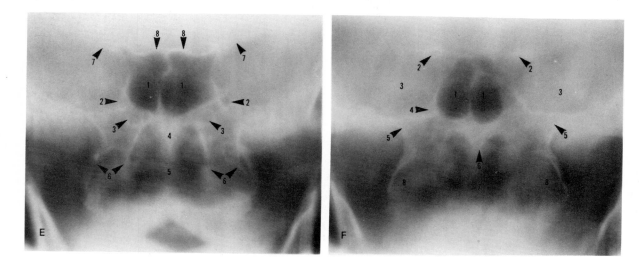

Figure 10.2. E–F.

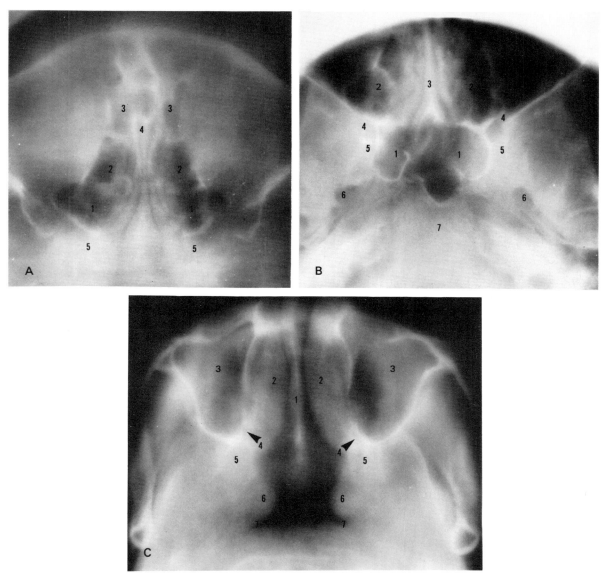

Figure 10.3. Base tomograms of the base of the skull (Polytome). *A*, Superior section: *1*, posterior ethmoid cells; *2*, middle ethmoid cells; *3*, anterior ethmoid cells; *4*, crista galli; *5*, pterygopalatine fossa. *B*, Middle section: *1*, sphenoid sinus; *2*, posterior ethmoid cells; *3*, vomer; *4*, inferior orbital fissure; *5*, superior orbital fissure; *6*, foramen ovale; *7*, clivus. *C*, Inferior section: *1*, vomer; *2*, inferior turbinate; *3*, maxillary sinus; *4*, pterygopalatine canal; *5*, pterygoid fossa; *6*, torus tubarius; *7*, fossa of Rosenmüller.

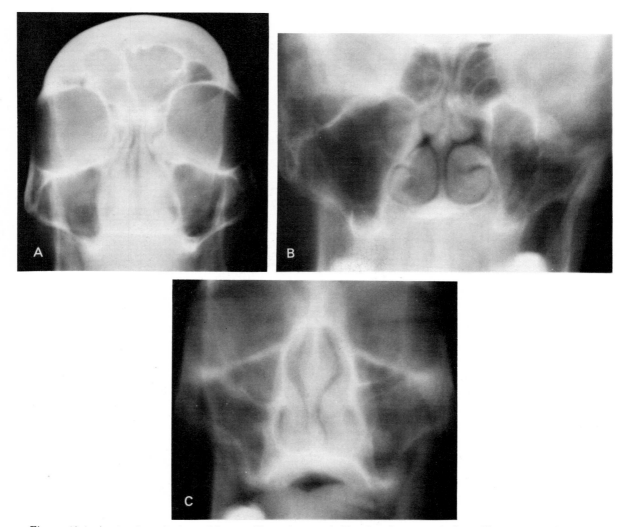

Figure 10.4. Anatomic variations of the maxillary sinuses. *A*, Equal and symmetrical maxillary sinuses; *B*, small left maxillary sinus; *C*, complete partition of both maxillary sinuses.

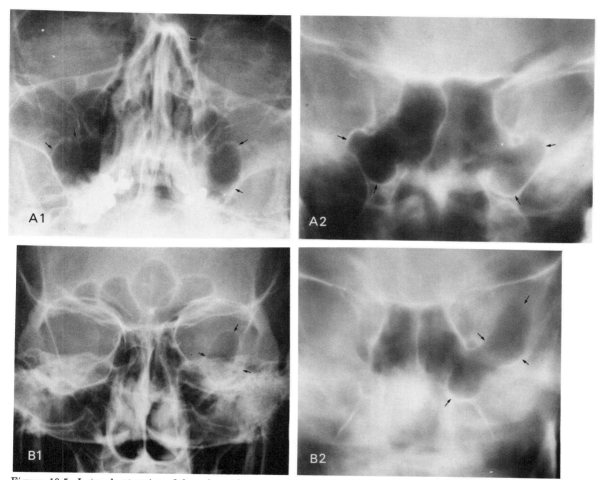

Figure 10.5. Lateral extension of the sphenoid sinuses. *A*, Lateral extension shown in the maxillary sinuses. *1*, Waters view: a well defined radiolucent area in the maxillary sinus on each side; *2*, posteroanterior tomogram: lateral extension of sphenoid sinuses is well demonstrated on each side. *B*, Lateral extension shown in the left orbit. *1*, posteroanterior view: an oval-shaped sharply defined radiolucency in the left orbit; *2*, posteroanterior tomogram: lateral extension of the left sphenoid sinus is demonstrated to good advantage.

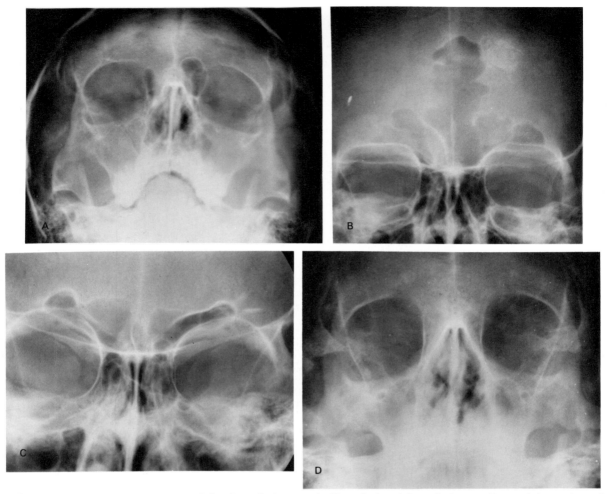

Figure 10.6. Anatomic variations of the frontal sinuses. *A*, Hypoplasia of frontal sinuses; *B*, serpiginous frontal sinuses; *C*, hypoplasia of vertical extension of frontal sinuses; *D*, agenesis of frontal sinuses.

shape, and extent of the frontal sinuses. A common variation is the formation of two cavities of unequal size divided by a bony septum located on the side of the small compartment. Occasionally, one sinus is so poorly developed that it approximates the size of the superior ethmoid cell while the other sinus may be of more or less normal size. Sometimes both frontal sinuses are poorly developed and remain at the ethmoidal level (Fig. 10.6*A*). Another normal variation is a marked enlargement of one chamber as a result of hyperpneumatization of the vertical and/or horizontal plates of the frontal bone. It is unusual to visualize lateral extensions of the sinuses beyond the lateral margins of the orbits. However, the frontal sinuses frequently extend superiorly above their usual level in the direction of the coronal suture (Fig. 10.6*B*).

Pneumatization of the horizontal plate of the frontal bone occurs in varying degrees. When the vertical plate is poorly pneumatized, aeration of the horizontal plate is apt to be manifest as slit-like or oval radiolucent zones at the level of the superior orbital margins (Fig. 10.6*C*). When the vertical plate is well pneumatized, this variation is not striking in the Waters or Caldwell views. For this reason, lateral views are important for determination of the extent of pneumatization of the horizontal plate.

Unilateral or bilateral agenesis is rare (Fig. 10.6*D*).

Extensive enlargement of the frontal sinuses often occurs as one of the manifestations of the enlargement of the skull and facial bones in acromegaly (Fig. 10.7). Occasionally, there may be a congenital or an acquired defect in the

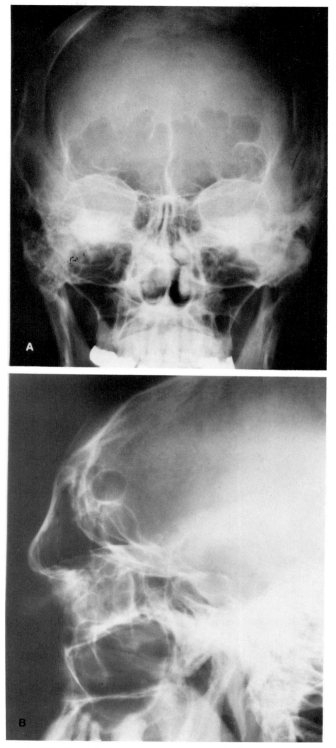

Figure 10.7. Enlargement of the frontal sinuses in acromegaly. *A*, Posteroanterior view; *B*, lateral view.

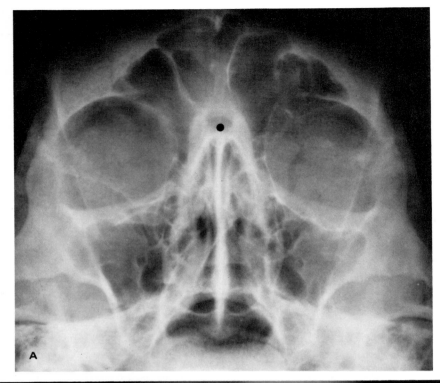

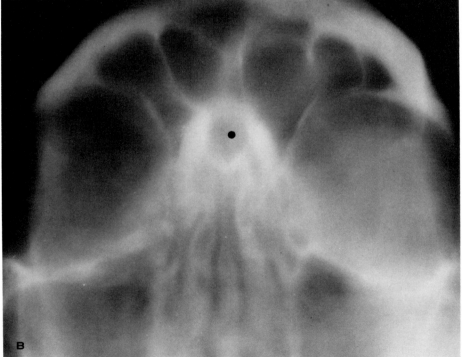

Figure 10.8. Crista galli cell. *A*, Waters view: agger ethmoid cell in the crista galli (*dot*); *B*, anteroposterior tomogram: the crista galli cell is shown to good advantage (*dot*).

septum between the two sinuses. Knowledge of this possibility has a practical implication in the interpretation of roentgenograms because unilateral disease may cause protrusion of the mucous membrane through the defect into the contralateral uninvolved sinus.

ETHMOID SINUSES

The normal roentgen anatomy of the ethmoid sinuses is shown in Figures 10.1, *A*, *B*, *C*, *D*, and *F*, 10.2, *B*, to *D*, and 10.3, *A* and *B*.

In each labyrinth the number of the ethmoid cells varies from three to 18. The fewer the number of the cells, the larger the individual cells.

The agger ethmoid cells are often found in the base of the frontal sinuses, the orbital plate of the frontal bone, the sphenoid, the maxillary, the lacrimal, the palatine, or the nasal bones. The most familiar cell in this group is the agger nasi cell, which is located in front of the nasofrontal duct and behind the frontal process of the maxilla. Occasionally, a larger maxillary agger cell may be mistaken for a supernumerary maxillary sinus. Not uncommonly, an ethmoid cell, the so-called crista galli cell (Fig. 10.8), is found in the crista galli.

The lateral view is useful in demonstrating the extent of ethmoid pneumatization, whereas the Caldwell and Waters views are of value for lateralizing disease in the ethmoids. The superior ethmoid cells are often well visualized in the Caldwell view, whereas the Waters view may reveal some of the posterior cells. Under certain conditions the oblique views may prove useful, but their value is limited because of the superimposed shadows of the various structures around the ethmoid cells.

REFERENCES

Etter, L. E.: *Atlas of Roentgen Anatomy*. 2nd printing. Charles C Thomas, Springfield, Illinois, 1964.

Law, F. M.: *Nasal Accessory Sinuses. Annals of Roentgenology XV*. Paul B. Hoeber, Inc. New York, 1933.

Shapire, R., and Janzen, A. H.: *The Normal Skull*. Paul B. Hoeber, Inc., Medical Division of Harper & Brothers, New York, 1960.

Young, B. R.: *The Skull, Sinuses and Mastoids. A Handbook of Roentgen Diagnosis*. Year Book Publishers, Chicago, Illinois, 1951.

Chapter 11

Roentgen Interpretation

A thorough knowledge of the normal roentgen anatomy is essential to the proper interpretation of both conventional and tomographic examinations of the paranasal sinuses. Careful examination of all available roentgenograms is also of importance; an apparent abnormality may be significant only when demonstrated in several different projections. Additionally, detailed information concerning the history and physical findings may be required to differentiate accurately between certain diseases which are similar in their gross appearance.

In examining roentgenograms of the paranasal sinuses one must observe the character of the walls as well as the appearance of the chamber. Sinuses are normally radiolucent owing to the air content, but their densities may vary according to the thickness of the walls. In a normal sinus the mucous membrane is not clearly visualized, but the bony walls are sharply defined by the adjacent air. Depending upon the degree of pneumatization, the wall thickness may vary from person to person. In the normal individual a small but thick walled sinus is more opaque than the large, well pneumatized variety. Different parts of the walls of the same sinus may also vary in thickness. In general, however, regardless of variations in thickness, the internal contours of a sinus in a normal adult are well defined and easily evaluated. In children the contours are frequently vague, as a result of active growth. With cessation of growth the sinus contours become sharply defined.

DIAGNOSTIC CRITERIA

The roentgen diagnosis of diseases of the paranasal sinuses is based on demonstration of (1) thickening of the lining mucous membrane, (2) opacity of the sinus cavity, (3) presence of a soft tissue mass in the sinus, (4) changes in the bony walls, and (5) invasion of adjacent structures. The nature and extent of the involvement depend on the type, severity, and duration of the disease.

Thickening of the Lining Mucous Membrane

The initial roentgen finding in inflammatory disease is usually an increase in the thickness of the lining mucous membrane; this varies considerably depending upon the type of infection. Thickening of the lining mucous membrane may be the only manifestation of an acute inflammatory process, but fluid may also be present. In chronic inflammatory or allergic sinusitis, the mucosa may become markedly hypertrophic and thickened, with complete opacification of the sinus cavity.

Opacification of the Sinus

Opacification of a sinus may be caused by changes in the lining mucous membrane or in the bony walls. It may be also produced by fluid accumulation or soft tissue mass in the sinus. The radiological demonstration of a cloudy sinus is of limited value in the differential diagnosis between inflammatory and malignant disease. Persistent opacity of one sinus, or unilateral opacity of several sinus chambers, should suggest the possibility of a malignant process, however.

Presence of a Soft Tissue Mass in the Sinus

Soft tissue masses arising in the paranasal sinuses may be of various kinds. Cysts and polyps are smooth in outline and are seen on the roof, the lateral wall, or the floor of the sinuses. Tumors and infectious granulomatous lesions may form soft tissue masses with smooth or irregular contours. A soft tissue mass arising

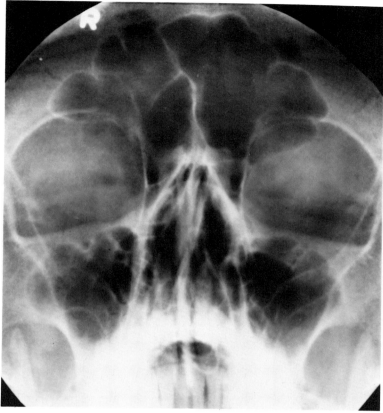

Figure 11.1. Waters view showing normal mucoperiosteal lines of the maxillary sinuses. The lateral walls are somewhat denser than the remainder of the sinus boundaries due to their obliquity.

from a sinus and extending beyond the bony confines is highly suggestive of a malignant process.

Changes in the Bony Walls

The changes in the bony walls of the paranasal sinuses are indicated by disappearance of the mucoperiosteal line and demineralization, thinning and expansion, sclerosis, or destruction of the bone. In a normal sinus, the mucoperiosteal line is seen as a sharp, thin, dense white line at the periphery of the sinus (Fig. 11.1). This line may disappear with inflammatory or neoplastic diseases. Demineralization of the bony walls is usually associated with acute infections and is most frequently seen in the ethmoid sinuses. When the sinuses are well pneumatized one can readily appreciate the loss of sharp definition of the cell septa owing to demineralization. This is the earliest sign of ethmoid disease. Later changes include disappearance or thickening of cell septa and opacification of the cells.

Thinning and expansion of the bony walls of a sinus is usually caused by pressure from an expanding soft tissue mass within the sinus. Generally associated with benign expansible tumors, it is occasionally found with adenocarcinomas and soft tissue sarcomas of the sinuses.

Sclerosis of the bony walls of a sinus may result from osteoblastic reaction to tumor invasion and/or from a chronic inflammatory process of the sinus. In the maxillary sinus, bony sclerosis is often attributable to osteitis caused by an inflammatory process rather than to osteoblastic neoplastic reaction. This change is best seen in the posterolateral walls of the sinus (Fig. 11.2). In the absence of specific evidence of an inflammatory process, localized sclerosis of the wall of the maxillary sinus should suggest the possibility of a malignant lesion (Fig. 11.3).

Destruction of the bony walls of the sinus indicates the invasive nature of a lesion and is usually reliable evidence of a malignant tumor. However, benign tumors and granulomatous lesions may, on occasion, erode and destroy and bony walls of the paranasal sinuses.

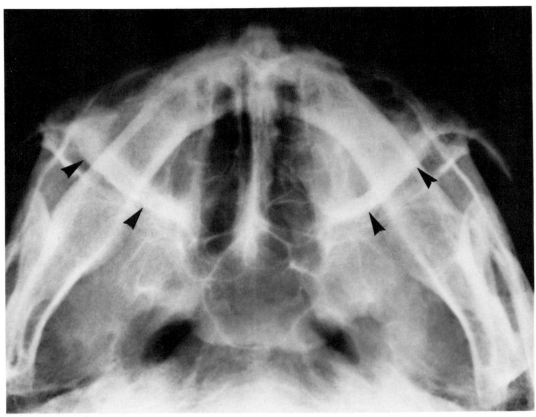

Figure 11.2. Sclerosis of posterolateral walls of the maxillary sinuses. Submentovertical view showing bony sclerosis of the posterolateral walls of the maxillary sinuses resulting from chronic sinusitis.

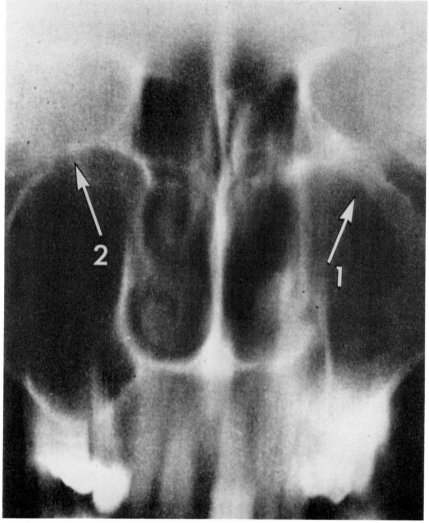

Figure 11.3. Localized sclerosis of the wall of the maxillary sinus. Posteroanterior tomogram showing localized sclerosis of the left maxillary sinus (*1*) secondary to squamous cell carcinoma. There is a normal-appearing bony wall of the right maxillary sinus (*2*).

Invasion of the Surrounding Structures

Invasion of adjacent structures is commonly seen in the advanced stage of malignant tumors of the paranasal sinuses. Occasionally, benign, slowly expanding lesions may also extend to surrounding structures, with erosion and destruction of bone.

PRECAUTIONS IN INTERPRETATION

It is worthy of note that comparison of one sinus with its fellow of the opposite side is often misleading and a frequent cause of error. The sinuses may be unequally developed, unequally opaque, or unequally diseased. The lining mucous membrane in one sinus may show marked thickening attributable to a chronic inflammatory process, which may divert the examiner's attention from the much less striking haziness attributable to an acute infection in the opposite chamber.

In the maxillary sinus, in the Waters view,

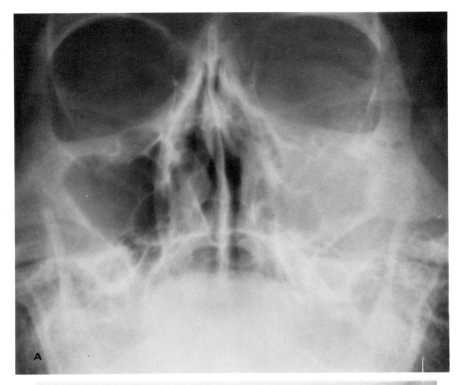

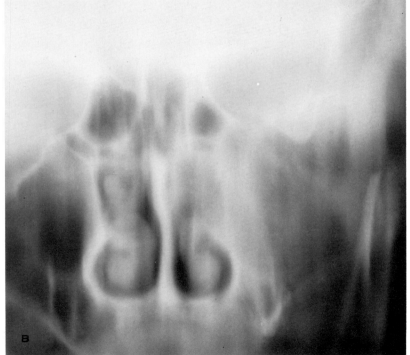

Figure 11.4. Intact medial wall interpreted as bony destruction. *A*, Waters view: opacity of the left maxillary sinus. The presence or absence of destruction of the medial wall cannot be determined. *B*, Posteroanterior tomogram: apparent destruction of upper portion of the medial wall of the left maxillary sinus. At surgery, the medial wall was intact. The error in interpretation was due to the thinness of the bony wall, an unusual obliquity, and thickening of the lining mucous membrane.

the soft tissue shadow of the upper lip is frequently superimposed on the floor of the sinus and may be mistaken for mucous membrane thickening. This can be avoided by following the shadow outside the maxillary sinus. Occasionally, a prominent nasal ala or a soft tissue swelling of the cheek overlies the floor of the maxillary sinus and simulates a polyp or cyst. Frequently, the lateral wall of the maxillary sinus appears somewhat denser than the remainder of the sinus. In the presence of normal aeration and a normal mucoperiosteal line, the density of the lateral wall is a function of the obliquity of the wall and is not caused by sinusitis (Fig. 10.1B). The vascular groove of the posterior superior alveolar artery lies midway along the lateral wall of the sinus and should not be mistaken for a fracture. The medial wall of the maxillary sinus is sometimes not visualized in normal people because of the thinness of the bony wall and an unusual degree of obliquity. For this reason, great care must be taken not to confuse this normal variation with actual bony destruction (Fig. 11.4). Lateral extensions of the sphenoid sinus are frequently projected over the maxillary sinus and should not be mistaken for a cyst or mucosal thickening (Fig. 10.5). The innominate line, or oblique orbital line, is produced by the depression on the temporal surface of the greater wing of the sphenoid bone, where it forms the medial wall of the temporal fossa. This line is often straight, but at times, its lower end shows a medial right angle turn. In the Waters view the lower end of the oblique line may appear just below the inferior rim of the orbit and may be confused with a blow-out

fracture of the floor of the orbit.

In incomplete or absent pneumatization, the frontal sinus is abnormally dense. This increased density of the frontal bone, in the area of the frontal sinus, may be erroneously attributed to osteitis caused by chronic infection. Correlation of the Caldwell projection with lateral and overangulated submentovertical views should lead to a correct interpretation and diagnosis.

The ethmoid cells are usually small and often obscured by regional soft tissue or bony structures. It is frequently difficult to demonstrate early disease of the ethmoid sinus and therefore painstaking study may be necessary. The misinterpretations of roentgenograms of the ethmoid sinuses are usually attributable to incomplete examination, poor technique, or lack of correlation with the clinical history and findings.

The roentgen examination of the sphenoid sinus is often inadequate and, consequently, minor manifestations of disease may be overlooked. For the detection of diseases of the sphenoid sinus, especially in the early stage, a complete study is imperative. Primary tumors of the sphenoid sinus are rare, but secondary invasion of the sphenoid sinus by tumors of the pituitary fossa, nasal cavity, ethmoid sinus, or nasopharynx is not uncommon. If the lesion is extensive, it may be impossible to distinguish secondary invasion from a primary growth roentgenographically. When the lesion is of limited size, the site of origin may be determined from the relative position of the soft tissue mass and of the bony destruction.

REFERENCES

De Lorimer, A. A., Hoehring, H. G., and Hannan, J. R.: *The Head, Neck and Spinal Column. Clinical Roentgenology,* Vol. 2, Charles C Thomas, Springfield, Illinois, 1954.

Dodd, G. D., Collins, L. C., Egan, R. L., and Herrera, J. R.: The systematic use of tomography in the diagnosis of carcinoma of the paranasal sinuses. Radiology, *72:* 379, 1959.

Jing, B. S.: Roentgen diagnosis of malignant disease of paranasal sinus and nasal cavity. Ann. Otol. Rhinol. Laryngol., *79:* 584, 1970.

Law, F. M.: *Nasal Accessory Sinuses. Annals of Roentgenology, Vol. 15.* Paul B. Hoeber, Inc. New York, 1933.

Merrell, R. A., Jr., and Yanagisawa, E.: Radiographic anatomy of the paranasal sinuses. I. Waters' view. Arch. Otolaryngol., *87:* 184, 1968.

Pendergrass, E. P., Schaeffer, J. P., and Hodes, P. J.: *The*

Head and Neck in Roentgen Diagnosis, 2nd ed. p. 545–702. Charles C Thomas, Springfield, Illinois, 1956.

Samuel, E.: The radiology of the paranasal sinuses. In *Textbook of X-ray Diagnosis by British Authors,* 4th ed., H. K. Lewis & Company Ltd., London, 1969.

Welin, S.: Roentgen diagnosis of paranasal sinuses. Minn. Med. *43:* 325, 1960.

Welin, S.: Paranasal sinuses. In *Roentgen Diagnosis.* 2nd American ed., edited by L. G. Rigler, Grune & Stratton, New York, 1969.

Yanagisawa, E., Smith, H. W., and Thaler, S.: Radiographic anatomy of the paranasal sinuses. II. Lateral view. Arch. Otolaryngol. *87:* 96, 1968.

Young, B. R.: *The Skull, Sinuses and Mastoids. A Handbook of Roentgen Diagnosis.* Year Book Publishers, Chicago, Illinois, 1951.

Chapter 12

Syndromes Involving the Paranasal Sinuses

A syndrome is a set of symptoms which occur together. Congenital or acquired syndromes may involve the skull and facial bones. In several, growth disturbances such as aplasia, hypo- plasia, or hyperplasia occur in the paranasal sinuses. In others, inflammatory processes involving the nose, paranasal sinuses, pharynx, and lungs may be present.

GARDNER'S SYNDROME

General Considerations

Gardner's syndrome consists of osteomas of the bone, cysts and benign tumors of the skin, and polyposis of the intestine. Although it was first reported in 1935 (Case Records of Massachusetts General Hospital), Gardner and his co-workers recognized the condition as a syndrome and extensively described it in 1953. Gardner's syndrome is inherited as an autosomal dominant trait with marked penetrance and variable expressivity. The syndrome may be present before puberty.

Cutaneous Manifestations

The most common skin lesions are epidermoid inclusion cysts. The cysts may appear anywhere on the face, trunk, or extremities. The time of the first appearance of the cysts is variable, but it is usually after puberty. New cysts appear periodically. Fibromas, desmoid tumors, lipomas, lipofibromatosis, and fibrosarcoma of the skin and of the mesentery may also be present. The desmoid tumors frequently arise from the abdominal scar after colonic surgery. They may, however, arise from the skin in the absence of any previous surgery.

Roentgen Findings

Polyposis of the Colon

Multiple polyps of the colon and rectum, with a marked tendency to malignant degeneration, are characteristic of this syndrome (Fig. 12.1*A*). The polyps may occur before puberty. Postoperative adhesions are usually severe, frequently resulting in intestinal obstruction.

Polyposis of the small intestine has been reported, but is rare.

Osteomas

Multiple osteomas may be scattered throughout the calvarium, the facial bones, and the long bones. In the skull and facial bones, the frontal bone, mandible, and maxilla are the most frequent sites (Fig. 12.1*C*). The paranasal sinuses can also be involved (Fig. 12.1*B*). When large, the osteoma may obliterate the sinus cavity. In most cases, osteomas appear at about the time of puberty and generally precede the appearance of polyposis of the colon.

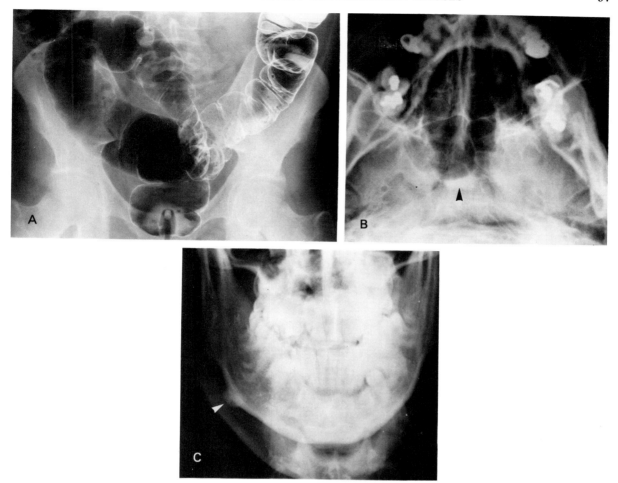

Figure 12.1. Gardner's syndrome. *A*, Double contrast study of the colon demonstrating multiple polyps; *B*, base view of the paranasal sinuses showing a small osteoma in the posterior wall of the sphenoid sinus; *C*, oblique view of the mandible showing an osteoma on the inferior aspect of the body of the right mandible.

KARTAGENER'S SYNDROME

General Considerations

As early as 1904, the relationship between bronchiectasis and situs inversus viscerum was reported by Siewert. In 1933 and 1935, Kartagener added chronic rhinosinusitis to the complex and the combination has received the appellation Kartagener's syndrome. The syndrome is inherited as an autosomal recessive trait with variable expressivity. However, it may occasionally be inherited as a dominant characteristic. There is probably some alteration of the immune mechanism in these patients. The syndrome has occurred in identical twins and in several individuals within a family.

Roentgen Findings

The characteristic findings are Kartagener's triad.

Chronic Rhinosinusitis

This is a frequent but inconstant finding. Both ethmoid and maxillary sinuses are usually involved. There is a diffuse thickening of the lining mucosa, with sclerosis of the walls of the involved sinuses (Fig. 12.2*A*).

Bronchiectasis

The bronchiectasis is generally of the cylindrical type (Fig. 12.2*B*) and may be congenital.

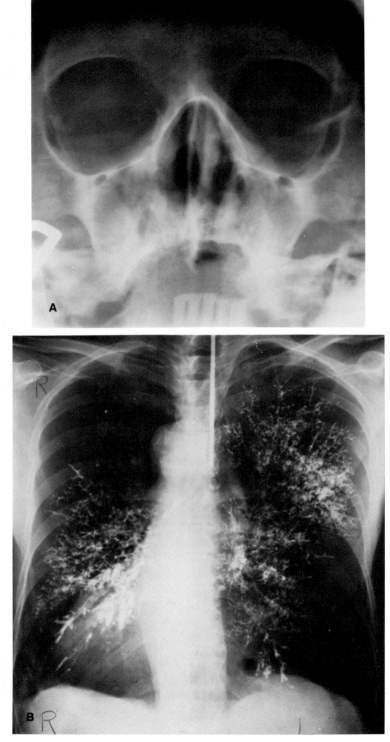

Figure 12.2. Kartagner's syndrome. *A*, Waters view: diffuse thickening of the lining mucosa and sclerosis of the bony walls of the maxillary sinuses. There is opacity of the ethmoid sinuses. Agenesis of the frontal sinuses is also noted. *B*, Bronchogram: cylindrical bronchiectasis, more on the right side. There is a dextrocardia.

Situs Inversus Viscerum

Situs inversus viscerum occurs once in about 8,000 to 10,000 births. Dextrocardia, without other visceral transposition, has been frequently reported (Fig. 12.2B). About 15 to 20% of cases of situs inversus viscerum are associated with bronchiectasis. The incidence is much higher than that in the general population, which is 0.25 to 0.50%. The relationship between situs inversus viscerum and bronchiectasis is not fully understood.

Other Associated Findings

Other findings associated with this syndrome are as follows: (1) agenesis or underdevelopment of frontal sinuses (Fig. 12.2A); (2) vascular anomalies, such as anomalous subclavian artery and renal artery; (3) bony anomalies such as absence of the xyphoid process and turricephaly.

NEVOID BASAL CELL CARCINOMA SYNDROME

General Considerations

The name nevoid basal cell carcinoma was originally suggested by Nomland in 1932. The major features of the disease were described by Brinkley and Johnson in 1951, but they considered the cutaneous tumors to be epitheliomata adenoides cysticum. Howell and Caro, in 1959, emphasized the distinction between epitheliomata adenoides cysticum and basal cell nevi. Gorlin and Goltz in 1960 added the anomaly of a broad nasal root (hypertelorism) to the complex and emphasized the concept of a syndrome. Ward, in 1960, added to the syndrome dyskeratotic lesions, i.e., pit-like defects, of the hands and feet. Approximately 350 cases of nevoid basal cell carcinoma syndrome have been reported.

The nevoid basal cell carcinoma syndrome is an inherited autosomal dominant disorder with high penetrance but variable expressivity. The components may occur in various combinations. Some individuals may not manifest the nevoid basal cell carcinomas and/or the jaw cysts at the time of the examination. Some of the variability is undoubtedly the consequence of age, because the frequencies of the nevoid basal cell carcinomas, pitting of the skin of the palms and feet, jaw cysts, and ectopic calcification increase at different rates with increasing age. Skeletal anomalies, however, are not age-dependent.

The affected parents will transmit the disorder to about 50% of their sons and daughters. Although some features of the syndrome may be present at birth, the majority of the features develop after birth.

Cutaneous Manifestations

Multiple Basal Cell Carcinomas

Nevoid basal cell carcinomas generally develop after birth. Any area of the body may be affected; however, lesions tend to develop generally in the central part of the face and cheeks. The chest and back are also frequent sites, as are the torso, scalp, and extremities, including the palms and soles. The number of lesions in any one patient is variable, ranging from a few to hundreds. The tumor may be papular, nodular, pigmented, erythematous, or ulcerated. Several varieties may be noted in the same patient. The lesions are benign during childhood, becoming variable in behavior after puberty, generally during the second decade, when one or more may begin to show activity and exhibit malignant traits. The activity is characterized by an increase in size and by ulceration, bleeding, and crusting (Fig. 12.3A).

Pit-like Defects of the Skin of the Palms and Feet

These lesions develop on the palms, including the thenar and hypothenar emiences, and on the sides and the dorsal surfaces of the hands. In the feet, the lesions appear on the soles, the medial and lateral borders of the feet, and the sides and the dorsal surfaces of the toes.

Other Manifestations

Cutaneous lesions include epithelial lined cysts, fibromas, and neurofibromas of the skin. The most common lesions of the eyes are ectropia and exophoria.

Roentgen Findings

The roentgen findings of this syndrome are often evident. The characteristic features are as follows:

Skull and Central Nervous System

Calcification of the Falx Cerebri and Tentorium Cerebelli. In normal persons, the calcifi-

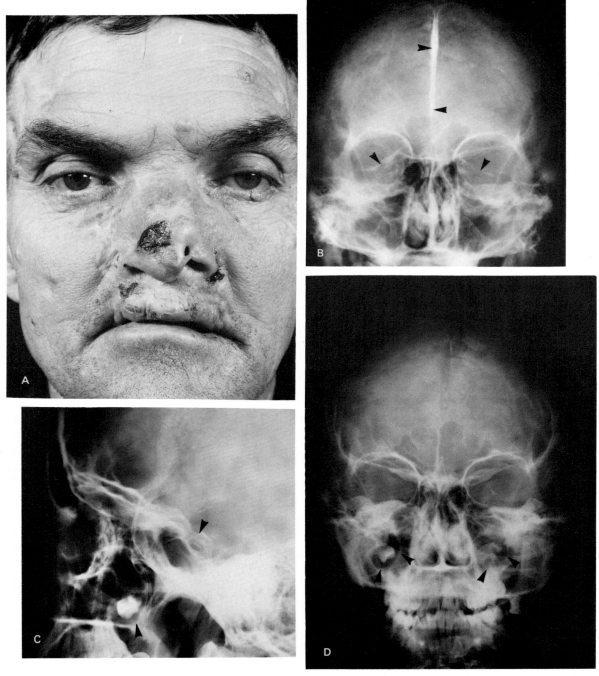

Figure 12.3. Nevoid basal cell carcinoma syndrome. *A*, Photograph showing nevoid basis cell carcinomas scattered over the face. Note hypertelorism and mild prognathism. There are scars over the face and deformity of the nose, results of previous surgical procedures. Strabismus and ptosis are noted on the left side. *B*, Posteroanterior view of the skull showing lamellar calcification of the falx cerebri and tentorium cerebelli. *C*, Lateral view of the skull showing sellar bridging. There is a dentigerous cyst located in the posterior aspect of the maxillary sinus. *D*, Posteroanterior view of the paranasal sinuses demonstrating dentigerous cyst protruding into the maxillary sinus on each side. *E*, Chest film showing anomalies of the ribs and dorsal spine. Sprengel's deformity of the left scapula is seen. *F*, Anteroposterior view of the right hand showing shortening of the fourth metacarpal and a subcortical cyst of the fifth metacarpal.

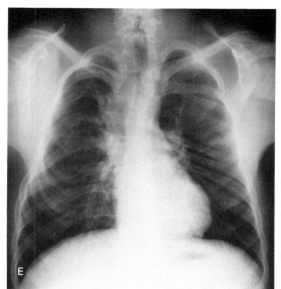

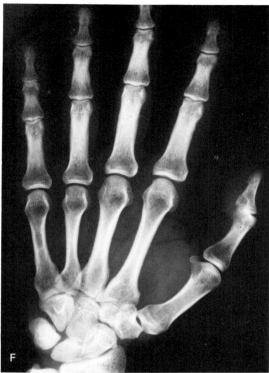

Figure 12.3. E–F.

cation of the falx cerebri and tentorium cerebelli appears to be button-like in configuration or plaque-like with an elliptical configuration. The calcifications measure a few millimeters in transverse diameter and one to several centimeters in vertical diameter, as seen in the frontal projection. Calcifications in the falx cerebri and tentorium cerebelli in patients with the nevoid basal cell syndrome show a distinctly lamellar appearance with one or more flat sheets, which usually are more extensive than those in normal persons (Fig. 12.3*B*).

Sellar Bridging. This is a common finding in patients with the syndrome (Fig. 12.3*C*).

Miscellaneous. Hypertelorism, frontal and biparietal bossing, and congenital hydrocephalus have been reported. Medulloblastoma and agenesis of corpus callosum have been observed in association with the syndrome.

Jaw Bones

Cysts of the Jaw Bones. The mandible is more frequently involved than the maxilla. In some instances the cysts in the maxilla protrude into the maxillary sinus (Fig. 12.3, *C* and *D*). The cysts are primordial or dentigerous in nature and lined with epithelium. Ameloblastic elements may be present. The lesions are usually multiple and vary in size from microscopic to several centimeters in diameter. Mild prognathism is often present.

Spine and Rib Cage Anomalies

Ribs. Rib anomalies include bifid, fused, dysplastic, and cervical ribs. Flattened anterior ends of the ribs have been frequently observed. Sprengel's deformity of the scapula, associated with spinal and rib anomalies, has been reported (Fig. 12.3*E*).

Spine. Spina bifida and scoliosis are the most frequent findings. Hemivertebra and fused vertebra are occasionally observed (Fig. 12.3*E*).

Sternum. Pectus excavatum and pectus carinatum are seen in some cases.

Extremities

Shortening of the fourth metacapal is often observed. Subcortical cystic changes have been reported (Fig. 12.3*F*). Soft tissue calcification, polydactylism, and syndactylism are occasionally noted.

Miscellaneous

Calcifications of the ovary and sacrotuberous ligament and cysts of mesentary and ovary have been reported.

VON RECKLINGHAUSEN'S DISEASE (NEUROFIBROMATOSIS)

General Considerations

Neurofibromatosis is a congenital disease of mesodermal and neuroectodermal origin, affecting principally the central and peripheral nervous systems. It is characterized by pigmentation of the skin, soft cutaneous fibromas (fibroma mollusca), and neurofibromas of the cranial and peripheral nerves. At times, the skeletal and endocrine systems are affected.

Neurofibromatosis is considered an inherited autosomal dominant, although less than 50% of the affected persons appear to have a family history of the disease. The incidence is about 1:2,000 in the general population and 1:200 in mental defectives. There is no sex preference.

Clinical Features

Cutaneous Manifestations

The skin lesions are the constant finding. These are of two types:

1. Pigmentation of the skin (cafe-au-lait spots). The pigmentation is caused by deposition of melanin. It usually appears within the first decade of life and precedes the appearance of the skin tumors. The color varies from yellowish to chocolate brown. The pigmentations vary in size from a freckle to an area several centimeters in diameter.

2. Skin tumors. Skin tumors consist of soft fibromas (fibroma mollusca), neurinomas, and neurofibromas. Such growths are present at birth or appear in early life, and increase in number and size at puberty. They may vary in size from the head of a pin to a huge pedunculated mass. They may be single, may be few in number, or may cover the entire body (Fig. 12.4*A*). In plexiform neurofibromas, elephantoid hypertrophy of the soft tissues may occur. They are usually unilateral, involving a lower extremity or the head and neck.

Central and Peripheral Nervous System

Neurilemmomas and neurofibromas of the cranial and spinal nerves occur frequently in neurofibromatosis. In addition, gliomas and meningiomas show an increased incidence. Neurofibromas of the cranial nerves (especially the optic, trigeminal, acoustic, and vagus) may cause a variety of symptoms depending upon which nerve is involved. Neurofibromas of the spinal nerves may lead to a multitude of motor symptoms. Neurofibromas of the peripheral nerves are usually discrete, but a diffuse enlargement of the nerves is also seen.

Other Manifestations

Pulsating exophthalmos may occur as a result of a bony defect in the posterosuperior wall of the orbit. The defect consists of absence of or failure of development of the membranous bone in the posterosuperior portion of the orbit, separating the cranial contents from those of the orbit. The resultant gap allows the temporal lobe of the brain to encroach upon orbital contents and leads to progressive pulsating exophthalmos with facial deformity and loss of vision.

Neurofibromatous involvement of the oral cavity is not common. It may involve any oral tissue, but appears to have some predilection for the tongue. Involvement of the maxilla and mandible is rare.

Roentgen Findings

The roentgen manifestations of neurofibromatosis of the skull and facial bones are as follows:

1. Bony defect of the posterosuperior wall of the orbit, resulting in a pulsating exophthalmos. This defect may be associated with an optic glioma. Radiologically on the Waters or Caldwell view, the orbit is generally larger on the affected side. The inferior margin may lie lower than normal. The orbit has a "blank" appearance, with loss of the usual landmarks; the oblique orbital line is absent and the sphenoid ridge is markedly elevated. There is often a pressure defect in the anterior ethmoid cells and the maxillary sinus (Fig. 12.4*B*). On the base view, anterior displacement of the middle fossa is well demonstrated on the affected side.

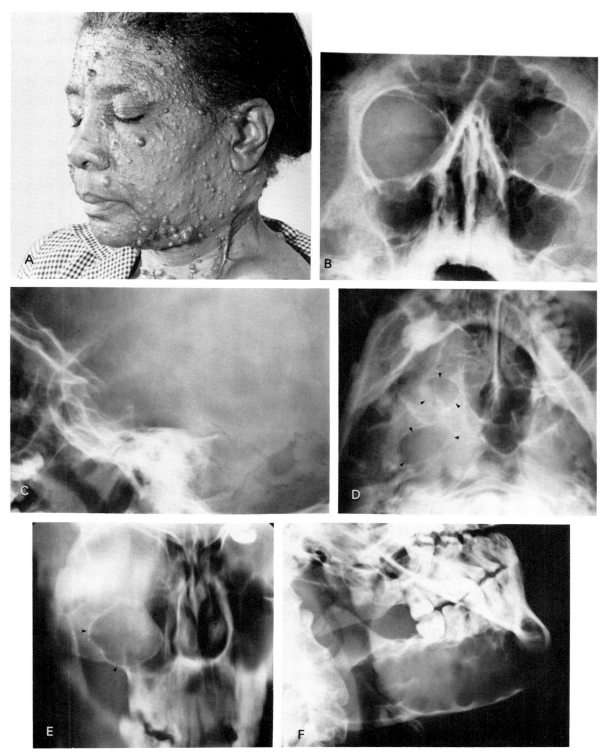

Figure 12.4. Neurofibromatosis. *A*, Photograph showing numerous neurofibromas over the face and the neck. There is a scar on the neck, a result of previous surgery. *B*, Bony defect in the posterosuperior aspect of the right orbit, with a blank appearance. *C*, J-shaped contour of the sella turcica and the bony defect in the left lambdoid suture. *D*, Enlargement of the foramina in the floor of the middle cranial fossa on the right side. *E*, Enlargement of the infraorbital foramen with cystic shadow within the right maxillary sinus. *F*, Enlargement of the mandibular canal with cystic change in the right mandible.

103

2. Bony defect in the region of the lower lambdoid suture (Fig. 12.4C).

3. J-shaped contour of the sella turcica, with deformity of the sphenoid sinus (Fig. 12.4C).

4. Erosion and enlargement of the foramina of the floor of the middle cranial fossa, especially the foramen ovale (Fig. 12.4D).

5. Erosion and enlargement of the infraorbital foramen, with a cystic shadow within the maxillary sinus (Fig. 12.4E).

6. Erosion and enlargement of the mandibular canal, with cystic changes in the mandible (Fig. 12.4F).

7. Facial dysplasia.

MUCOVISCIDOSIS (CYSTIC FIBROSIS OF THE PANCREAS)

Cystic fibrosis of the pancreas is a relatively recently recognized disease of heriditary origin. The disease was first described in 1936 by Fanconi, and Anderson, in 1938, defined the disease as an entity. In 1958, Di Sant'Agnese demonstrated the increase in the salinity of the sweat.

Cystic fibrosis of the pancreas is a systemic disorder of childhood caused by an unknown defect in the mucus-producing exocrine glands of the intestine, pancreas, liver, and upper respiratory tract and in the non-mucus producing salivary and sweat glands. The abnormal secretions in the pancreas, liver, intestine, and upper respiratory tract may lead to pancreatic achylia with malabsorption and meconium ileus, biliary cirrhosis with portal hypertension and esophageal varices, recurrent pulmonary infection, obstructive emphysema, and bronchiectasis. There is salt depletion, with increased electrolyte concentration in the sweat glands.

The mode of inheritance is an autosomal recessive with variable expressivity. The disease occurs in infants and during childhood. The incidence has been estimated to be between 1:1,000 and 1:2,000 live births. The disease appears to be less frequent in blacks and orientals. There has been no relationship between cystic fibrosis of the pancreas and adult chronic pulmonary disease. Several so-called instances of cystic fibrosis of the pancreas in the adults have been reported; however, after careful review, these were probably attributable, in large part, to uncritical use of the sweat test in adults.

Clinical Features

In cystic fibrosis of the pancreas the involvement of different organs is variable with variable clinical features. Most notable symptoms are progressive obstructive emphysema and recurrent pulmonary infection, steatorrhea, malnutrition, and growth retardation. Meconium ileus is pathognomonic of the condition. In about 10 to 20% of patients, small bowel obstruction of this nature occurs soon after birth. The patients may present with pancreatic insufficiency or with chronic pulmonary disease without intestinal symptoms. Less frequently, biliary cirrhosis with portal hypertension and esophageal varices may dominate the clinical picture. Rarely, acute salt depletion, especially during hot weather, is the first manifestation.

Nasal and paranasal sinus polyposis and chronic sinusitis may occur, often with increased incidence in older children. In one series of 742 cases of cystic fibrosis of the pancreas in which 80 had roentgen examination of the paranasal sinuses, 93% showed involvement of the paranasal sinuses. Histological examination of mucosal specimens from the sinuses revealed changes characteristic of those seen in the bronchi of patients with cystic fibrosis of the pancreas, including dilation of the mucus-secreting glands and the discharge of viscid secretions, often with evidence of secondary infection (Shwachman et al., 1962).

Rates of linear growth and osseous maturation are almost always depressed and puberty is frequently delayed. The disease may progress more slowly during adolescence. The ultimate prognosis is generally poor.

Roentgen Findings

The roentgen features of cystic fibrosis of the pancreas are many. The sinopulmonary changes are as follows.

Lungs

Pulmonary changes almost always occur in patients who have survived infancy. The abnormal, tenacious secretions occlude parts of the bronchial tree. Bronchiolar obstruction leads to subsequent obstructive emphysema, atelectasis, recurrent pulmonary infection, bronchiectasis, and pulmonary fibrosis (Fig. 12.5A).

Nasal Cavity and Paranasal Sinuses

1. Nasal polyps (Fig. 12.5B).

2. Chronic sinusitis—usually there is in-

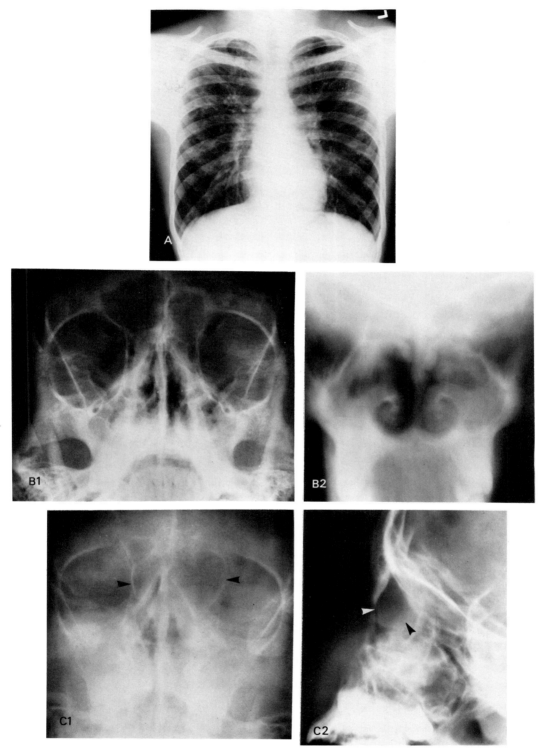

Figure 12.5. Cystic fibrosis of the pancreas. *A*, Pulmonary changes: posteroanterior view of the chest showing chronic infiltrations, mainly in both upper lung fields, associated with mild emphysema. There is prominence of pulmonary arteries. *B*, Sinusitis and polyposis of the nasal cavity and maxillary sinuses. *1*, Waters view of the paranasal sinuses: marked opacity of the maxillary sinuses, more on the right. There is cloudiness of the ethmoid and frontal sinuses. *2*, Anteroposterior tomogram of the paranasal sinuses: thickening of the lining mucosa with polyp formation in both maxillary sinuses. There are also polyps in the nasal cavity. *C*, Mucocele of the ethmoid sinuses. *1*, Waters view: expansion of the ethmoid sinuses with lateral bowing of the lamina papyracea. There is loss of cell structures of the ethmoid sinuses. Both maxillary sinuses are opaque. There is hypoplasia of the frontal sinuses. *2*, Lateral view: expanding lesion in the anterior portion of the ethmoid sinus with downward extension to the nasal cavity. (Courtesy of Dr. Milton Wagner and Dr. E. B. Singleton.)

volvement of all of the paranasal sinuses (Fig. 12.5B).

3. Polyposis of the paranasal sinuses (Fig. 12.5B). In advanced cases the changes in the sinuses may simulate mucocele and benign expanding tumor. In the diagnosis of polyposis, routine projections of the paranasal sinuses are often inconclusive and may be misleading. Tomograms are imperative in delineating the location and extent of the lesion.

4. Mucocele of the paranasal sinuses. This may occur in the maxillary and ethmoid sinuses. The roentgen manifestations are very similar to that described in mucocele after infectious sinusitis (Fig. 12.5C).

5. Agenesis or hypoplasia of the frontal sinuses (Fig. 12.5C).

LOUIS-BAR SYNDROME

General Considerations

The complex of ataxia, oculocutaneous telangiectasia, and sinopulmonary infections was first described by Mme. Louis-Bar in 1941 and bears her name. Because of the occurrence in siblings and the absence of parental manifestations, the syndrome appears to be inherited as an autosomal recessive trait. An affected person rarely lives to be over 20 years of age.

The ataxia is of a cerebellar type and usually becomes manifest when the child begins to walk. The speech is low and slurred. Movement of the eyes is slow and often halted midway between a lateral and upward gaze. Nystagmus is present in the majority of the cases. The affected child usually is 4 to 6 years of age when fine, symmetrical, bright red telangiectases appear in the temporal and nasal sides of the conjunctiva of the eyes. These gradually extend to the face, neck, ears, antecubital and popliteal areas, and dorsa of the hands and feet. Recurrent sinopulmonary infections occur in the majority of the patients. Mental deficiency is noted in about one-third of cases. Growth is markedly retarded in over two-thirds of cases.

Roentgen Findings

Paranasal Sinuses

Chronic sinusitic is a frequent finding. The maxillary and ethmoid chambers are often affected.

Lungs

Chronic inflammatory disease with bronchiectasis and pulmonary fibrosis is a common finding.

WEGENER'S GRANULOMATOSIS

General Considerations

Wegener's granulomatosis is a disease consisting of the pathological triad of necrotizing granulomas in the respiratory tract, vasculities of the small arteries and veins, and glomerulitis.

The etiology is unknown, but most authors agree that it is related to hypersensitivity and autoimmune reactions. The disease is encountered most frequently in the fourth and fifth decades, but may occur earlier. The duration is short; many patients die of uremia within a few months after onset of the disease.

Clinically, in the upper respiratory tract, there are inflammatory changes, with thickening of the mucous membrane in the sinuses and nasal cavity in the early state. Later, there is necrosis and bony destruction of the sinuses and nasal cavity; occasionally perforation of the hard palate occurs. Pulmonary involvement is frequent. There are usually granulomatous lesions with cough and hemoptysis. Glomerulitis is usually a late manifestation. In terminal cases, nearly any organ may be involved.

Roentgen Findings

Nasal Cavity and Paranasal Sinuses

In the nasal cavity there is mucosal thickening with granulation and ulceration; later in the disease, there may be destruction of bony walls and septum. The paranasal sinuses, especially the maxillary sinuses, show opacification in the early stage (Fig. 12.6A). Later, there is destruction of the bony walls of the sinuses, which may simulate malignant disease.

In the advanced stage of the lesion, there is

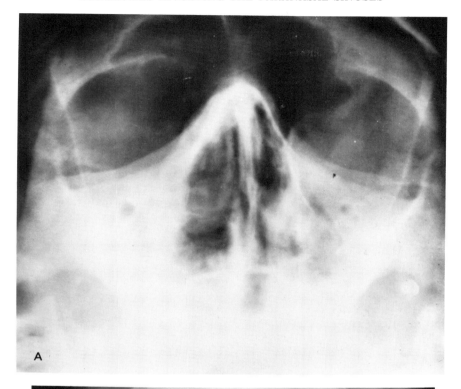

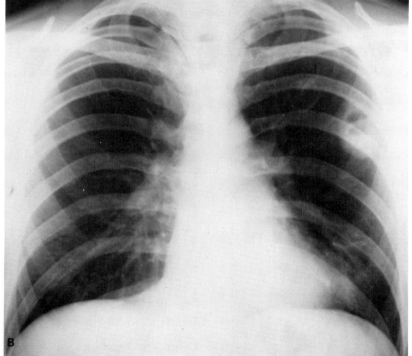

Figure 12.6. Wegener's granuloma. *A*, Waters view showing opacification with thickening of the lining mucosa of the maxillary sinuses, more on the right side. *B*, Posteroanterior chest film demonstrating thin-walled cavities in the left lung. One of the cavities shows a fluid level.

often involvement of the hard palate, with bony destruction and perforation.

Lungs

In the lungs, there are sharply defined rounded nodules or rather well demarcated areas of parenchymal consolidation in both lungs. In one-third of the cases there is cavity formation. The cavity may be unilocular or multilocular, with thin or thick walls (Fig. 12.6*B*). The lesion may be either stationary or progressive.

MIDLINE LETHAL GRANULOMA

General Considerations

Midline lethal granuloma is a progressive ulcerative gangrenous lesion of the face, nose, sinuses, palate, and pharynx. Pulmonary, renal, and generalized vascular granulomas have been reported.

The etiology of midline lethal granuloma remains unknown, but it has been considered to be one form of an allergic vasculitis related to polyarteritis nodosa or the results of an altered immune mechanism. Spear and Walker (1956) believe midline lethal granuloma to be a disease entity. However, Burston (1959) stated that midline lethal granuloma represents little more than a general term for a number of conditions causing tissue ulceration and destruction.

This disease is rather rare, and is usually fatal. There is a progressive, massive ulceration and destruction of the involved soft tissues and bones eventually ending in septis, hemorrhage, and death.

Roentgen Findings

The roentgen findings in the nasal cavity, paranasal sinuses, and palate are as follows: (1) thickening and ulceration of the mucosa and destruction of the bony walls and septum of the nasal cavity (Fig. 12.7*B*); (2) opacification of the maxillary sinuses with destruction of the bony walls (Fig. 12.7*A*); (3) destruction of the hard palate associated with soft tissue mass (Fig. 12.7*B*); (4) ulcerative lesions in the pharynx.

OTHER SYNDROMES

There are a number of syndromes of the head and neck area which affect the paranasal sinuses primarily on the basis of aplasia or hypoplasia. The following are a few examples.

Cleidocranial Dysostosis (Scheuthauer-Marie-Sainton Syndrome)

Cleidocranial dysostosis is characterized by defective ossification of the cranial bones and complete or partial absence of the clavicles. The paranasal sinuses are often absent or underdeveloped.

Mandibulofacial Dysostosis (Treacher Collins Syndrome)

Mandibulofacial dysostosis is defined by a downward sloping of the palpebral fissures, often with colobomas, hypoplasia of the facial bones with poor malar and mandibular develop-ment, deformity of the ears, and malocclusion. The paranasal sinuses are often small and may be completely absent.

Progeria (Hutchinson-Gilford Syndrome)

Progeria is manifested by dwarfism, immaturity, and pseudosenility. There are often no frontal sinuses.

Mongolism (Trisomy 21 Syndrome)

In mongolism, frontal and sphenoid sinuses are often absent or small in size.

Occulomandibulodyscephaly with Hypotrichosis (Hallerman-Streiff Syndrome)

The syndrome consists of dyscephaly, parrot nose, mandibular hypoplasia, proportionate nanism, hypotrichosis, and blue sclera. The paranasal sinuses are often small in size.

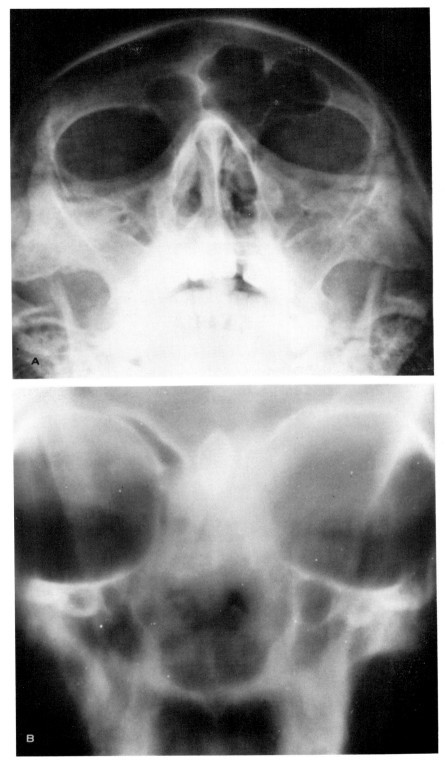

Figure 12.7. Midline lethal granuloma. *A*, Waters view showing opacification of both maxillary sinuses. *B*, Posteroanterior tomogram of paranasal sinuses demonstrating granulation and ulceration of the nasal cavity with destruction of the septum and lateral walls. There is also destruction of the hard palate. Both maxillary and ethmoid sinuses are cloudy.

REFERENCES

Ahmoo, E. W.: Lethal midline granuloma of the face. Oral Surg., *23:* 578, 1967.

Anderson, D. E.: Linkage analysis of the nevoid basal cell carcinoma syndrome. Ann. Hum. Genet., *32:* 113, 1968.

Anderson, D. E., Taylor, W. B., Falls, H. F., and Davidson, R. T.: The nevoid basal cell carcinoma syndrome. Am. J. Hum. Genet., *19:* 12, 1967.

Anderson, D. H.: Cystic fibrosis of the pancreas and its relation to celiac disease. Am. J. Dis. Child., *56:* 344, 1938.

Andrews, B. F., Kopack, F. M., and Bruton, O. C.: A syndrome of ataxia, oculocutaneous telangiectasia and sinopulmonary infections. U.S. Armed Forces Med. J., *11:* 587, 1960.

Antenucci, A. J., Jacobs, T. P., and Garret, R.: Sinusitis, nodules in lungs and hard prostate. N. Y. J. Med., *65:* 2808, 1965.

Bechner, R. E., and Lamppin, D. W.: Midline malignant reticulosis. Arch. Otolaryngol., *95:* 467, 1972.

Becker, M. H., Kopf, A. W., and Lande, A.: Basal cell nevus syndrome. Its roentgenologic significance. Am. J. Roentgenol. Radium Ther. Nucl. Med., *99:* 817, 1967.

Border, E., and Sedgwick, R. P.: Ataxia-telangiectasia. A familial syndrome of progressive cerebellar ataxia, oculocutaneous telangiectasia and frequent pulmonary infection. Pediatrics, *21:* 526, 1958.

Brown, H. A., and Woolner, L. B.: Findings referrable to the upper part of the respiratory tract in Wegener's granulomatosis. Ann. Otol. *69:* 810, 1960.

Bruner, A. J. and Kierland, R. R.: Neurofibromatosis and congenital unilateral pulsating and nonpulsating exophthalmos. Arch. Ophthalmol., *53:* 2, 1955.

Burston, H. H.: Lethal midline granuloma *Laryngoscope,* *69:* 1, 1959.

Carrington, C. B., and Liebow, A. A.: Limited forms of angiitis and granulomatosis of Wegener's type. Am. J. Med., *41:* 497, 1966.

Case Records of Massachusetts General Hospital. Case 2,061. N. Engl. J. Med., *212:* 263, 1935.

Davidson, K. C.: Cranial and intracranial lesions in neurofibromatosis. Am. J. Roentgenol., *98:* 550, 1966.

Di Sant'Agnese, P. A.: Cystic fibrosis of the pancreas. Am. J. Med., *21:* 406, 1956.

Edelken, J., and Hodes, P. J.: *Roentgen Diagnosis of Disease of Bone.* Williams & Wilkins, Baltimore, 1966.

Evans, R. A., Schwartz, J. F., and Chutorian, A. M.: Radiologic diagnosis in pediatric ophthalmology. Radiol. Clin. North Am., *1:* 459, 1963.

Fanconi, G., Ühlinger, E., and Knauer, C.: Das Coeliaksyndrom bei angeborener zystischer Pankreasfibromatose und Bronchiektasen. Wien Med. Wochenschr., *86:* 753, 1936.

Felson, B., and Braunstein, H.: Non-infectious necrotizing granulomatosis, Wegener's syndrome, lethal granuloma, and allergic angitis and granulomatosis. Radiology, *70:* 326, 1958.

Gardner, E. J., and Richards, R. C.: Multiple cutaneous and subcutaneous lesions occurring simultaneously with hereditary polyposis and osteomatosis. Am. J. Hum. Genet., *5:* 139, 1953.

Gonzales, C. and Van Ordstrand, N. S.: Wegener's granulomatosis. Radiology, *107:* 295, 1973.

Gorlin, R. J., and Chaudhry, A. P.: Multiple osteomatosis, fibromas, lipomas and fibrosarcomas of the skin and mesentery, epidermoid inclusion cysts of the skin,

leiomyomas and multiple intestinal polyposis. N. Engl. J. Med., *63:* 1151, 1960.

Gorlin, R. J., and Pindborg, J. J.: *Syndromes of the Head and Neck.* McGraw-Hill Book Company, New York, 1964.

Gorlin, R. J., and Sedano, H. O.: Syndromes involving the sinuses — congenital and acquired. Semin. Roentgenol., *3:* 33, 1968.

Gorlin, R. J., and Goltz, R. W.: Multiple nevoid basal cell epithelioma, jaw cysts and bifid rib. N. Engl. J. Med., *262:* 908, 1960.

Greenberger, N. J.: The gastrointestinal system. In *Genetic Disorders of Man,* edited by R. M. Goodman, Little, Brown and Company, Boston, 1970.

Hanafee, W. N., and Dayton, G. O.: The roentgen diagnosis of orbital tumors. Radiol. Clin. North Am., *8:* 403, 1970.

Holt, J. F., and Wright, E. M.: The radiologic features of neurofibromatosis. Radiology, *51:* 647, 1948.

Howell, J. B., Anderson, D. E., and McClendon, J. L.: The basal cell nevus syndrome. J. A. M. A., *190:* 274, 1964.

Hunt, J. C., and Pugh, D. G.: Skeletal lesions in neurofibromatosis. Radiology, *76:* 1, 1961.

Jacobs, M. H.: Oral manifestations in Von Recklinghausen's disease (neurofibromatosis). Am. J. Orthod. Oral Surg., *32:* 28, 1946.

Kartagener, M.: Zur Pathogenese der Bronkiektasien, Bronkietasien bei Situs Viscerum inversus. Beitr. Klin. Tuberk. *83:* 489, 1933.

Kartagener, M. and Horlacher, A.: Zur Pathogenese der Bronkietasien Situs Viscerum inversus und Polyposis nasi in einem Falle familiar Bronchiektasien. Beitr. Klin. Tuberk. *87:* 331, 1935.

Lobeck, C. C.: Cystic fibrosis of the pancreas. In *The Metabolic Basis of Inherited Disease,* 2nd ed., edited by J. B. Stanburg, J. B. Wyngaarden, and D. S. Fredrickson, McGraw-Hill Book Company, New York, 1960.

Louis-Bar, M.: Sur un syndrome progressif comprenant des télangiectasies capillaires cutanées et conjonctivales symétriques, a disposition naevoïd et des troubles cérébelleux. Confinia Neurol. *4:* 32, 1941.

Luri, M. H.: Cystoc fibrosis of the pancreas and nasal mucosa. Ann. Otol. Rhinol. Laryngol., *68:* 478, 1959.

McGregor, M. B., and Sander, G.: Wegener's granulomatosis. Br. J. Radiol., *37:* 430, 1964.

Marshak, R. H., and Linder, A. E.: *Radiology of the Small Intestine.* W. B. Saunders Company, Philadelphia, 1970.

Meszaros, W. T., Guzzo, F., and Schorsch, H.: Neurofibromatosis. Am. J. Roentgenol., *98:* 557, 1966.

Mills, J., and Foulker, J.: Gorlin's syndrome. Radiological and cytogenetic study of nine cases. Br. J. Radiol., *40:* 366, 1967.

Neely, J. G., Harrison, G. M., Jerger, J. F., Greenberg, S. D. and Presberg, H.: The otolaryngologic aspects of cystic fibrosis. Trans. Am. Acad. Ophthalmol. Otolaryngol. *76:* 313, 1972.

Nomland, R.: Multiple basal cell epitheliomas originating from congenital pigmented basal cell nevi. Arch. Dermatol., *25:* 1002, 1932.

Overholt, E. L., and Bauman, D. F.: Variants of Kartagener's syndrome in same family. Ann. Intern. Med., *48:* 574, 1958.

Pastore, P. N., and Olsen, A. M.: Absence of the frontal sinuses and bronchiectasis in identical twins. Proc. Mayo Clin., *16:* 593, 1941.

Pennington, C. L.: Paranasal sinus changes in fibrocystic

disease of the pancreas. Arch. Otolaryngol., *63:* 576, 1956.

Peterson, R. D. A., Kelly, W. D., and Good, R. A.: Ataxia-telangiectasia. Its association with a defective thymus, immunological deficiency disease, and malignancy. Lancet, *1:* 1189, 1964.

Raskowski, H. J., and Hufner, R. F.: Neurofibromatosis of the colon. A unique manifestation of Von Recklinghausen's disease. Cancer, *27:* 134, 1971.

Rittersma, J., Ten Kate, L. P. and Westerink, P.: Neurofibromatosis with mandibular deformities. Oral Surg. Oral Med. Oral Pathol., *33:* 718, 1972.

Rosedale, R. S.: Massive fibroma of the maxillary antrum as part of multiple neurofibromatosis in siblings. Arch. Otolaryngol., *42:* 208, 1945.

Ross, P.: Gardner's syndrome. Am. J. Roentgenol., *96:* 298, 1966.

Schwachman, H., Kulczycki, L. L., Meuller, H. L., and Flake, C. G.: Nasal polyposis in patients with cystic fibrosis. Pediatrics *30:* 389, 1962.

Siewart, A. K.: Über einen Fall van bronchiextasie bei einen patienten mit situs inversus viscerum. Berl. Klin. Wochenschr. *41:* 139, 1904.

Spear, G. S., and Walker, W. G.: Lethal midline granuloma (granuloma gangrenescens) at autopsy. Bull. Johns Hopkins Hosp., *99:* 313, 1956.

Tadjoedin, M. K., and Fraser, F. C.: Hereditary of ataxia-telangiectasis (Louis-Bar syndrome). Am. J. Dis. Child., *110:* 64, 1965.

Thomaidis, T. S., and Arey, J. B.: The intestinal lesions in cystic fibrosis of the pancreas. J. Pediatr., *63:* 444, 1963.

Von Recklinghausen, F.: *Über die Multiplen Fibroma der Haut und ihre Beziehung zu den Multiplen Neuromen.* A. Hirschwald, Berlin, 1882.

Ward, W. H.: Nevoid basal celled carcinoma associated with a dyskeratosis of the palms and soles. A new entity. Aust. J. Dermatol., *5:* 204, 1960.

Williams, H. L.: Lethal granulomatous ulceration involving the midline facial tissues. Ann. Otol. Rhinol. Laryngol., *58:* 1013, 1949.

Winter, S. E., et al.: Neurofibromatosis (Von Recklinghausen's disease) with involvement of the mandible. Oral Surg. *13:* 76, 1960.

Yarington, C. T., Abbot, J., and Raines, D.: Wegener's granulomatosis. Laryngoscope, *75:* 259, 1965.

Chapter 13

Inflammatory and Allergic Diseases of the Paranasal Sinuses

ACUTE SINUSITIS

Infectious Sinusitis

General Considerations

Acute infectious sinusitis is usually caused by the extension of an inflammatory process from the nasal cavity after an upper respiratory infection, or, more rarely, from the oral cavity along the root of a tooth. The responsible organisms are usually streptococci, staphylococci, or pneumococci. Bacterial infections may also be secondary to a viral infection of the upper respiratory tract.

The primary site of reaction is the lining mucosa. The bony walls of the sinus are unaffected. The major gross pathological change is edematous swelling of the lining mucous membrane with outpouring of secretions into the sinus cavity. The secretions may become mucopurulent or frankly purulent.

Roentgen Findings

Acute infectious sinusitis may be present and produce symptoms before roentgen findings become evident. The roentgen features are as follows.

Opacity of the Sinus. Thickening of the lining mucous membrane of the sinus and the accumulation of secretions give a hazy or cloudy appearance to the sinus. If the air space is completely obliterated, the entire cavity may become uniformly opaque (Fig. 13.1).

Thickening of the Lining Mucosa. When sufficient air remains in the sinus cavity, the thickened mucosa can be well delineated as a smooth, and somewhat curved, hazy rim around the periphery of the sinus. The rim follows the contour of, and is parallel to, the walls of the sinus. These changes are well seen in the maxillary and the sphenoid sinuses, but are difficult to visualize in the frontal sinus. In the ethmoid sinus they appears as a diffuse increase in density.

Care must be taken in interpreting thickening of the lining mucosa of the lateral aspect of the maxillary sinus. Normally, the lateral wall of the maxillary sinus appears somewhat denser than the rest of the chamber because of its obliquity. This may simulate thickening of the lining mucosa (Fig. 11.1). Poor pneumatization of the frontal sinus with relative thickening of the bony walls also results in a loss of radiolucency (Fig. 13.2).

Fluid Levels. Radiologically, it is impossible to determine whether fluid levels in the sinuses are attributable to serous fluid, pus, mucus, blood, or fluid left behind from diagnostic or therapeutic procedures. The important point is that if fluid is present, it can be recognized as such provided that the patient is radiographed in the proper position, usually the upright. Patency of the ostium is also necessary. The presence of the fluid can be demonstrated as an air-fluid level if sufficient air remains to layer out above the fluid.

The presence of a fluid level is readily demonstrated in the maxillary (Fig. 13.3A) and frontal sinuses; they can usually be seen in the sphenoid sinus, but rarely in the ethmoid cells because the amount of the fluid is too small to form a recognizable interface. If the horizontal shadow as seen in the upright position is questionable, tilting of the head to either side will lead to the formation of a new horizontal level

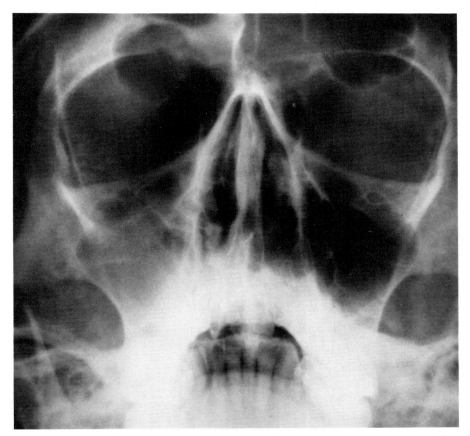

Figure 13.1. Acute infectious maxillary sinusitis. Waters view shows complete opacity of the right maxillary sinus caused by thickening of the lining mucosa and/or fluid accumulation. The mucoperiosteal line is preserved. The left maxillary sinus is normal in appearance.

and will establish the presence of fluid (Fig. 13.3B). Sometimes thick, viscid fluid may require a few moments to assume a new level when the head is tilted. In addition to testing by tilting the head, a roentgenogram may be taken in the prone position. This usually results in disappearance of the fluid level.

In the frontal sinus, the presence of a small amount of fluid is best demonstrated by a roentgenogram made in the posteroanterior projection using a horizontal beam with the patient lying on his side. In this position, the fluid gravitates to the lateral part of the sinus and forms a fluid level. In the sphenoid sinus, a fluid level is best demonstrated in the brow-up lateral projection of the skull made with a horizontal beam (see also Fig. 16.10).

Disappearance of the Mucoperiosteal Line. This change is usually seen in the late phase of

an acute infection. It is due to a thickening of the lining mucous membrane which diminishes air-bone contrast.

Changes in the Bony Walls of the Sinuses. The relatively thick bony walls of the maxillary, frontal and sphenoid sinuses do not show any change in the acute stage of infectious sinusitis, but the delicate ethmoid cells may be indistinct as a result of demineralization secondary to hyperemia.

Allergic Sinusitis

General Considerations

Allergic sinusitis is a local manifestation of an allergic reaction in the upper respiratory tract. It causes edema of the lining mucous membrane of the sinuses and nasal cavity, with

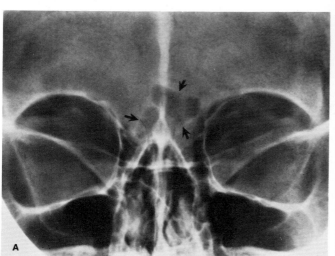
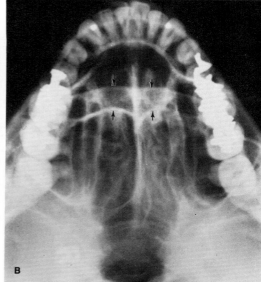
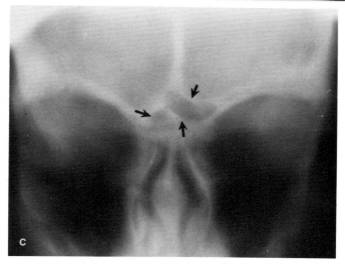

Figure 13.2. Poor pneumatization of the frontal sinuses. *A*, Caldwell view: there is a small area of pneumatization of the frontal sinuses (*arrows*). *B*, Overangulated submentovertical view: the anterior and posterior walls of the frontal sinuses are shown to good advantage (*arrows*). There is poor pneumatization of the frontal sinuses. *C*, Anteroposterior tomogram: there are two small areas of pneumatization of the frontal sinuses (*arrows*). The mucoperiosteal line is well preserved.

an outpouring of secretions. The secretions are usually clear or rarely mucopurplent.

Allergic and infectious sinusitis may coexist; not infrequently a bacterial sinusitis will be superimposed upon the allergic variety as the result of impaired drainage and stasis. The differential diagnosis may be impossible from the roentgen standpoint unless the presence of bony changes indicates the infectious component. In the early stages the presence or absence of fever and a purulent discharge is more reliable.

Roentgen Findings

The roentgen manifestations of allergic sinusitis may develop within a few minutes. The chief findings are as follows (Fig. 13.4).

Involvement of All of the Paranasal Sinuses. Allergic sinusitis often involves all of the paranasal sinuses. The maxillary sinuses are the most commonly affected.

Marked Thickening of the Nasal Turbinates. In the vast majority of cases the nasal

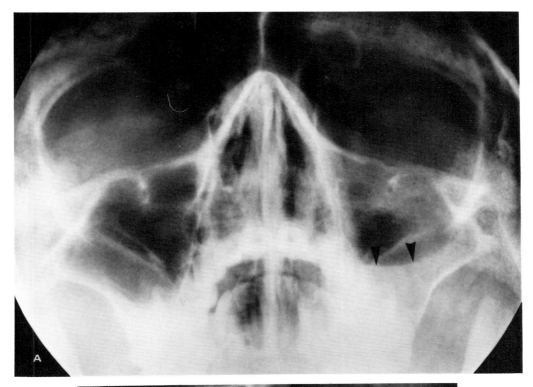

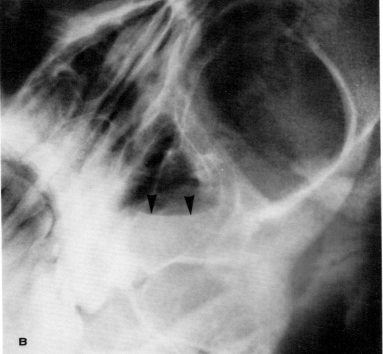

Figure 13.3. Acute infectious maxillary sinusitis with fluid level. *A*, Waters view: there is considerable thickening of the lining mucosa with a fluid level in the left maxillary sinus. The mucoperiosteal line is preserved. *B*, Tilted Waters view: shifting of the fluid level has occurred, confirming the presence of fluid.

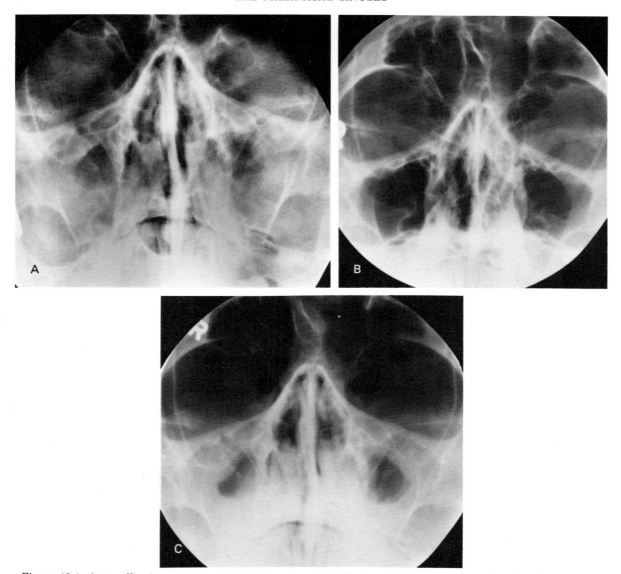

Figure 13.4. Acute allergic sinusitis. *A*, Waters view during attack: there is lobulated thickening of the lining mucosa of both maxillary sinuses. Cloudiness of the ethmoid and frontal sinuses is also present. There is marked enlargement of the nasal turbinates. The mucoperiosteal line of the maxillary sinuses is preserved. *B*, Waters view after recovery: complete regression of sinusitis with normal radiolucency of the sinuses. There is still some enlargement of the nasal turbinates. *C*, Waters view during reattack: recurrence of sinusitis with marked thickening of lining mucosa of the maxillary sinuses.

turbinates are involved, with marked thickening and edema. When the turbinates are of normal size and contour, an allergic basis for the thickening of the lining mucosal membrane of the sinus is unlikely.

Thickening of the Lining Mucosal Membrane of the Sinus. If there is a sufficient amount of air in the sinus cavity, the thickened mucosa can be visualized. Frequently the mu-

cosal margins exhibit a scalloped appearance attributable to localized thickening, cyst formation, or polypoid changes.

Opacity of the Entire Sinus. If the sinus cavity is completely filled by thickened mucosa and secretions, the entire sinus may become opaque. Under such circumstances, an allergic sinusitis cannot be distinguished from an infectious one.

Fluid Level. A fluid level is seldom seen because of the mucoid nature of the secretion.

Differential Diagnosis of Acute Infectious and Allergic Sinusitis

In most cases of acute sinusitis, the differentiation between allergic and infectious sinusitis is difficult; however, certain roentgen features may help in the differentiation.

Allergic

1. The thickened mucosa has a scalloped appearance, notably in the maxillary chambers.
2. Fluid levels are rare.
3. The nasal turbinates are usually swollen and thickened.

4. Polyp formation is common.
5. Involvement of all of the sinuses is common.
6. Involvement is usually bilateral.

Infectious

1. The thickened mucosa usually has a smooth and somewhat curved outline, paralleling the walls of the sinus.
2. Fluid levels are common.
3. The nasal turbinates may not be affected. Occasionally a nearby turbinate may be involved.
4. Polyp formation is rare.
5. Involvement of one or a few sinuses is the rule.
6. Involvement is generally unilateral.

CHRONIC SINUSITIS

General Considerations

Chronic sinusitis is usually a sequela of an acute infection which fails to resolve. Occasionally it is caused by repeated allergic attacks, or a combination of infection and allergy.

The pathological changes are mucosal thickening and edema, with accumulation of mucus, pus or mucopus within the sinus cavity.

Roentgen Findings

The roentgen features of chronic sinusitis, both of the infectious and allergic varieties, are as follows (Figs. 13.5 to 13.7).

Thickening of the Lining Mucous Membrane

The thickened mucosa is very dense and is almost as opaque as the surrounding bone. The inner outline of the sinus walls is hazy and ill defined. The thickening of the mucosal membrane is variable, often because of the presence of cysts or polyps. Polyp formation is thought to be the result of allergy.

Opacity of the Entire Sinus

The entire sinus may become opaque because of gross thickening of the mucosa, the overlapping of polyps or cysts, or the filling of the central part of the sinus with the fluid.

Fluid Level

A small fluid level may be present.

Changes in the Bony Walls

1. Disappearance of the mucoperiosteal line indicates extension of the infection from the mucous membrane into the bony margin of the sinus.
2. When the infection extends beyond the mucoperiosteal line, there will be a change in the bony density. In chronic infectious disease the bony walls may become sclerotic. This feature is usually more readily seen in the frontal sinus (Fig. 13.8). In the maxillary sinus, when the infectious process is not severe, the demonstrable sclerotic bony change is best seen in the posterolateral wall. In advanced disease all of the walls may show sclerotic changes (Fig. 13.5). In chronic allergic sinusitis, the sinus walls are generally rarefied.

Differential Diagnosis of Chronic Infectious and Allergic Sinusitis

There is no radiological means for differentiating the types of chronic sinusitis with certainty. Generally speaking, polyps are thought to be the result of allergy. In chronic infectious sinusitis the bony walls of the affected sinus frequently become sclerotic, whereas in chronic allergic sinusitis, the bony walls of the affected sinus often show rarefying osteitis.

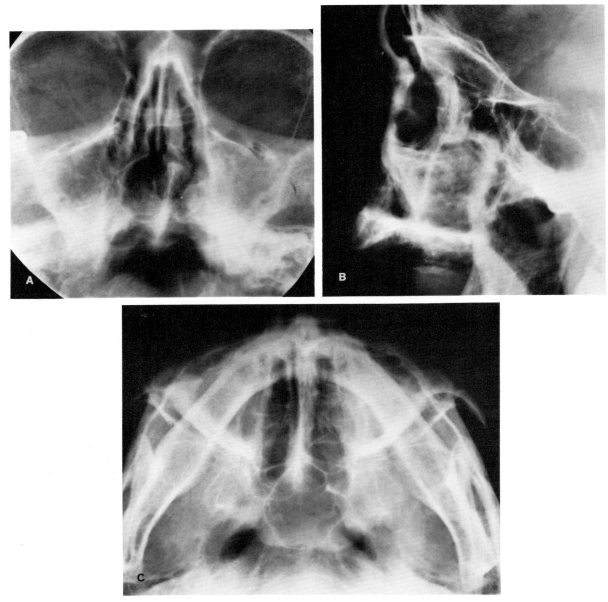

Figure 13.5. Chronic infectious maxillary sinusitis. *A*, Waters view: marked opacity of the left maxillary sinus and, to a lesser degree, of the right maxillary sinus, with disappearance of mucoperiosteal lines and sclerosis of the bony walls. *B*, Lateral view: marked bony sclerosis of all of the walls of the maxillary sinuses. *C*, Submentovertical view: marked sclerosis of posterolateral walls of the maxillary sinuses.

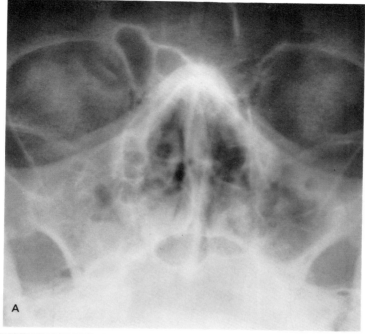

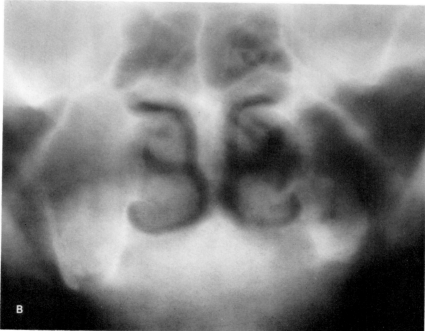

Figure 13.6. Chronic infectious sinusitis, probably dental in origin, of the maxillary sinuses. *A*, Waters view: marked opacity of the right maxillary sinus and, to a lesser degree, of the left maxillary sinus. There is disappearance of the mucoperiosteal lines and sclerosis of the bony walls. *B*, Posteroanterior tomogram: erosion and destruction of the alveolar ridge adjacent to the socket of the molar tooth in association with marked thickening of the lining mucosa in the inferior aspect of the sinus on each side. This change is more marked on the right side.

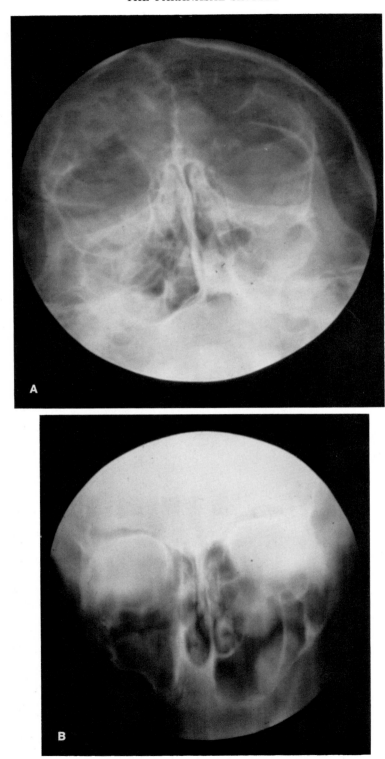

Figure 13.7. Chronic allergic sinusitis. *A*, Waters view: opacity of the maxillary, ethmoid, and frontal sinuses. The thickened lining mucosa has a scalloped appearance in the maxillary sinuses. *B*, Posteroanterior tomogram: multiple polyps in the maxillary sinuses. The mucoperiosteal lines are still preserved, but not well-defined. There is no evidence of sclerosis of the bony walls.

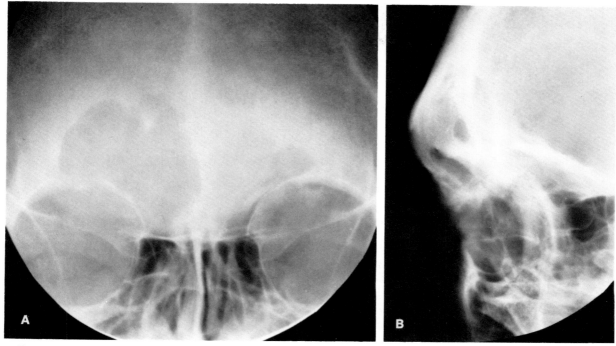

Figure 13.8. Chronic sinusitis with marked sclerosis of the frontal sinuses. *A,* Posteroanterior view: opacity of the frontal sinuses, more marked on the left side. There is disappearance of mucoperiosteal line and marked bony sclerosis. *B,* Lateral view of the frontal sinus: cloudiness of the frontal sinuses with diffuse bony sclerosis. There is no evidence of expansion and erosion of the walls. Pathological diagnosis: chronic sinusitis with mucous cyst.

COMPLICATIONS OF SINUSITIS

Polyps

Polyps of the paranasal sinuses arise from the lining membrane as a result of hyperplasia of the mucosa. It is the opinion of most observers that the great majority are the sequelae to allergic changes in the lining mucosa. Polyps may be seen in any of the paranasal sinuses, but are most common in the maxillary antra.

Radiologically, the polyp appears as a smooth, rounded soft tissue shadow projecting from the sinus walls. The shadow may be faint or fairly dense, according to the size and consistency of the polyp. Polyps are difficult to observe in the ethmoid and sphenoid sinuses, more readily seen in the frontal sinuses, and easily demonstrated in the maxillary sinuses (Fig. 13.7). They may undergo cystic degeneration and change shape in different positions.

The polyp arising from the maxillary sinus may project into the nasal cavity, occluding the nasal passage. A large polyp may project anteriorly into the nostril or posteriorly through the choana into the nasopharynx, forming an antro-choanal polyp. When the polyp protrudes into the nasopharynx and appears as a round soft tissue mass, it must be differentiated from a nasopharyngeal juvenile angiofibroma in the adolescent male patient and a nasopharyngeal carcinoma in an elderly patient. With an antrochoanal polyp, the antrum on the affected side is usually opaque. Between the roof of the nasopharynx and nasopharyngeal aspect of the polyp, there is often a radiolucent curvilinear zone, indicating that the polyp does not arise from the roof of the nasopharynx (Fig. 13.9).

Mucous Cysts (Secreting or Retention Cysts)

Mucous cysts are lined with epithelium and are the result of obstruction of the ducts of a mucous gland after sinus infections. The vast majority of cysts are small and seldom give rise to any symptoms. They may resolve spontaneously. Rarely, the cyst may enlarge and fill the sinus cavity, producing pressure deformity of the wall of the sinus. On transillumination, there is a loss of translucency of the affected

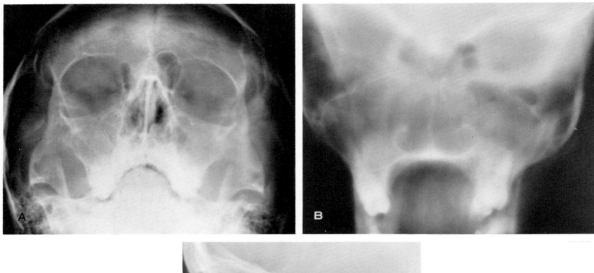

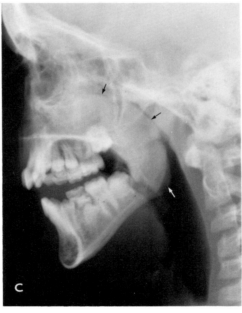

Figure 13.9. Multiple polyps in the nasal cavity and maxillary sinuses with protrusion into the nasopharynx. *A*, Waters view: opacity of both maxillary sinuses and partial opacity of the nasal cavity. *B*, Anteroposterior tomogram: multiple polyps in the nasal cavity and maxillary sinuses. *C*, Lateral soft tissue roentgenogram: a large, elongated polyp protrudes into the nasopharynx. There is a radiolucent, curvilinear zone separating the roof of the nasopharynx from the polyp.

sinus. Mucous cysts commonly occur in the maxillary sinus; occasionally they are found in the frontal and the sphenoid sinuses. Retention cysts are more commonplace than polyps, but the differential diagnosis may be difficult if change in shape cannot be demonstrated.

Radiologically, the main features of mucous cyst are as follows: (1) the cyst appears as a homogenous, dome-shaped, sharply defined mass, often along the floor of the maxillary

sinus; (2) there is generally associated thickening of the mucous membrane; (3) cysts are broadly based and do not move or change in shape with changes in position (Fig. 13.10).

Mucocele

A mucocele is an expanding secreting cyst lined by sinus mucous membrane. It occurs when an infectious sinusitis subsides but the

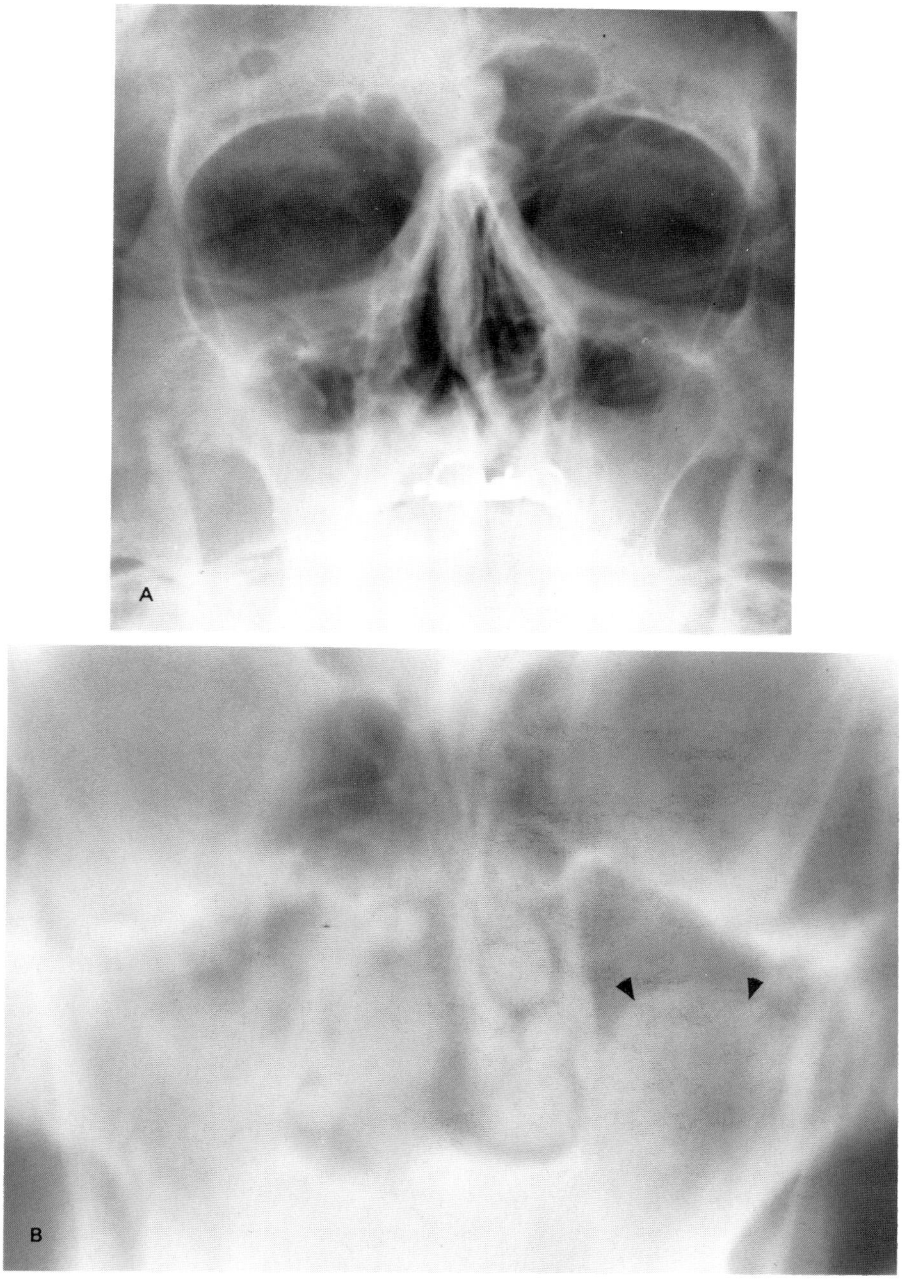

Figure 13.10. Chronic maxillary sinusitis and mucous cyst of the left maxillary sinus. *A*, Waters view: thickening of the lining mucosa and opacity in the lower aspect of both maxillary sinuses, more on the left side. *B*, Posteroanterior tomogram: a homogeneous, dome-shaped, sharply defined soft tissue density in the floor of the left maxillary sinus (*arrow*). A polyp is also noted in the nasal cavity.

ostium remains occluded. There is a continuous accumulation of secretions, consisting mostly of mucus, within the sinus. It is destructive in nature and causes gradual expansion of the sinus and pressure erosion of the bony walls. In advanced cases, it encroaches on, and displaces, adjacent structures. Mucoceles most often occur in the frontal and ethmoid sinuses, less frequently in the sphenoid, and rarely in the maxillary antra.

Radiologically the classical roentgen feature is a smooth, expanding area of bony erosion with an abnormal radiolucency of the affected sinus. There is frequently a zone of bone sclerosis around the lesion. Calcification of the wall of mucocele may occasionally be seen. Other roentgen manifestations depend on the location and extent of the lesion.

In the frontal sinus, the bony erosion is first recognized by disappearance of the sinus ridge, then by the decalcification of the bony outline and later by expansion of the sinus. The expansion is in the direction of least resistance, usually toward the orbital roof. The orbital roof may be destroyed and the globe may be displaced downward and outward, producing proptosis and diplopia. On occasion, the mucocele may extend through the anterior wall of the frontal sinus, causing an external deformity. Mucoceles may also extend through the posterior wall into the anterior cranial fossa or into the ethmoid sinus. The affected frontal sinus may appear more radiolucent than normal because of the loss of bony density. In the absence of air within the mucocele no fluid level is seen (Fig. 13.11).

In mucocele of the ethmoid sinus, the direction of expansion is usually through the thin lamina papyracea; consequently, the globe is displaced laterally and downward (Fig. 13.12).

In the sphenoid sinus, the mucocele expands anteriorly into the ethmoid region, but it may expand upward into the pituitary fossa and the middle cranial fossa (Fig. 13.13).

In the maxillary sinus a mucocele can cause slow expansion with pressure atrophy of the lining mucous membrane and erosion of the bony walls. The clinical signs and symptoms which result from a mucocele in this location are: (1) upward displacement of the globe, resulting in proptosis and diplopia; (2) enophthalmus, caused by a loss of the roof of the maxillary sinus; (3) swelling of the cheek over the affected side; and (4) a palpable defect in the anterior wall of the maxillary sinus or in the infraorbital rim.

Pyocele or Empyema

Pyocele of the sinus is an infected mucocele; it occurs when the infection of the sinus persists and the ostium remains occluded. The sinus is then filled with pus. Pyoceles most often occur in the frontal sinus but may be found in other sinuses. Pyoceles of the frontal and sphenoid sinuses are potentially dangerous because of the possibility of extension of the infection into the cranial cavity.

Initially the affected sinus becomes opaque, so that pyocele cannot be radiologically differentiated from sinusitis. Later, demineralization of the bony wall of the sinus and adjacent bone secondary to hyperemia may suggest the diagnosis of pyocele. In the vast majority of cases, the severe constitutional symptoms and physical findings often lead to the correct diagnosis. If the pyocele enters a chronic stage, the roentgen manifestations may be similar to that of the mucocele, and the final diagnosis depends upon the histological findings (Fig. 13.14).

Osteomyelitis

Osteomyelitis is the result of spread of the sinus infection into the bony wall of the sinus and the surrounding bones. It may also follow surgical intervention in sinus disease. Osteomyelitis may occur in any of the paranasal sinuses, but is most commonly seen in the frontal cells. Occasionally, it may occur in the maxillary sinuses as a result of dental sepsis or as a sequel to dental extraction.

Radiologically, the findings depend upon the stage of the lesion.

Acute stage findings include: (1) opacity of the sinus; (2) demineralization of the bony wall of the sinus, with loss of normal sharp outlines; (3) irregular patchy radiolucency of the bony wall of the sinus; (4) loss of trabecular pattern, with moth-eaten appearance of the surrounding bone.

Chronic stage findings (Fig. 13.15) include: (1) irregular thickening and sclerosis of the bony wall of the sinus and the surrounding bones; (2) sequestration—a dense, amorphous, fragment of bone is visualized, surrounded by an area of bone destruction.

Superior Orbital Fissure Syndrome

The lateral wall of the sphenoid sinus, if well pneumatized, is in close proximity to the superior orbital fissure through which the third,

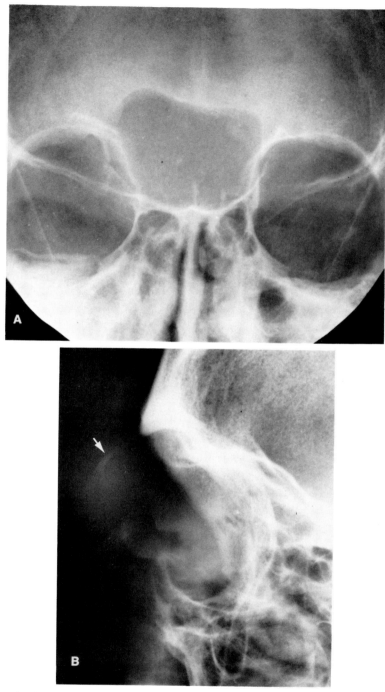

Figure 13.11. Mucocele of the frontal sinuses. *A*, Posteroanterior view: expansion of the sinuses with disappearance of the bony septum and ridges. There is a zone of bony sclerosis around the sinuses. Erosion and displacement of the superomedial wall of the right orbit is noted. *B*, Lateral view: expansion of the sinus with thinning and anterior displacement of the anterior wall (*arrow*).

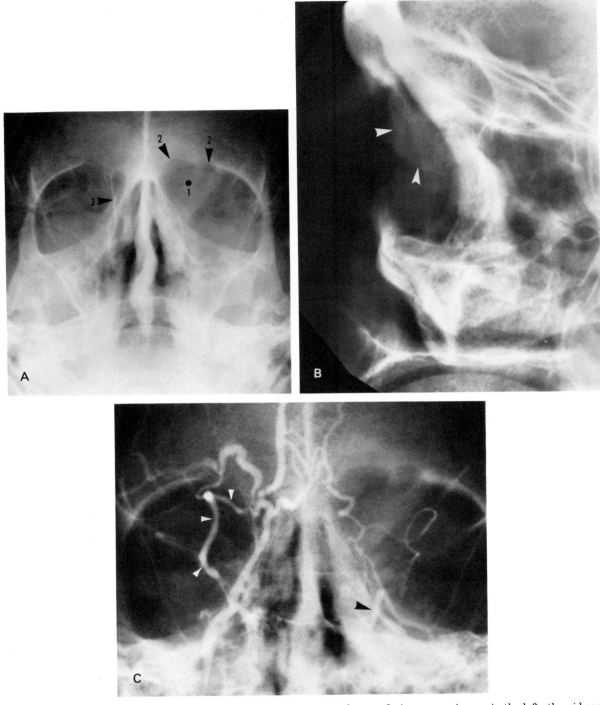

Figure 13.12. Mucocele of left ethmoid sinus. *A*, Waters view: *1*, a large soft tissue mass is seen in the left ethmoid and medial aspect of the left orbit: *2*, erosion of the superomedial margin of the left orbit; *3*, destruction of the left lamina papyracea but the right one is normal. There is opacity of both maxillary sinuses resulting from chronic sinusitis. *B*, Lateral view: a well defined soft tissue mass in the superoposterior aspect of the orbit (*arrows*). *C*, Orbital venogram (anteroposterior projection): complete obstruction of the first and second segments of the left superior ophthalmic vein. There is visualization of the distal end of the third segment of the vein (*arrow*). The right superior ophthalmic vein is normal (*arrows*).

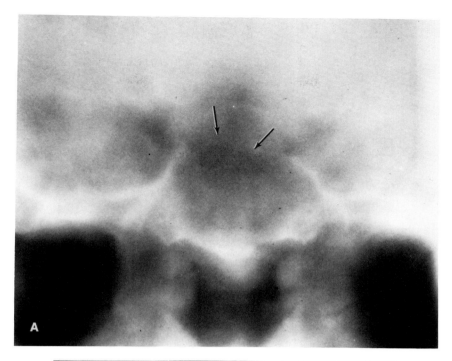

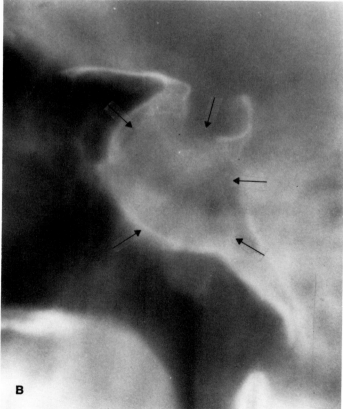

Figure 13.13.   Mucocele of the sphenoid sinus. *A*, Anteroposterior tomogram: a large soft tissue mass occupies the entire sphenoid sinuses with thinning and expansion of the lateral walls, especially on the left side. There is extension of the lesion to the floor of the sella turcica (*arrows*). The floor of the sphenoid sinuses is slightly depressed and somewhat ragged. *B*, Lateral tomogram: the large soft tissue mass is shown to good advantage (*arrows*). There is expansion of the sinuses with ballooned appearance. A bony defect is seen in the floor of the sella turcica.

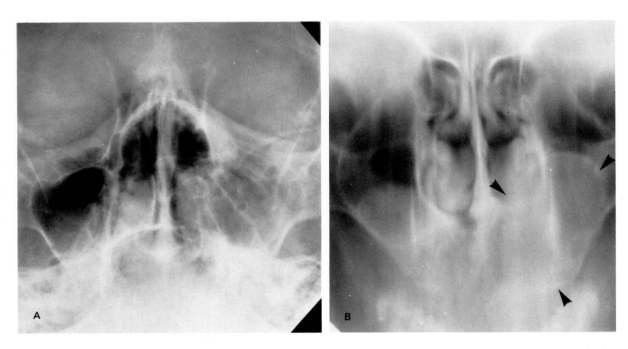

Figure 13.14. Pyocele or infected mucocele of the left maxillary sinus. *A*, Waters view: a large soft tissue mass in the left maxillary sinus destroying the medial wall of the sinus and protruding into the left nasal fossa. There is thickening of the lining mucosa. A mucous cyst is noted in the floor of the right maxillary sinus. There is soft tissue swelling in the left ala nasi. *B*, Posteroanterior tomogram: an expanding cystic mass with a well demarcated capsule arising in the alveolar recess of the left maxillary sinus and extending upward into the sinus cavity, medially into the floor of the left nasal fossa, and inferiorly to the alveolar ridge. There is erosion of the floor of the left nasal fossa. A mucous cyst is again noted in the floor of the right maxillary sinus.

An 80-year-old female complained of pain and swelling over the left maxillary area for about 4 weeks. Examination of the face revealed swelling involving the tissues of the left side of the floor of the nose and adjacent left cheek area. There was also swelling of the left upper alveolar ridge, extending to the hard palate. Aspiration of the mass revealed necrotic and purulent material.

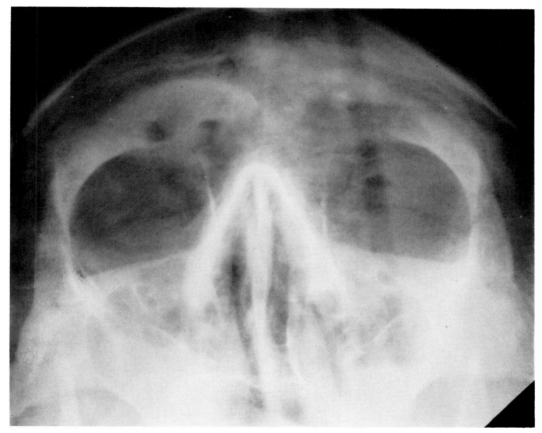

Figure 13.15. Chronic sinusitis with osteomyelitis of the frontal sinus. Waters view: opacity of the frontal sinuses with disappearance of the mucoperiosteal line. There is marked irregular bony sclerosis. Focal areas of bony destruction are noted in the inferior aspect of the sinuses, more marked on the left. The superomedial wall of the orbit is destroyed on each side.

fourth, and sixth cranial nerves, the first division of the fifth cranial nerve, the ophthalmic vein, and sympathetic nerves from the cavernous plexus exit. An acute or chronic inflammatory process in the sphenoid sinus may extend to the superior orbital fissure and any or all of the structures passing through the fissure may be affected. The sixth cranial nerve is usually involved first, with subsequent involvement of the third and fourth nerves. As the disease progresses, the fifth cranial nerve is involved, with resultant pain in the eye and forehead. This stage is followed by exophthalmos and total ophthalmoplegia.

Radiologically, there is erosion or destruction of the medial wall of the superior orbital fissure. Demineralization or spotty destruction of the bone surrounding the fissure may also be demonstrable.

Optic Nerve-Retrobulbar Neuritis

Approximately 15 to 20% of retrobulbar neuritis is said to be caused by sinusitis. This is not surprising because the optic nerve may be in close relationship to the sphenoid, ethmoid, and maxillary sinuses, depending upon the degree of pneumatization. Inflammatory processes involving the sinuses may spread directly through the sinus wall or by venous channels to the retrobulbar space.

Radiologically, in retrobulbar neuritis, conventional views of the paranasal sinuses may fail to reveal any pathological changes. Tomograms of the paranasal sinuses, especially at the posterior levels, are of extreme value in detecting infection in the sinuses as a possible cause of retrobulbar neuritis.

Intracranial Complications

Possible intracranial complications from inflammatory processes in the paranasal sinuses are meningitis, extradural and subdural abscess, brain abscess, and septic thrombosis of the cavernous or superior longitudinal sinus.

Acute sinusitis is more apt to result in intracranial complications than is chronic sinusitis. Any infected paranasal sinus may give rise to an intracranial complication, but extension from the maxillary sinus is rare.

REFERENCES

Ballenger, J. J.: *Diseases of the Nose, Throat and Ear,* 11th ed., Lea and Febiger, Philadelphia, 1969.

Belal, A.: Mucocele of the maxillary sinus. J. Laryngol. Otol., *65:* 286, 1951.

Ceurville, C. B., and Rosenvold, L. K.: Intracranial complications of infections of nasal cavities and accessory sinuses. Arch. Otolaryngol., *27:* 629, 1938.

Christensen, Lt. J. R., and Houck, L.: Mucocele of the maxillary sinus. Arch. Otolaryngol., *59:* 147, 1954.

Cone, A. J., More, S. and Dean, L. W.: Relationship of paranasal sinus disease to ocular disorders. Laryngoscopy, *49:* 374, 1934.

Hansel, F. K.: *Allergy of the Nose and Paranasal Sinuses.* C. V. Mosby, St. Louis, 1936.

Hansel, F. K.: *Clinical Allergy.* C. V. Mosby, St. Louis, 1953.

Isben, B.: Basal convex shadows in maxillary sinus. Nord. Med., *27:* 1487, 1945.

James, A. E., Jr.: Destroyed sphenoidal sinus (x-ray seminar). J.A.M.A., *206:* 2732, 1968.

Kern, R. A., and Schenck, H. P.: Allergy a constant factor in the etiology of so-called mucous nasal polyps. J. Allergy, *4:* 485, 1933.

Lundgren, A. and Olin, T.: Muco-pyocele of sphenoid sinus or posterior ethmoid cells with special reference to the apex orbital syndrome. Acta Otolaryngol., *53:* 61, 1961.

Nugent, G. R., Sprinkle, P., and Bloor, B. M.: Sphenoid sinus mucocele. J. Neurol., *32:* 443, 1970.

Reinecke, R. D., and Montgomery, W. W.: Oculomotor nerve palsy associated with mucocele of the sphenoid sinus. Brde. Epletbal., *71:* 50, 1964.

Samuel, E.: Inflammatory disease of the nose and paranasal sinuses. Semin. Roentgenol., *3:* 148, 1968.

Selden, H. S., August, D. S.: Maxillary sinus involvement—an endodontic complication. Oral Surg., *30:* 117, 1970.

Skillern, S. R.: Obliterative frontal sinusitis. Arch. Otolaryngol., *23:* 267, 1938.

Welin, S.: Paranasal sinuses. In *Roentgen Diagnosis,* 2nd American ed., edited by L. G. Rigler, Grune and Stratton, New York, 1969.

Zizmor, J., Noyek, A. M. and Chapnik, J. S.: Mucocele of the paranasal sinuses. Can. J. Otolaryngol. (suppl.), 3:1, 1974.

Chapter 14

Benign Cysts of the Paranasal Sinuses

CLASSIFICATION

Benign cysts usually arise from the lining mucosa or submucosa of the paranasal sinuses. The sinuses may also be involved by cysts of extrinsic origin. Cysts, arising from or involving the paranasal sinuses, may be classified as follows:

Cysts of Intrinsic Origin

1. Secreting (Mucous or retention) Cyst
2. Nonsecreting (serous) Cyst
3. Mucocele
4. Cholesteatoma

Cysts of Extrinsic Origin

1. Odontogenic Origin
   a. Primordial cyst
   b. Dentigerous cyst
   c. Radicular (root or dental) cyst
   d. Tuberosity cyst
   e. Others
2. Neurogenic Origin
   a. Meningocele
   b. Encephalocele
3. Miscellaneous
   a. Globulomaxillary cyst
   b. Dermoid cyst

CYSTS OF INTRINSIC ORIGIN

Mucous Cyst (Secreting or Retention Cyst)

The mucous cyst is described in Part II, Chapter 13.

Serous Cyst (Nonsecreting Cyst)

Serous cysts are thought to be the end result of a degenerative process after a localized edematous distention of the loose connective tissue underlying the sinus mucosa. The cysts are inflammatory or allergic in origin and nonsecretory in nature. They are lined by connective tissue and contain a clear amber-colored fluid with a high percentage of cholesterol. They are generally unassociated with changes in the lining mucosa in other parts of the affected sinus and often show no abnormality on transillumination, presumably because of the double refractory property of cholesterol. The cysts usually do not give rise to symptoms and require no treatment.

Serous cysts are the most common type found in the maxillary sinuses, but they are occasionally found in other paranasal chambers.

Roentgen Findings

The following radiological features may be seen in serous cysts:

1. A smooth walled and dome-shaped density is often found in the affected sinus, notably in the floor of the maxillary sinus (Fig. 14.1A).
2. A slight alteration of the contour of such a cyst may occur with change in position (Fig. 14.1B).
3. The lining mucosa of the remainder of the affected sinus is usually normal in appearance (Fig. 14.1A).

Mucocele

Mucoceles are described in Part II, Chapter 13.

Cholesteatoma

General Considerations

Cholesteatomas are benign, expansile cystic lesions which are lined by keratin-producing

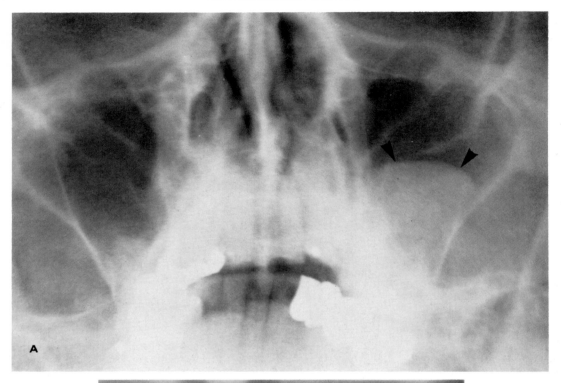

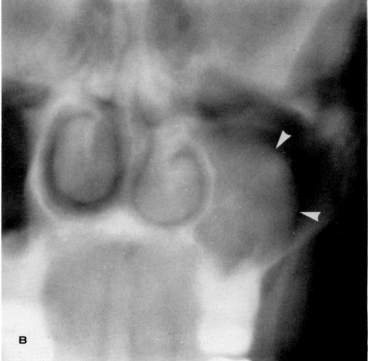

Figure 14.1. Serous cyst of the maxillary sinus. *A*, Waters view: a smooth, well defined soft tissue density is present in the floor of the left maxillary sinus. There is no associated thickening of the lining mucosa. *B*, Tomogram in the supine position: an alteration of the contour of the soft tissue density is apparent with the change in position.

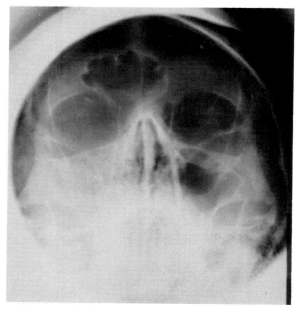

Figure 14.2. Cholesteatoma of the maxillary sinus. Waters view: Opacification of the right maxillary sinus without erosion or destruction of the walls. The findings are nonspecific. There is diffuse haziness in the right orbit, probably due to soft tissue swelling. (Reprinted with permission from Archives of Otolaryngology *82:* 532, 1965.)

mous epithelium then develops into the cholesteatoma. Whereas cholesteatomas do occur as primary lesions of congenital origin, the majority are secondary to chronic sinusitis.

Cholesteatomas are quite rare in the paranasal sinuses. They are usually found in the maxillary chambers but may occur in other sinuses. Clinically, the patient may have pain in the cheek, deformity of the face, and displacement of the eyeball.

Roentgen Findings

The roentgen findings are as follows (Fig. 14.2): (1) opacification of the affected sinus; (2) thinning and erosion of the bony walls of the sinus; (3) destruction of the bony walls of the sinus.

Differential Diagnosis

The roentgen findings are nonspecific and should be differentiated from:

1. Pseudocholesteatoma — a pseudocholesteatoma may arise when chronic infection blocks the ostium of a maxillary sinus resulting in the accumulation of an inflammatory exudate. This may occur even when the infective organism is no longer active. Breakdown products of the purulent exudate contain certain insoluble fats, including cholesterol. The crystallized cholesterols are surrounded by foreign body giant cells and fibrous tissue, giving the appearance of cholesteatomatous debris. The cavity is usually lined by a fibrous capsule.

2. Mucocele — this is discussed in Chapter 13.

3. Malignant lesions of the maxillary sinus — these are discussed in Chapter 18.

stratified squamous epithelium and which contain desquamated material.

The tumors may form as the result of: (1) inclusion of epidermis in the sinus cavity during fetal life; (2) invasion of the sinus cavity by buccal epithelium via an oroantral fistula; and (3) squamous metaplasia of the lining mucosa attributable to a chronic sinusitis. The squa-

CYSTS OF EXTRINSIC ORIGIN

Cysts of Odontogenic Origin

Primordial Cysts

General Considerations. A primordial cyst is derived from the primordial enamel cells and formed in the tooth bud before the enamel and dentin have developed. It therefore may arise in portions of the jaws where there have been no teeth or are no teeth. The most common site is the anterior border of the ascending ramus of the mandible above the level of the alveolar crest. It may also be found at the angle, in the coronoid process, and in the central portion of the ascending ramus. It is rarely found in other parts of the mandible. In the maxilla the canine

area is the most common site of occurrence and the sinus may be invaded. The cysts may be found at any age, ranging from 4 to 84 years in one series, but all probably develop in early life.

Roentgen Findings. These are as follows:

1. Primordial cysts appear as oval-shaped or rounded areas of radiolucency with well-defined sclerotic margins. They are usually located posterior to the third mandibular molar and are not attached to the root or crown of the tooth.

2. The cysts almost always have a single cavity although roentgenograms may show a lobulated appearance.

3. The cysts may occur between the roots of the teeth with separation of the adjacent teeth.

4. When the cysts arise in the maxilla, they may bulge into the maxillary sinus.

Primordial cysts should be differentiated from unilocular ameloblastomas.

Dentigerous Cysts

General Considerations. Dentigerous cysts develop in or from some part of the tooth follicle after the crown of the tooth has been formed, but before the tooth has erupted. They are found more often in the mandible than in the maxilla. The most common sites in the mandible are the canine and the molar regions. In the maxilla the cysts often involve the canine teeth. The cysts generally are much larger than the primordial or radicular variety and consequently often cause considerable expansion of the bone. When the cysts occur in the maxilla, the maxillary sinus may be involved. Dentigerous cysts are usually found in young persons.

Roentgen Findings. These are as follows.

1. Dentigerous cysts are generally classified into three types: (1) central type—the entire crown of the tooth is included in the cyst; (2) lateral type—only a part of the crown is involved in the cyst; (3) circumferential type—the entire enamel organ around the neck of the tooth becomes cystic and allows the tooth to erupt through the cyst (this type is extremely rare).

2. The central or lateral cyst shows a well defined oval or rounded area of radiolucency, attached to the crown of the tooth (Fig. 14.3A).

3. In the maxilla, dentigerous cysts may extend upward into the maxillary sinus, with displacement of the floor. The cysts, including the involved tooth, may fill the sinus (Fig. 14.3B). With increase in size, pressure erosion of the bony walls of the sinus may ensue. This bony erosion must be differentiated from destruction caused by malignant tumors.

4. Dentigerous cysts are usually solitary. When multiple, the possibility of an association with the nevoid basal cell carcinoma syndrome should be considered.

Radicular Cysts (Dental Cysts, Periodontal Cysts, or Dental Root Cysts)

General Considerations. A radicular cyst occurs in a tooth which has matured and erupted. It is primarily an infected granuloma about the root of a nonvital tooth. The root of the tooth may have been partially absorbed. There is probably little difference in the relative involvement of the upper and lower jaws, al-though some observers state that the cysts occur more often in the maxilla than in the mandible. The molar and incisor teeth are most frequently involved and cysts of the upper jaw may extend into the maxillary sinus. Radicular cysts are more frequently seen in the fourth and fifth decades.

Roentgen Findings. These are as follows:

1. A radicular cyst appears as a rounded or oval radiolucency at the root of a tooth, often demarcated by marginal sclerosis.

2. The root of the tooth may be partially destroyed.

3. The cysts may be found in edentulous areas of the jaws or may persist after extraction of a tooth or root (Fig. 14.4).

4. With increase in size the cyst may extend to the maxillary sinus, completely filling it and causing expansion and erosion of the wall of the sinus. However, the degree of involvement is usually not as extensive as that produced by dentigerous cysts.

The cysts should be differentiated from rarefying osteitis caused by an abscess.

Tuberosity Cysts

A cyst of the maxillary tuberosity is of dental origin and quite rare. Radiologically it appears as a small radiolucent area involving the tuberosity, with expansion of the cortex. It should be differentiated from a normal maxillary tuberosity. Careful roentgen examination should include dental films and tomograms. A lattice-like appearance of the cancellous bone of the tuberosity and the absence of bony expansion will differentiate the normal tuberosity from a cyst.

Cysts of Neurogenic Origin

Meningocele and Encephalocele

Rarely, neurogenic cysts, usually meningoceles or encephaloceles, may involve the paranasal sinuses. The majority are of congenital origin but they may also be traumatic in nature. The origin and nature of meningoceles and encephaloceles have been described in Part I, Chapter 4.

In the sincipital type, the frontal and ethmoid sinuses may be involved. Occasionally, in the basal type, the cyst may appear in the sphenoid sinus as a well defined soft tissue mass. When encountered in the frontal, ethmoid, and sphenoid sinus regions, mucoceles and cholesteatomas should be considered in the differential diagnosis.

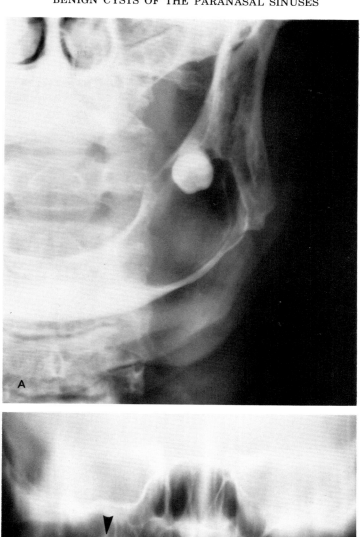

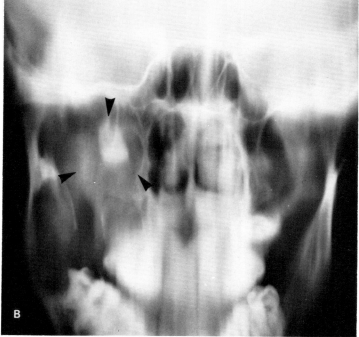

Figure 14.3. Dentigerous cyst. *A*, Typical dentigerous cyst of the mandible. Anteroposterior view of the mandible: a well defined oval-shaped area of radiolucency is attached to the crown of the left lower third molar. *B*, Posteroanterior tomogram of the paranasal sinuses: a large dentigerous cyst in the right maxillary sinus with upward displacement of the molar tooth. The cyst occupies most part of the sinus and its wall can be fairly well visualized.

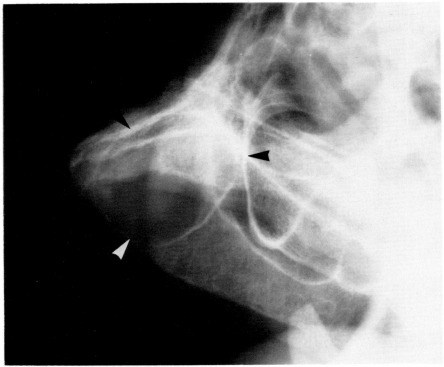

Figure 14.4. Radicular cyst of the maxilla. Oblique view of the maxilla: a well defined, oval-shaped radiolucency with sharp margination is present in the right upper canine region in an edentulous patient. There is no involvement of the maxillary sinus.

Miscellaneous Cysts

Globulomaxillary Cysts

General Considerations. Globulomaxillary cysts are the most common type of lateral fissural inclusion cysts of the maxilla. They represent cystic involutions from a failure of complete fusion of the globular and maxillary processes of the maxilla. The cysts develop in the alveolar ridge between the roots of the lateral incisor and canine teeth and tend to displace and separate the involved teeth. They may encroach upon the maxillary sinus.

Roentgen Findings. These are as follows:

1. If a globulomaxillary cyst is small, it usually appears as a unilocular, radiolucent teardrop, with a smooth, well defined margin, in the canine fossa.

2. As it grows larger, the cyst extends beyond the apices of the teeth and may destroy the globular portion of the palate. This will cause separation of the roots of the canine and incisor teeth.

3. The cyst may grow upward into the maxillary sinus, with displacement of the floor of the sinus.

4. Globulomaxillary cysts may be confused with a large alveolar recess of the maxillary sinus. With Waters and lateral views it is often possible to differentiate between the two, but in doubtful cases a tomogram may be required to establish the diagnosis.

Dermoid Cysts

General Considerations. Dermoid cysts involving the paranasal sinuses are quite rare. They may occur in the nose, with extension into the frontal, ethmoid, or maxillary sinuses. Those arising in the hard palate or orbit may also involve the maxillary antra. Dermoid cysts are of the inclusion variety and contain the normal structures of the skin, such as glands, hair, and teeth. The cysts are present at birth but may not become manifest until adolescence.

Roentgen Findings. The roentgen findings are variable, depending upon the location of the cyst and the extent of involvement of the paranasal sinuses. The chief roentgen features are: (1) expanding cystic lesion with little or no marginal sclerosis at the site of origin; (2) displacement and erosion of the bony wall of the involved paranasal sinus and adjacent facial bones.

REFERENCES

Baxter, J. S. R.: Cholesteatoma of the maxillary antrum. J. Laryngol., *80:* 1059, 1966.

Beekhuis, G. J., and Watson, T. H.: Mid-facial cysts. Arch. Otolaryngol., *85:* 65, 1967.

Blumenfeld, R., and Skolnik, E. M.: Intranasal encephaloceles. Arch. Otolaryngol., *82:* 527, 1965.

Hardy, G.: Benign cysts of the antrum. Ann. Otol. Rhinol. Laryngol., *48:* 649, 1939.

McGavran, M. H.: Neurogenic nasal neoplasms. Ann. Otol. Rhinol. Laryngol., *79:* 547, 1970.

Pogorel, B. S., and Budd, E. G.: Cholesteatoma of the maxillary sinus. A case report. Arch. Otolaryngol., *82:* 532, 1965.

Paparella, M. M.: Mucosal cyst of the maxillary sinus. Diagnosis and treatment. Arch. Otolaryngol., *77:* 650, 1963.

Sonesson, A.: Odontogenic cysts and cystic tumors of the jaws. Acta Radiol. (Suppl.), *81:* 50, 1950.

Walsh, F. V.: *Clinical Neuro-Ophthalmology,* 2nd. ed., p. 952. Williams & Wilkins, Baltimore, 1957.

Zizmer, J., and Noyek, A. M.: Cysts and benign tumors of the paranasal sinuses. Semin. Roentgenol., *3:* 172, 1968.

Chapter 15

Chronic Granulomatous Diseases and Mycotic Infections of the Paranasal Sinuses

TUBERCULOSIS OF THE PARANASAL SINUSES

General Considerations

Tuberculous infection of the paranasal sinuses is rare. Not more than 60 cases have been reported in the literature. The disease is nearly always secondary to pulmonary or extrapulmonary tuberculosis, with involvement of the paranasal sinuses by the hematogenous route or by direct extension.

In the early stage the disease involves the lining mucosa. The sinuses are filled with polypoid masses and thickened mucosa, but purulent discharge is scanty. As the disease progresses there is often bony involvement, with the formation of fistulae. At this stage, purulent discharge is abundant and tubercle bacilli are readily demonstrable.

The symptoms of tuberculous sinusitis are nonspecific. Local symptoms include pain in the region of the affected sinus, nasal discharge and obstruction, epistaxis, and swelling of the eyelids. Pain is a late symptom and progresses in severity as the bony wall is involved. Constitutional symptoms are insomnia, vertigo, and weakness.

Any sinus may be affected, but the maxillary and ethmoid sinuses are the most susceptible.

Roentgen Findings

The common roentgen findings are as follows (Fig. 15.1): (1) opacity of the affected sinus caused by thickening of the lining mucosa and polyp formation; (2) erosion and destruction of the bony walls of the affected sinus in advanced lesions; (3) association with pulmonary or osseous tuberculosis.

Differential Diagnosis

There are no characteristic roentgen findings which allow a radiological diagnosis of tuberculosis of the paranasal sinuses. In the early stage, with only mucosal involvement, it must be differentiated from chronic nonspecific sinusitis. When the disease becomes advanced, with bone destruction, malignant tumors of the paranasal sinus must be excluded.

SYPHILIS OF THE PARANASAL SINUSES

General Considerations

Syphilitic infection of the paranasal sinuses is extremely rare. As in the case of tuberculosis, it is usually secondary in nature. The tertiary variety is the form which affects the paranasal sinuses. This is always periosteal and/or perichondral. Syphilis has a predilection for the intermaxillary bones and the lamina papyracea of the ethmoid cells; syphilitic sinus infections are generally secondary to bone lesions in these regions. The vomer, as well as the hard and soft palates, is also involved in tertiary syphilis. Ulcerations of the palate may perforate into the nasal cavity or maxillary sinus. Syphilis may cause suppuration in any one or all of the paranasal chambers.

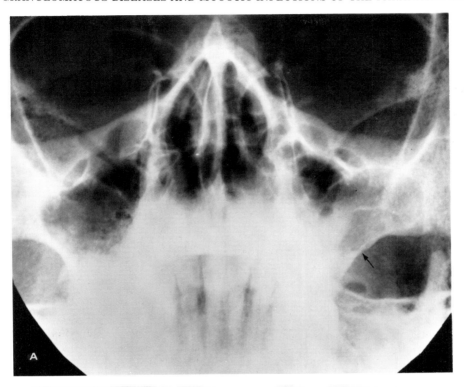

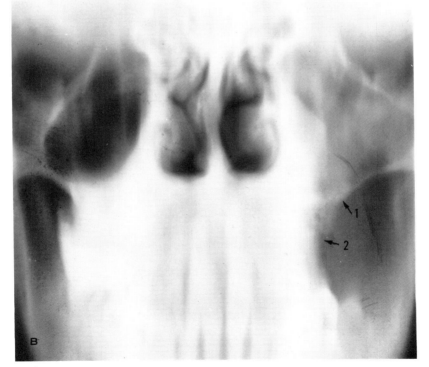

Figure 15.1. Tuberculosis of the hard palate and left maxillary sinus. *A*, Waters view: opacity of the lower part of the left maxillary sinus with slight irregularity and thinning of the lateral wall of the sinus (*arrow*). *B*, Posteroanterior tomogram: irregular densities in the left maxillary sinus with destruction of the inferolateral wall and alveolar recess of the sinus (*arrow 1*). There is also destruction of the maxilla and hard palate on the left side (*arrow 2*).

Headache is a common symptom; mild pain and discomfort are often present. Nasal obstruction together with a foul smelling discharge is characteristic but not pathognomonic.

Roentgen Findings

The chief roentgen features of syphilis of the paranasal sinuses are: (1) marked irregular thickening of the lining mucosa of the affected sinus; (2) disappearance of the mucoperiosteal line with loss of the underlying bone density; (3) sclerotic, moth-eaten destruction of the maxilla and zygoma.

The roentgen diagnosis of syphilitic infection is possible, especially if supported by clinical evidence. An infected sinus, with a foul nasal discharge and the presence of necrotic bone, should suggest syphilis. The differential diagnosis includes granulomatous and malignant lesions, both of which thicken the mucous membranes and destroy bone.

SARCOIDOSIS OF THE PARANASAL SINUSES

General Considerations

Sarcoidosis (Boeck's sarcoid) is recognized as a systemic disease with lesions occurring in the skin, lymph nodes, lungs, bones, spleen, liver, eye, parotid gland, and other organs. The lesions consist of granulomatous nodules resembling noncaseating tubercles.

Involvement of the paranasal sinuses is infrequently observed and rarely reported. The majority of patients are between 20 and 40 years of age and are usually asymptomatic. Some obstruction and a sparse nasal discharge may be present.

Roentgen Findings

The roentgen findings are nonspecific and are as follows (Fig. 15.2): (1) opacity of the affected sinus; (2) soft tissue mass in the sinus; (3) destruction of the walls of the affected sinus; (4) associated changes in the lungs.

When changes in the paranasal sinuses and lung parenchyma are considered together they may simulate malignant disease of the sinuses with metastasis to the lungs. The final diagnosis often depends upon tissue biopsy.

WEGENER'S GRANULOMA

Wegener's granuloma has been discussed in detail in Part II, Chapter 12.

MIDLINE LETHAL GRANULOMA

Midline lethal granuloma has been discussed in Part II, Chapter 12.

MYCOTIC INFECTIONS OF THE PARANASAL SINUSES

General Considerations

Mycotic infections of the paranasal sinuses are rare. They usually occur as part of a generalized secondary invasion and only rarely as a primary process without predisposing factors. The incidence of mycotic infections has increased considerably in recent years. This is probably related to alterations of the normal flora balance as a result of the extensive use of antibiotics and also to the immunosuppressive effects of chemotherapeutic agents such as folic acid antagonists and nitrogen mustards. Patients with debilitating diseases, diabetes, and other metabolic disturbances are also susceptible to mycotic infection. The infections occur more frequently in cancer patients and, more specifically, in patients with malignant lymphoma than in any other type of disease. Patients with chronic, debilitating illnesses who are receiving steroids, antibiotics, and antimetabolites may die from an unsuspected mycotic infection rather than the primary disease.

Mycotic infections may be cutaneous or vis-

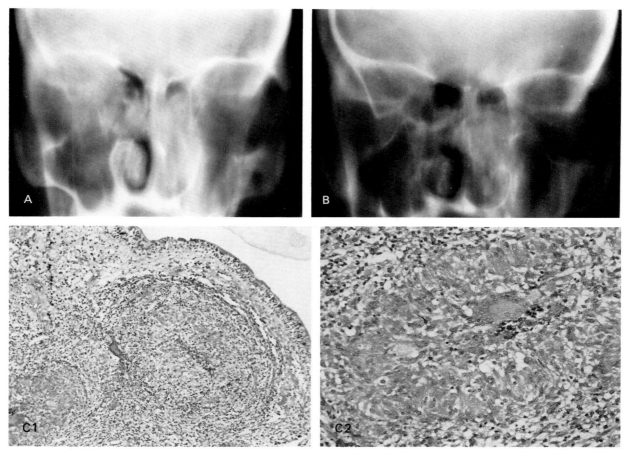

Figure 15.2. Boeck's sarcoid of the nasal cavity and paranasal sinuses. *A*, Anteroposterior tomogram — anterior section: there is opacification of left nasal cavity, left ethmoid, and left maxillary sinus. The right ethmoid and right nasal cavity are also partially opacified. *B*, Anteroposterior tomograms — posterior section: the soft tissue mass is well shown. It involves both sides of the nasal cavity, both ethmoid sinuses, and the left maxillary sinus. There is destruction of the medial wall of the left maxillary sinus and nasal septum. *C*, Histological findings: *1*, noncaseating granuloma in submucosa of maxillary antrum (H&E, × 80); *2*, a tubercle-like granuloma composed of epithelioid cells. Note the absence of caseous necrosis (H&E, × 180).

This 25-year-old woman had had generalized sarcoidosis for 10 years and symptoms of maxillary sinusitis for 4 years. Biopsy of the left nasal and left maxillary mucosa confirmed the presence of sarcoid. (Reprinted with permission from Radiology *113:* 619, 1974.)

ceral. There are three main types of visceral infection: (1) pulmonary, (2) gastrointestinal, and (3) head and neck. In the latter the mycotic organisms enter through the upper respiratory tract and may infect the nose, the paranasal sinuses, the pharynx, or the palate.

Among the mycotic organisms involved, Aspergillus and Mucormyces are the most common; Actinomyces, Nocardia, and other fungi are rare.

The maxillary sinuses are most often involved; the ethmoid, frontal, and sphenoid sinuses are rarely affected.

The symptoms of mycotic infection of the paranasal sinuses are variable, but do not differ significantly from that of bacterial infection. In addition, there is frequently a foul smelling, purulent nasal discharge. Most mycotic infections tend to be chronic, whereas mucomycotic infections tend to be acute and progress rapidly.

Roentgen Findings

The radiological findings in mycotic infection are nonspecific and consist of the following (Fig. 15.3): (1) thickening of the lining mucosa of the affected sinus; (2) opacity of the affected sinus; (3) sclerosis of the bony walls of the affected sinus; (4) erosion and destruction of the bony walls of the affected sinus.

The roentgen changes are not diagnostic and may simulate bacterial infection or malignant disease. The final diagnosis depends upon biopsy and culture of the suspected organisms.

AMYLOIDOSIS OF THE PARANASAL SINUSES

General Considerations

Amyloidosis is a degenerative disorder characterized by the deposition of amyloid in the intercellular spaces of various organs. The exact chemical composition of amyloid is unknown, but a protein-chondroitin-sulfuric acid complex has been identified. Amyloid is probably not a single chemical substance, but a series of closely related protein compounds. The cause of amyloid deposition remains in question, although much is known about the incidence, histological appearance, and clinical course of amyloidosis. Recent thinking tends to include it among diseases of the immunologic system. Amyloidosis is customarily classified into four types: (1) generalized primary amyloidosis with absence of predisposing disease; (2) generalized secondary amyloidosis accompanying a wide variety of chronic inflammatory processes such as tuberculosis, Mediterranean fever, and rheumatoid arthritis; (3) amyloidosis associated with myeloma; and (4) localized primary amyloidosis.

In generalized amyloidosis, amyloid deposition in the upper digestive tract, especially the tongue, is common. Localized primary amyloidosis of the upper respiratory tract is uncommon; the usual site of involvement is the larynx. Amyloidosis of the paranasal sinuses is extremely rare. Trible and Loewenheim (1963) reported an example of maxillary sinus amyloidosis, Brown (1966) reported a case involving the frontal sinus, and Garrett (1968) reported the details of a third patient with maxillary sinus and nasal septum involvement.

The clinical behavior of amyloidosis of the paranasal sinuses may be that of an actively growing benign tumor. The process is expansile and destructive and often recurs.

Roentgen Findings

The roentgen manifestations are nonspecific and include: (1) opacity of the sinus; (2) soft tissue mass in the sinus; (3) bony erosion and destruction of the wall of the sinus (Fig. 15.4).

Differential Diagnosis

The diseases to be differentiated are benign expanding tumors and chronic granulomatous diseases of the paranasal sinuses.

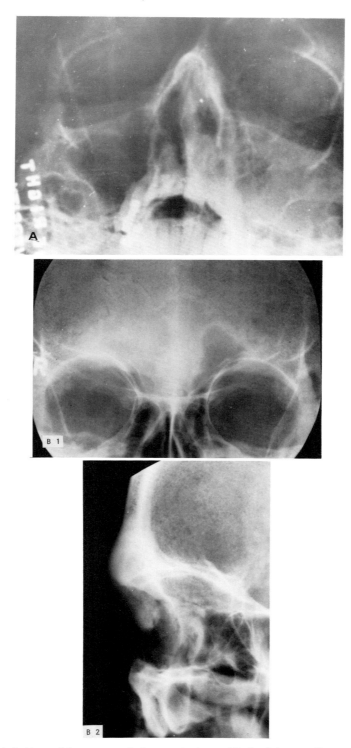

Figure 15.3. Mycotic infections of the paranasal sinuses. *A*, Aspergillosis of the maxillary sinus. Waters view: opacity of the left maxillary sinus, with bony erosion of the lateral and medial walls. The change is nonspecific, simulating chronic sinusitis.

B, Nocardiosis of the frontal sinus. *1*, posteroanterior view: opacity of the frontal sinuses, especially on the right, with disappearance of mucoperiosteal line and bony sclerosis; *2*, lateral view: opacity of the frontal sinuses with marked sclerosis of the anterior and posterior walls and adjacent frontal bone. (Reprinted with permission from Radiology 90: 49, 1968.)

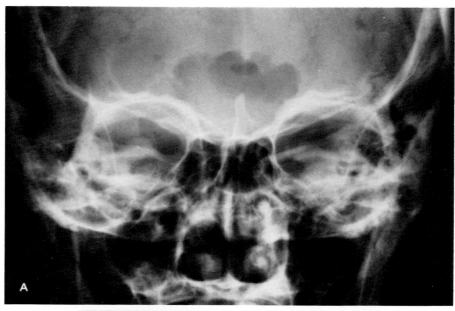

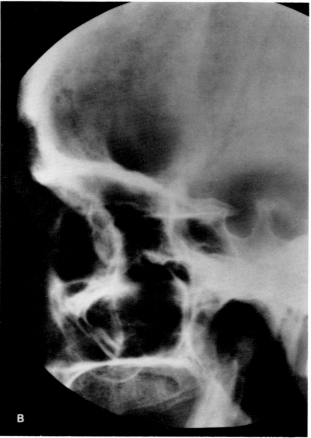

Figure 15.4. Amyloidosis of the frontal sinus. *A*, Posteroanterior view: opacity of the frontal sinuses with erosion and destruction of the bony wall. There is disappearance of the septum and marginal ridges. *B*, Lateral view: Erosion and destruction of both the anterior and posterior walls of the frontal sinuses. (Reprinted with permission from the Journal of Laryngology and Otology 80: 571, 1966.)

GIANT CELL REPARATIVE GRANULOMA

For convenience, giant cell reparative granuloma is discussed together with true giant cell tumors in Part II, Chapter 17.

REFERENCES

Becker, M. H., Ngo, N., and Beranbaum, S. L.: Mycotic infection of the paranasal sinuses: radiographic manifestations. Radiology, *90:* 49, 1968.

Bordley, J. E., and Proclor, D. E.: Destructive lesion in the paranasal sinuses associated with Boeck's sarcoid. Arch. Otol., *36:* 740, 1942.

Brown, B.: Amyloidosis of frontal sinus. J. Laryngol. Otol. *80:* 571, 1966.

Cabriele, O. F.: Mucormycosis. Am. J. Roentgenol., *80:* 337, 1960.

Chodosh, P. L., and Willis, W. W.: Tuberculosis of the upper respiratory tract. Laryngoscope, *80:* 679, 1970.

Dowie, L. N.: A short review of sarcoidosis with a report of three cases with involvement of the nasal mucosa. J. Laryngol. Otol. *81:* 931, 1967.

Garrett, J. A.: Amyloid deposition of the nose and maxillary sinuses. Arch. Otolaryngol., *87:* 103, 1068.

Graham, H. B.: Frequent but neglected evidence of syphilis from the side of the nose, accessory sinuses and ear. Am. J. Syphilis, *3:* 26, 1919.

Hara, H. J., and Crane, W. E.: Tuberculosis of the maxillary sinus. Ann. Otol. Rhinol. Laryngol., *57:* 1077, 1948.

Hersch, J. H.: Tuberculosis of the maxillary sinus. Arch. Otolaryngol., *28:* 987, 1938.

Hersch, J. H.: Primary infection of maxillary sinus by *Actinomyces necrophorus*. Arch. Otolaryngol., *41:* 204, 1945.

Hora, J. F.: Primary aspergillosis of the paranasal sinuses and associated areas. Laryngoscope, *75:* 768, 1954.

Hutter, R. V.: Phycomycetous infection (mucomycosis) in cancer patients: complication of therapy. Cancer, *12:* 330, 1959.

Ingberg, B.: Tuberculous sinusitis. Acta Otolaryngol. Supp., *140:* 163, 1957.

Juselius, H.: Tuberculosis of the maxillary sinus. Acta Otolaryngol., *53:* 424, 1961.

McAlpine, J. C., Radcliffe, A. and Friedmann, I.: Primary amyloidosis of the upper air passages. J. Laryngol. Otol., *77:* 1, 1963.

Mittermaies, R.: *Otorhinolaryngologic Radiology*, translated by P. W. Hoffman, Grune and Stratton, New York, 1970.

Meng, C. M., and Wu, Y. K.: Tuberculosis of the flat bones of vault of the skull. A study of forty cases. Chinese Med. J., *61:* 155, 1943.

Pantazopoulos, P. E.: Tuberculosis of the frontal sinus. Ann. Otol. Rhinol. Laryngol., *73:* 173, 1967.

Pendergrass, E. P., Schaeffer, J. P., and Hodes, P. J.: *The Head and Neck in Roentgen Diagnosis*, 2nd ed., p. 545. Charles C Thomas, Springfield, Illinois, 1956.

Spencer, F. R.: Chronic granulomas of the nose and nasal accessory sinuses. In *Otolaryngology*, edited by G. M. Coates, H. P. Schenk and V. Miller, Chap. 19, p. 1–24. W. F. Prior Publishers, Baltimore, 1960.

Trachtenberg, S. B., Wilkinson, E. E., and Jackson, G.: Sarcoidosis of the nose and paranasal sinuses. Radiology, *113:* 619–620, 1974.

Trible, W. M. and Loewenheim, H.: Amyloid tumor of the nose and sinuses. Laryngoscope, *73:* 424, 1963.

Wilson, G. E., and Stern, W. K.: Tuberculosis of the nose and paranasal sinuses. In *Otolaryngology*, edited by G. M. Coates, H. P. Schenk, and V. Miller, Chap. 20, p. 1–7. W. F. Prior Publishers, Baltimore, 1960.

Chapter 16

Traumatic Lesions Involving the Paranasal Sinuses

Traumatic lesions involving the paranasal sinuses may be classified into three major groups depending upon the underlying etiological factors:

1. Lesions of the paranasal sinuses associated with fractures of the facial bones or skull resulting from direct physical violence.

2. Lesions of the paranasal sinuses resulting from a sudden change in pressure within the sinuses or orbital cavity. These pressure changes may be caused by rapid alterations of barometric pressure or direct blows to the eye.

3. Lesions of the paranasal sinuses caused by foreign bodies.

FRACTURES INVOLVING THE PARANASAL SINUSES

Fractures of the Maxillary Sinuses

Classifications

The maxilla, being the most prominent of the facial bones, is frequently involved in facial fractures; injuries of the maxillary sinus, therefore, are relatively common.

Fractures of the maxilla range from a simple fracture of the maxillary bone to extensive injuries including the bones of the nose, orbit, palate, or any other bones of the face and skull with which the maxilla is intimately associated. These may be classified as follows:

Isolated Fractures of the Maxillary Sinus. Isolated fractures of the maxillary sinus resulting from direct impact most commonly involve the anterolateral wall of the maxillary sinus. A less frequent site is the inferior rim of the orbit. Fractures of the alveolar recess of the maxillary sinus are rare.

Multiple Fractures of the Facial Bones with Involvement of the Maxillary Sinuses. Multiple fractures are usually caused by motor vehicle accidents or by violent blows.

LATERAL FACE (ZYGOMATICOMAXILLARY) FRACTURES. These are classified as follows.

Fracture of the Zygoma. These fractures primarily affect the zygomatic bone, but frequently involve the maxilla, the zygomatic process of the temporal bone, and the zygomatic process of the frontal bone. The damaging force is usually directed to the lateral aspect of the face.

Tripod Fracture. These fractures usually occur simultaneously in three directions: medially there are fractures of the infraorbital rim and lateral wall of the maxillary sinus; superoposteriorly there is separation of the frontozygomatic suture with fracture above or below the suture; and laterally there is fracture of the zygomatic arch.

MIDDLE FACE (NASOMAXILLARY) FRACTURES. For practical purposes, middle face fractures can be classified according to the level at which they occur. According to LeFort, these levels are determined by the natural lines and areas of weakness of the facial bones and the type and location of force applied (Fig. 16.1).

Transverse Maxillary, LeFort Type I, Fracture. This fracture occurs when the traumatic force is applied to the lower maxillary region. It passes transversely through the maxillae above the level of the teeth and extends posteriorly through the lower parts of the maxillary sinuses to the lower parts of the pterygoid plates of the sphenoid bones.

Pyramidal, LeFort Type II, Fracture. This is caused by a violent trauma to the upper maxil-

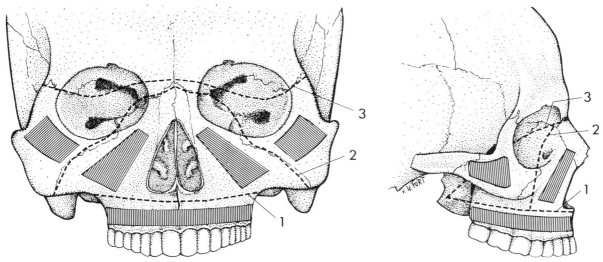

Figure 16.1. Line drawing of types of the middle face fracture (after Le Fort): *1*, transverse maxillary, Le Fort type I, fracture; *2*, pyramidal, Le Fort type II, fracture; *3*, craniofacial, Le Fort type III, dysjunction.

lary region which produce fractures of the nasal bones and of the frontal processes of the maxillae. The fractures pass laterally through the lacrimal bones and the floors of the orbits and near or through the zygomaticomaxillary sutures. The fractures then continue posteriorly along the lateral walls of the maxillae through the pterygoid plates and into the pterygomaxillary fossae. This has been designated as the pyramidal fracture because of the somewhat pyramidal shape of the fragments when viewed in the face. In severe comminuted types there may also be involvement of the ethmoid and frontal sinuses.

Craniofacial, LeFort Type III, Dysjunction. Craniofacial dysjunction occurs when the traumatic force is sufficiently violent to cause complete separation of the facial bones from their cranial attachments. The fractures usually occur through the zygomaticofrontal, maxillofrontal, and nasofrontal sutures, through the floors of the orbits, and through the ethmoids and the sphenoids with complete separation of all structures of the middle facial skeleton from their attachments. These injuries are almost always associated with multiple fractures of the facial bones.

Roentgen Findings

The roentgen examination is indispensible in the diagnosis of facial fractures with involvement of the maxillary sinuses. Waters and lateral views of the sinuses are most useful for revealing fractures of the maxilla and associated structures. Because of superimposition of the various structures, conventional roentgen examinations may be insufficient and tomograms may be needed to delineate the injuries accurately. Multiple projections are often advisable to demonstrate associated fractures of other facial bones.

Isolated Fractures of the Maxillary Sinus. Fracture of the anterolateral wall may be either linear or comminuted with posterior displacement of the fragments (Fig. 16.2). There is often subcutaneous emphysema of the cheek on the affected side. Sensory disturbance may occur in fractures of the inferior rim of the orbit with involvement of the infraorbital nerve.

Multiple Fractures of the Facial Bones. A lateral face (zygomaticomaxillary) fracture may be accompanied by: (1) separation of the zygomaticofrontal suture with or without fracture; (2) step-like irregularity of the inferior orbital rim; (3) irregularity or a break in continuity of the lateral wall of the maxillary sinus; (4) deformity or discontinuity of the zygoma; (5) disruption of the zygomatic arch.

The most common type of zygomaticomaxillary fracture is known as the "tripod fracture", i.e., fracture of the zygomatic arch, the inferior orbital rim, and the lateral wall of the maxillary sinus. The fracture may be undisplaced or displaced with or without rotation (Fig. 16.3).

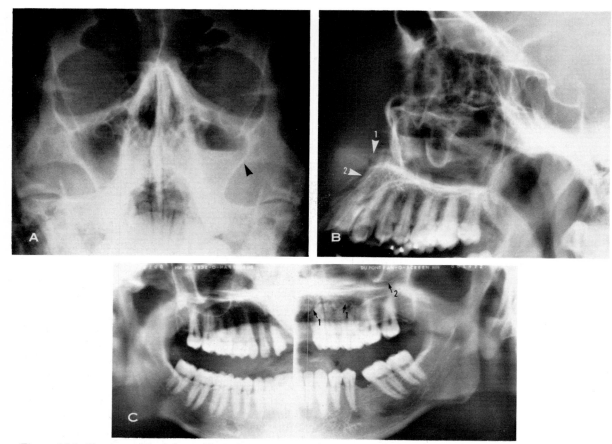

Figure 16.2. Fracture of the left maxilla including the alveolar ridge and anterior nasal spine. *A*, Waters view: fracture of the lateral wall of the left maxillary sinus (*arrow*). There is a fluid level in the left maxillary sinus. *B*, Lateral view: fracture of the anterior nasal spine of the maxilla (*1*) and of the alveolar ridge (*2*). A fluid level is seen in the maxillary sinus. *C*, Panorex view: comminuted fracture of the alveolar ridge (*1*) and floor of the left maxillary sinus (*2*).

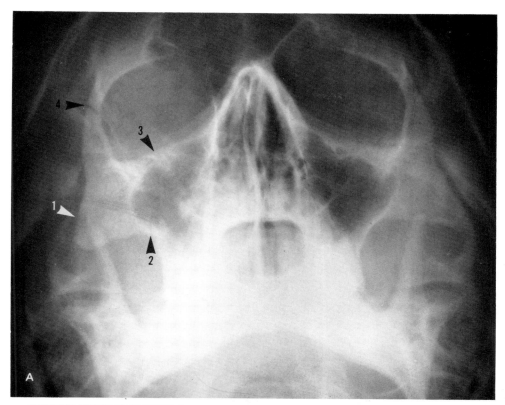

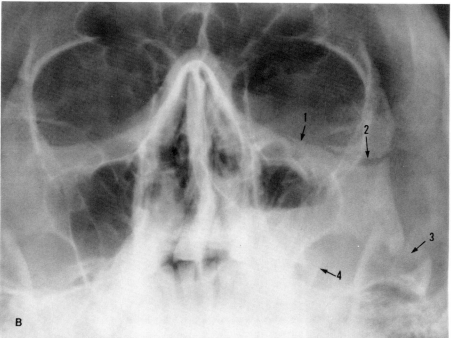

Figure 16.3. Tripod fractures. *A*, Displaced tripod fracture. Waters view showing fractures of the zygomatic arch (*1*), of the lateral wall of the maxillary sinus (*2*), and of the inferior rim of the orbit (*3*). There is also separation of the zygomaticofrontal suture (*4*). There is no rotation of the fragments of the fracture.

B, Displaced tripod fracture with lateral rotation. Waters view showing fractures of the inferior rim of the orbit (*1*), of the frontal process of the zygoma (*2*), of the zygomatic arch (*3*), and of the lateral wall of the maxillary sinus (*4*). There is medial displacement of the zygoma, with wide separation of the fragments of the lateral wall of the maxillary sinus. There is no definite separation of the frontozygomatic suture. A fluid level is seen in the left maxillary sinus. (Courtesy of Dr. George Campbell.)

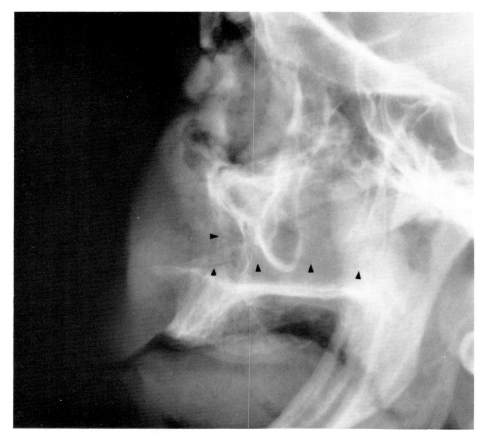

Figure 16.4. Transverse maxillary fracture and fracture of the anterior wall of the maxillay sinus. Lateral view showing a transverse fracture just above the alveolar ridge from the anterior wall of the maxillary sinus to the pterygoid plate (*vertical arrows*). There is a comminuted fracture with inward displacement of the anterior wall of the maxillary sinus (*horizontal arrow*) associated with regional soft tissue swelling. (Courtesy of Dr. George Campbell.)

Middle face (nasomaxillary) fractures have the following roentgen findings.

1. Transverse maxillary fractures are bilateral, with posterior displacement of the lower parts of the maxillary sinuses and transverse fractures of the pterygoid plates (Fig. 16.4).

2. In pyramidal fractures, the roentgen findings are as follows: (1) fractures of the frontal process of the maxilla on each side; (2) fractures of the inferior orbital rim on each side; (3) separation with or without fracture of the zygomaticomaxillary sutures on each side; (4) fractures of the lateral wall of the maxillary sinus on each side; (5) fractures through the pterygoid plate on each side; (6) coincident fractures of the ethmoid and frontal sinuses and other bones (Figs. 16.5 and 16.6).

3. In craniofacial dysjunction, the chief roentgen findings are: (1) separation of the zygomati-cofrontal, maxillofrontal, and nasofrontal sutures; (2) fracture of the nasal bones, of the orbital floors, and of the pterygoid plates; (3) fracture of the sphenoid and ethmoid sinuses; (4) associated fractures of other facial bones (Figs. 16.7 and 16.8).

Associated Changes in the Maxillary Sinuses

The roentgen changes in the maxillary sinuses vary and usually include the following.

Generalized Opacity of the Sinus. This is attributable to hemorrhage into the sinus or to edema of the mucosa of the sinus. A fracture of the maxillary sinus is often complicated by bleeding into the sinus cavity, with resultant opacity (Figs. 16.5 and 16.6). For this reason, an opaque maxillary sinus on the side of injury

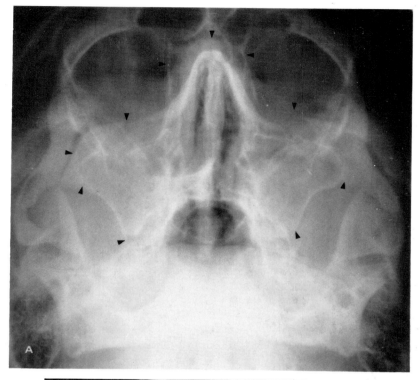

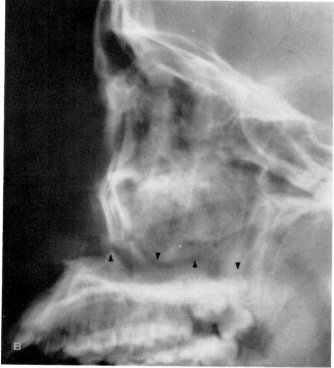

Figure 16.5. Transverse maxillary and pyramidal fractures. *A*, Waters view showing pyramidal fracture (*arrows*). There is a diffuse opacity of both maxillary sinuses. *B*, Lateral view showing a complete transverse maxillary fracture above the level of the alveolar ridge from the anterior wall of the maxillary sinus to the pterygoid plate (*arrows*).

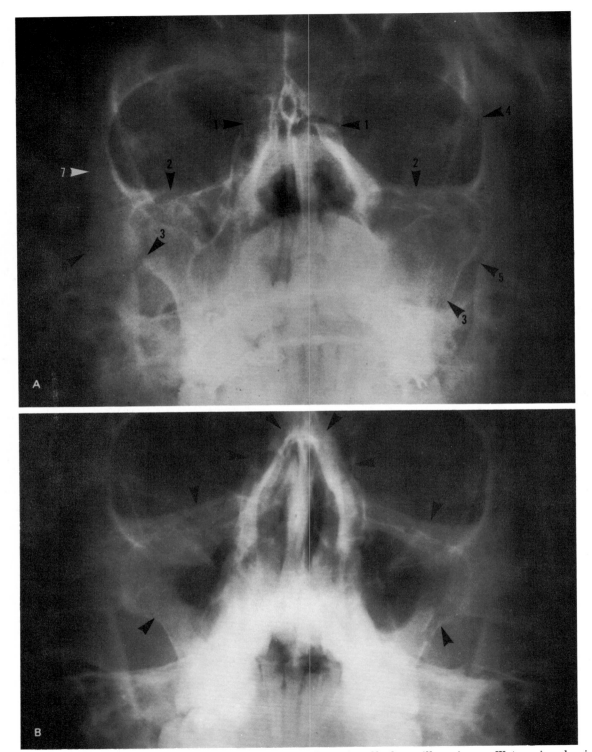

Figure 16.6. Pyramidal fractures. *A*, Pyramidal fracture with opacity of both maxillary sinuses. Waters view showing fracture of the ethmoid sinus (*1*), of the inferior rim of the orbit (*2*), and of the lateral wall of the maxillary sinus (*3*) on each side. There is separation of the zygomaticofrontal and zygomaticomaxillary sutures (*4, 5*) on the left side. Fracture of the zygomatic arch (*6*) and of the frontal process of the zygoma (*7*) is noted on the right. There is generalized opacity of both maxillary sinuses, more marked on the right.

B, Pyramidal fracture with submucosal hemorrhage of the maxillary sinuses (*arrows*). There is considerable thickening of the lining mucosa, especially on the right side, due to submucosal hemorrhage. (Courtesy of Department of Radiology, St. Joseph Hospital, Houston, Texas.)

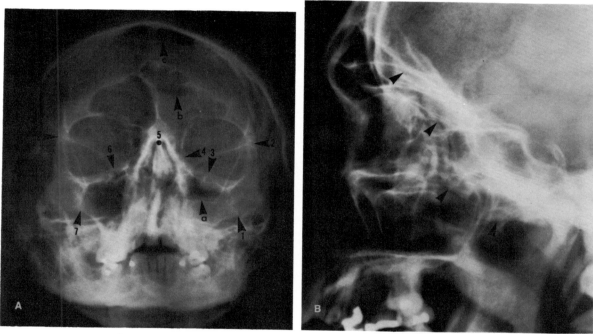

Figure 16.7. Craniofacial dysjunction. *A,* Waters view showing fluid levels in the left maxillary sinus (*a*), left frontal sinus (*b*), and pneumocephalus (*c*). Multiple fractures are present: left zygomaticomaxillary region (*1*), left zygomatico-frontal region (*2*), left inferior rim of the orbit (*3*), left lamina papyracea (*4*), nasal bones (*5*), right inferior rim of the orbit (*6*), right zygomaticomaxillary region (*7*), and right zygomaticofrontal region (*8*). There is vertical shortening of the face. *B,* Lateral view showing an oblique fracture from the frontal sinus through the ethmoid and maxillary sinuses down to the pterygoid plates. (Courtesy of Dr. Warren McFarland.)

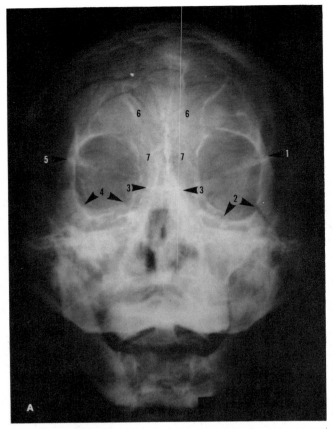

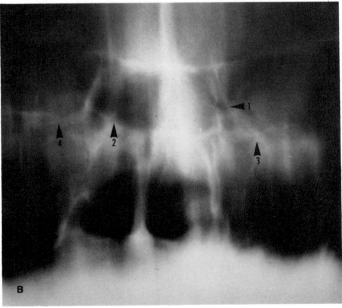

Figure 16.8. Craniofacial dysjunction. *A*, Posteroanterior view showing multiple fractures crossing from left to right (*arrows*): left zygomaticofrontal suture (*1*), left inferior rim and floor of the orbit with downward displacement of the fragment (*2*), entire bony complex between the two orbits (*3*), right inferior rim and the floor of the orbit (*4*), and right zygomaticofrontal suture (*5*). There is involvement of the frontal and ethmoid sinuses (*6 & 7*). Both maxillary sinuses are opacified. Comminuted fractures are seen in the frontal region. *B*, Anteroposterior tomogram showing fracture of the floor on the right side (*2*) and of the lateral wall on the left side of the sphenoid sinuses (*1*). There is opacification of the sphenoid sinuses, especially on the left side. Fracture of the floor of the middle cranial fossa is seen on each side (*3 & 4*). (Courtesy of Department of Radiology, St. Joseph Hospital, Houston, Texas.)

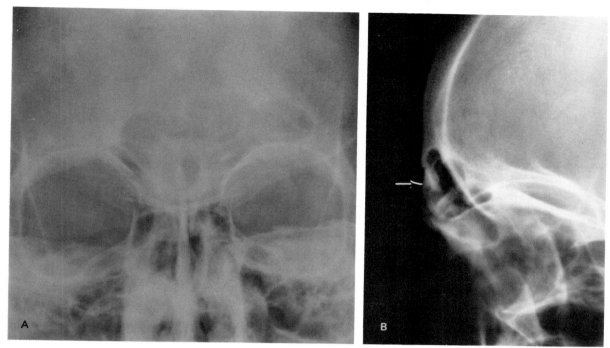

Figure 16.9. Fracture of the frontal sinus. *A*, Posteroanterior view showing cloudiness of the frontal sinuses. There is a large bony density with an irregular margin in the middle and lower aspect of the sinuses. *B*, Lateral view showing comminuted fracture of the anterior wall of the frontal sinuses with slight posterior displacement of the fragments of the fracture.

should be presumptive evidence of fracture until proven otherwise.

Localized Thickening of the Mucosa of the Sinus. This is a more common finding and represents a mucosal edema or submucosal hemorrhage. It is usually located at the site of the fracture line. It appears as a rounded density projecting into the air space of the sinus; the appearance may closely simulate a polyp or a cyst of the sinus (Fig. 16.6).

Fluid Level in the Sinus. A fluid level associated with mucosal thickening may appear in the maxillary sinus. The fluid is generally blood or serosanguinous material (Figs. 16.2, 16.3, and 16.7).

Radiolucency of the Maxillary Sinus. Occasionally the radiolucency of the maxillary sinus may be normal despite the presence of a fracture.

Fractures of the Frontal Sinuses

General Considerations

Fractures of the frontal sinuses are usually due to direct violence, but they may be indi-rectly caused by the extension of a fracture of the skull. One or both walls may be involved. Fractures of the posterior wall are extremely rare and are generally an extension of a fracture of the base or the vertex of the skull. Fractures that involve the anterior wall are more common and are usually the result of direct violence. In any fracture of the frontal sinus it is of importance to determine whether or not the posterior wall is involved. If it is involved, the danger of cerebrospinal fluid rhinorrhea, meningitis, pneumocephalus, and other intracranial complications should be taken into consideration.

Roentgen Findings

The demonstration of a fracture of the frontal sinus requires delineation of the walls of the chambers. This may at times be difficult. Lateral, Caldwell, and tangential views are of value and an overangulated view can be of help. Tomograms in the posteroanterior or lateral plane may also be of assistance in revealing the extent of a fracture and the degree of depression of the fragments (Figs. 16.9 and 16.10).

Anterior Wall of the Frontal Sinus. Frac-

tures of the anterior wall can be linear or mo-
saic-like. These are prone to be depressed and
may easily be overlooked.

Posterior Wall of the Frontal Sinus. Fractures of the posterior wall usually assume a
linear configuration.

*Associated Changes in the Frontal Sinuses
(Figs. 16.9 and 16.10)*

These include: (1) generalized opacity of the
sinus attributable to edema of the mucosa or
hemorrhage into the sinus; (2) fluid level in the
sinus—this may be formed by the accumulation
of cerebrospinal fluid or blood in the chamber.

Other Findings (Fig. 16.10)

These include: (1) emphysema of the orbit
resulting from disruption of the superior orbital
wall; (2) pneumocephalus caused by laceration
of the dura.

Fractures of the Ethmoid Sinuses

General Considerations

An isolated fracture of the ethmoid sinus
caused by direct violence is very rare. Fractures
may occur, however, in association with multiple fractures of the facial bones. A fracture of
the skull with extension to the roof of the ethmoid is not uncommon and has a sinister significance because it is often accompanied by laceration of the dura with subsequent risk of meningitis, hemorrhage, etc.

Roentgen Findings

By using Caldwell, Waters, base, and oblique
views, a fracture through the ethmoid cells or
an extension into this part of the ethmoid bone
may be diagnosed indirectly, even though the
fracture line is not always visible. There often is
evidence of fracture of the skull. Tomograms are
of value in the delineation of the extent and
location of the fractures (Fig. 16.11).

Changes in the Ethmoid Sinus. Opacity of
the ethmoid sinus is seen on the affected side.
This is caused by hemorrhage into the sinus or
edema of the ethmoid cells.

Emphysema of the Orbit. Even in the absence of a demonstrable ethmoid fracture line,
the presence of air in the orbit is unequivocable
evidence of a fracture through air-containing
sinuses.

Fractures of the Sphenoid Sinuses

General Considerations

An isolated fracture of the sphenoid sinus is
extremely rare. They usually accompany fractures of the base of the skull or multiple fractures of the facial bones.

Roentgen Findings

In LeFort type III fractures of the facial
bones, involvement of the sphenoid sinus usually can be demonstrated on multiple views of
the skull and the facial bones.

A fracture of the skull with extension into the
sphenoid sinus often can be recognized on the
lateral and base views of the skull and on tomograms of the skull and the sphenoid sinus (Figs.
16.8 and 16.12).

Changes in the Sphenoid Sinus

Direct visualization of the fracture line may
be difficult and, therefore, the following indirect
signs are particularly important: (1) opacity of
the sinus; (2) air in the orbital cavity; (3) fluid
level in the sphenoid sinus. If hemorrhage or a
flow of cerebrospinal fluid into the sinus cavity
occurs, an increased density in the sphenoid
sinus with a fluid level may be visualized. The
fluid level is best demonstrated in the brow-up
lateral projection of the skull with a horizontal
beam (Fig. 16.13); (4) pneumocephalus. This is
due to dural laceration with leakage of the air
into the intracranial cavity. It is a rare condition.

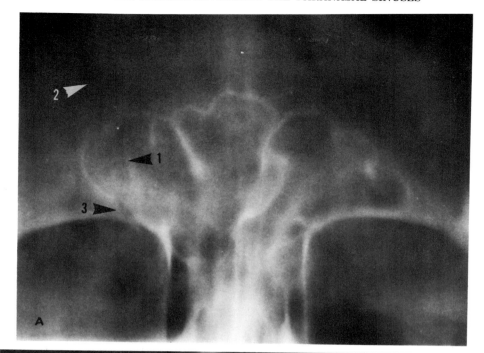

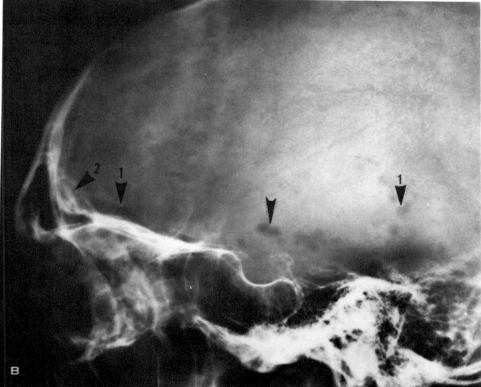

Figure 16.10. Fracture of the frontal sinus with pneumocephalus. *A*, Anteroposterior tomogram: linear fractures through the wall of the right frontal sinus (*1*) extending superiorly to the right frontal bone (*2*) and inferiorly to the superior rim of the right orbit (*3*). There is emphysema of the right orbit. *B*, Lateral view of the skull: small collections of air at the base of the skull (*1*). There is a linear fracture of the posterior wall of the frontal sinus (*2*). (Courtesy of D~ Thomas Harle.)

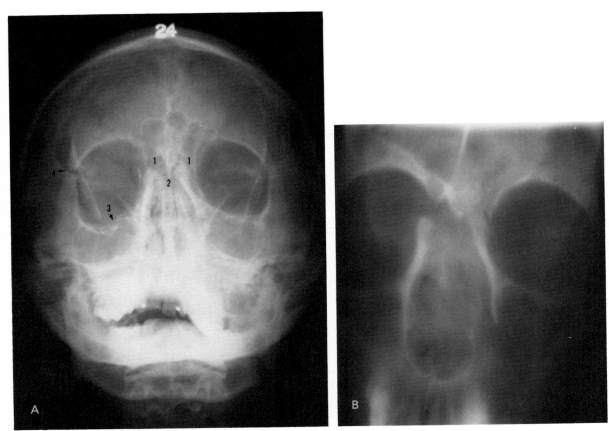

Figure 16.11. Fracture of the ethmoid sinus—pyramidal fracture. *A*, Waters view: there are fractures of the ethmoid sinuses (*1*), of the nasal bones (*2*), and of the right inferior rim of the orbit (*3*). There is separation of the right zygomaticofrontal suture (*4*). *B*, Posteroanterior tomogram: fractures of the ethmoid sinuses and nasal bones are shown to good advantage.

Figure. 16.12. Fracture of the skull with involvement of the sella turcica and sphenoid sinus. Lateral view of the skull showing fractures of the temporal bone, the floor of the sella turcica, and the wall of the sphenoid sinus (*arrows*).

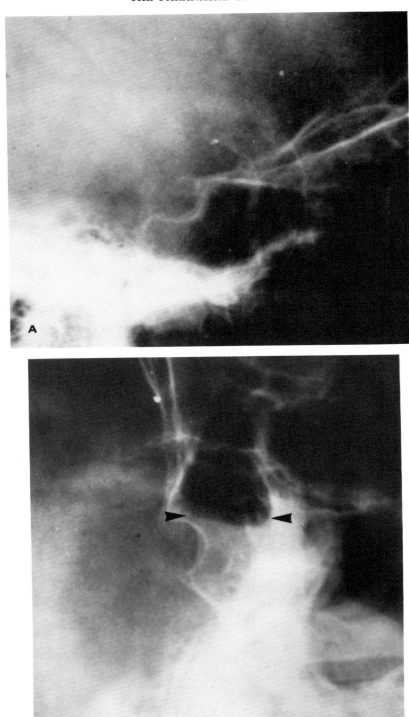

Figure 16.13. Fracture of the base of the skull with extension to the sphenoid sinus. *A*, Lateral view in sitting position showing no fluid level in the sphenoid sinus; *B*, Brow-up lateral view with horizontal beam showing air-fluid level in the sphenoid sinus.

BAROTRAUMA OF THE PARANASAL SINUSES (AEROSINUSITIS)

General Considerations

High altitude flying can produce a physical strain and pathological changes in the paranasal sinuses. Under pressurized conditions aerosinusitis is uncommon, but it may occur when pressurization fails. The rapid alterations in atmospheric pressure may force infected material from the nose into the sinuses, with a resultant sinusitis. If the ostium of the sinus or frontonasal duct becomes blocked because of a flap of thickened mucosa or polypoid mass, the alterations in atmospheric pressure may produce obstructive pathological changes within the sinus. During the ascent, when the ostium is blocked, the pressure within the sinus rises above the atmospheric pressure and severe pain and headache result. On the other hand, during the descent, if the ostium is obstructed, the pressure within the sinus is lower than the atmospheric pressure, and outpouring of fluid with thickening of the mucosa and hemorrhage may occur.

The frontal and maxillary sinuses are most often affected. Aerosinusitis is characterized clinically by a sudden severe pain over the affected sinuses and by bleeding from the nose during the descent from high altitude.

Roentgen Findings

The roentgen findings are nonspecific and shoud be correlated with the clinical history (Fig. 16.14).

Polypoid Mass in a Sinus

This is, in essence, a hematoma resulting from bleeding. It may be single or multiple. It may involve one or more of the sinuses, and is most commonly found in the frontal sinuses. The hematoma may assume an appearance of a round or ovoid density, of varying size, which may be indistinguishable from a retention cyst.

Generalized Opacity of the Sinus

This is attributable to edema with generalized mucosal thickening. In some cases, it may be caused by a transudate into the sinus.

Localized Thickening of the Mucosa of the Sinus

This may be brought about by localized edema of the mucosa.

Fluid Level in the Sinus

Occasionally, as a result of an outpouring of fluid into the maxillary or frontal sinus, a fluid level may appear.

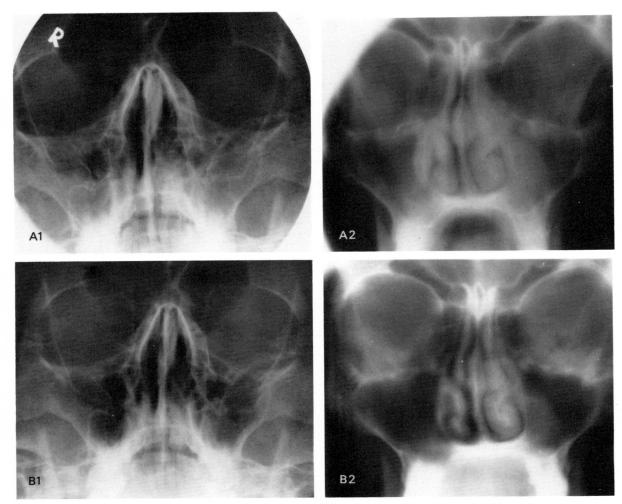

Figure 16.14. Aerosinusitis of the maxillary sinuses. *A1*, Waters view: generalized opacity of the left maxillary sinus and, to the lesser degree, of the right maxillary sinus. *A2*, Anteroposterior tomogram: thickening of the lining mucosa of both maxillary sinuses, marked on the left side. A polypoid soft tissue mass is noted in the medial aspect of the left maxillary sinus. There is opacity of the ethmoid sinuses. Enlargement of the nasal turbinates is seen, especially on the left side. *B1*, Waters view: almost complete clearing of the opacity of the right maxillary sinus. There is minimal mucosal thickening in the left maxillary sinus. *B2*, Anteroposterior tomogram: minimal mucosal thickening of both maxillary sinuses. Polypoid mass in the medial aspect of the left maxillary sinus appears smaller in size. There is minimal opacity of the ethmoid sinuses.

This patient traveled by air to a business appointment. Symptoms of "flu" had been present for 2 days before flying. Pain over the cheeks and impairment of hearing occurred during the descent. The same symptoms developed during the return flight the following day. Roentgen examination of the paranasal sinuses 4 days after the last flight revealed opacity and mucosal thickening of the maxillary sinuses (Fig. 16.14A). The pain gradually subsided and hearing returned to normal with treatment. Follow-up roentgen examination 16 days later revealed remarkable regression of opacity and mucosal thickening of the sinuses (Fig. 16.14B).

BLOW-OUT FRACTURES OF THE ORBIT

General Considerations

A blow-out fracture of the orbit generally is a fracture of the orbital floor caused by a sudden increase in the intraorbital pressure from a blow to the eye, usually a fist or a round, hard object. To cause this injury the object should have a diameter larger than that of the bony orbit, e.g., a baseball, which ensures that it is stopped by the bony rim of the orbit and does not rupture the eyeball. The orbital soft tissues are suddenly compressed and the increased pressure is transmitted to the orbital walls in all directions. The thin floor of the orbit, which is least resistant to sudden increases in pressure, is the usual site of fracture. Bony fragments and orbital soft tissues are forced downward into the maxillary sinus. This may be accompanied by hemorrhage. The mechanism of a blow-out fracture of the floor of the orbit is shown in Figure 16.15. Rarely, the blow-out fracture may involve the medial wall of the orbit in the region of the lamina papyracea. Medial displacement of the bony fragments and orbital soft tissues into the ethmoid sinus may occur.

The clinical manifestations of a blow-out fracture of the orbit are enophthalmos, downward displacement of the eyeball, impaired rotation of the eyeball, and diplopia, particularly on upward gaze. Sensory disturbances of the face on the affected side attributable to involvement of the infraorbital nerve may be present.

Roentgen Findings

In interpretation of the roentgenograms of the blow-out fracture of the orbit, one should be aware that a fracture may be mistaken for superimposed bony septa, or obscured by pathological changes in the underlying maxillary sinus. The chief roentgen findings are as follows (Figs. 16.16 to 16.18).

Bony Fragments in Superior Aspect of Maxillary Sinus

In an imcomplete fracture, the fracture is visualized as a depressed regular or irregular curvilinear bone density superimposed to a variable extent on the superior aspect of the maxillary sinus. The inferior orbital rim is often intact.

Soft Tissue Mass in Superior Aspect of Maxillary Sinus

This is caused by prolapse of orbital soft tissue into the maxillary sinus. The herniated soft tissue is visualized as a crescentic opaque soft tissue density on the roentgenograms. If the herniated soft tissue is mainly of orbital fat, its appearance may mimic a radiolucent cyst in the superior aspect of the sinus.

Depressed fractures of the orbital floor and prolapse of the orbital soft tissues into the maxillary sinus usually can be detected by Waters and lateral views. However, tomograms of the orbits and maxillary sinuses are imperative to delineate completely the location and extent of the fracture of the orbital floor and the prolapse of orbital soft tissue, especially when the sinus is opaque from hemorrhage or from preexisting sinus disease.

Opacity of Maxillary Sinus, with or without Fluid Level

This results from hemorrhage and edema of the mucosa.

Emphysema of Orbit

Occasionally, emphysema of the orbit is seen. This is caused by passage of air from the maxillary antrum into the orbital tissues via the fracture.

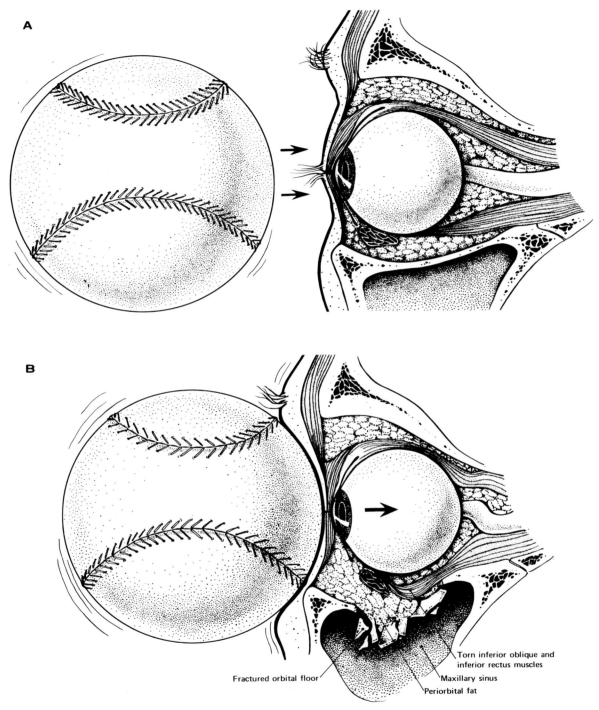

Figure 16.15. Mechanism of blow-out fracture of the floor of the orbit. *A*, Normal cross-section of the orbit and a hurled ball; *B*, External pressure causing a blow-out fracture of the floor of the orbit.

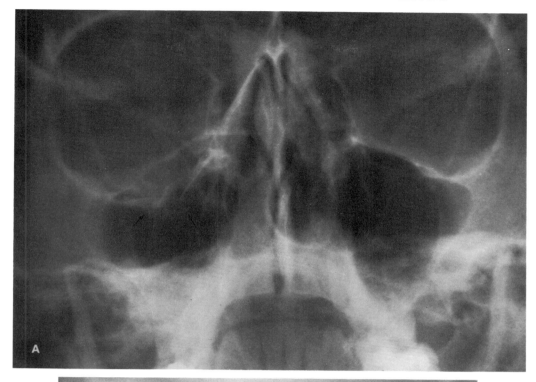

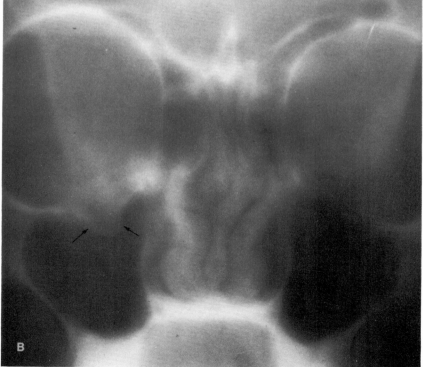

Figure 16.16. Blow-out fracture of the floor of the right orbit. *A*, Waters view: depressed fracture of the floor of the right orbit and orbital soft tissues protruding into the superior aspect of the maxillary sinus (*arrows*); *B*, Posteroanterior tomogram: the depressed fracture and prolapsed orbital soft tissue in the superior aspect of the sinus are shown to good advantage (*arrows*).

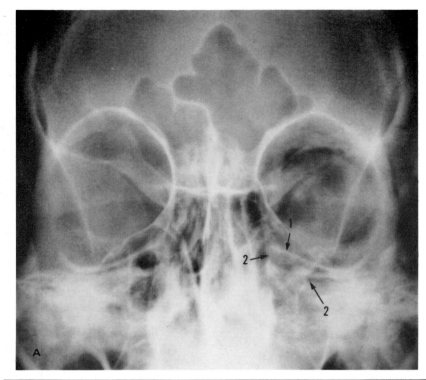

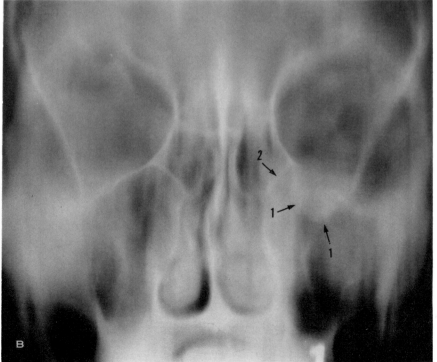

Figure 16.17. Blow-out fracture of the floor of the left orbit with emphysema of the orbit. *A*, Waters view: depressed fracture of the medial aspect of the floor of the left orbit (*1*). Orbital soft tissue is protruded into the superior aspect of the left maxillary sinus (*2*). There is air in the left orbital cavity. *B*, Posteroanterior tomogram: the fragments of the depressed fracture and protruded orbital soft tissue are clearly demonstrated (*1*). The ethmoidomaxillary plate is intact (*2*). There is opacity of the superomedial aspect of the left maxillary sinus, probably due to edema and hemorrhage.

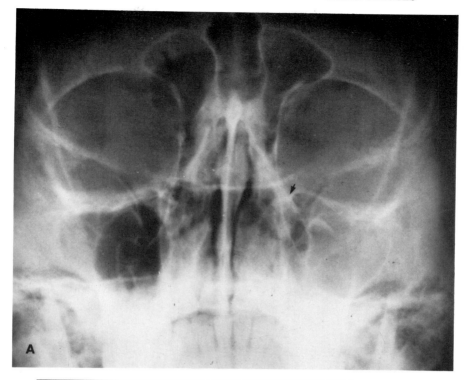

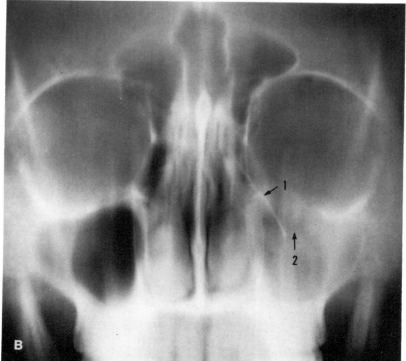

Figure 16.18. Blow-out fracture of the floor and medial wall of the left orbit. *A*, Waters view: fracture of inner aspect of the floor and medial wall of the left orbit (*arrow*). There is opacification of the left maxillary sinus. *B*, Posteroanterior tomogram: blow-out fracture is shown to good advantage. There is a downward and medial displacement of the fragments of the fracture (*1*). Orbital soft tissue is protruding into the superior aspect of the maxillary sinus (*2*). There is opacification of the left maxillary sinus and, to a lesser degree, of the left ethmoid sinus, probably due to hemorrhage and edema.

FOREIGN BODIES IN THE PARANASAL SINUSES

Foreign bodies may be found in the paranasal sinuses, most commonly in the maxillary, occasionally in the frontal or ethmoid, and rarely in the sphenoid chamber. The foreign bodies may gain entry through the wall of the sinus by a direct trauma or a gunshot wound. They may be accompanied by acute infections which often persist as long as the foreign bodies remain in the sinus (Fig. 16.19). The acute infection may be followed by a low grade inflammatory process and eventually by formation of granulomas simulating tumor masses (Fig. 16.20). At times, a foreign body may act as a nucleus and become encrusted with calcium, forming a rhinolith.

In the maxillary sinus, most of foreign bodies are dental in origin, either misplaced or displaced teeth. Frequently the displacement of the root of the tooth into the maxillary sinus occurs during dental extraction. An antro-oral fistula may develop if the antro-oral opening resulting from extraction of the tooth becomes infected (see also Fig. 17.26). A misplaced tooth is usually developmental in origin and the tooth is often surrounded by fibrous capsule. Occasionally, a dentigerous cyst may force the involved tooth into the maxillary sinus (Fig. 14.3B).

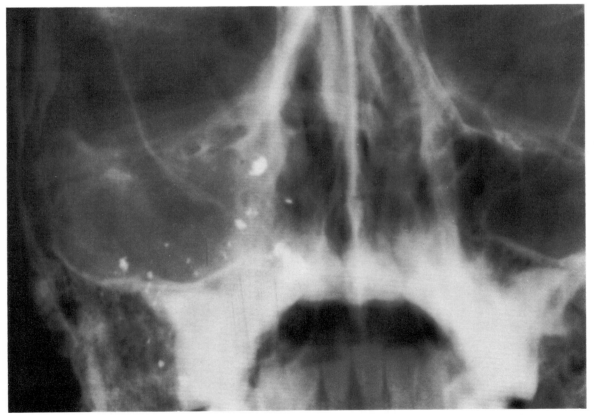

Figure 16.19. Gunshot wound of the right maxillary sinus with acute sinusitis or hematoma. Waters view: fragments of foreign bodies are seen in the right maxillary sinus. There is cloudiness of the right maxillary sinus, probably due to an acute inflammatory process or hematoma. Fragments are also seen in soft tissues of the cheek.

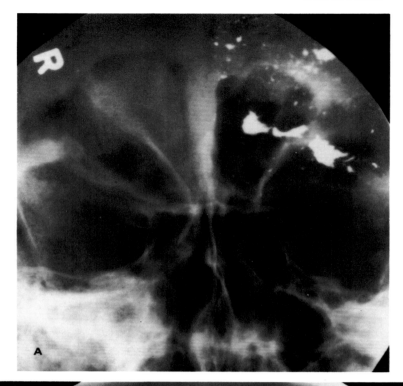

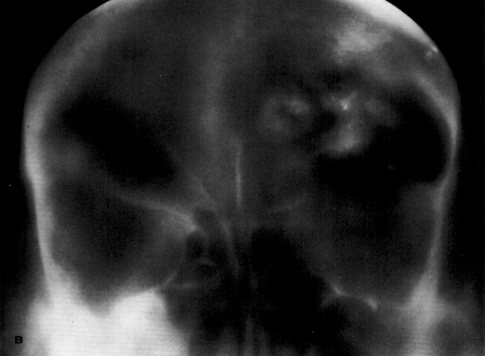

Figure 16.20. Gunshot wound with formation of chronic granuloma in the left frontal sinus. (This patient had a gunshot wound in the left frontal region 25 years ago). *A*, Waters view: fragments of foreign bodies in the left frontal sinus and left frontal region. There is sclerosing osteitis around the sinus. Marked bony sclerosis is noted in the right frontal sinus. *B*, Posteroanterior tomogram: fragments of the foreign body have become encrusted with granulation tissue, forming a well defined soft tissue mass.

REFERENCES

Campbell, P. A.: Aerosinusitis. Arch. Otolaryngol., *35:* 107, 1942.

Dingman, R. A. and Natvig, P.: *Surgery of Facial Fractures.* p. 43–95, 211–266, 295–310, 329–337. W. B. Saunders, Philadelphia.

Dolan, K. D. and Hayden, J., Jr.: Maxillary "pseudofracture" line, Radiology, *107:* 321, 1973.

Feore, D. R.: Blow-out fractures of the orbit, Clin. Radiol. *16:* 347, 1965.

Freimanis, A.K.: Fractures of the facial bones. Radiol. Clin. North Am., *4:* 341, 1966.

Fueger, G. F., Milauskas, A. T., and Britton, W.: The roentgenologic evaluation of orbital blow-out injuries. Am. J. Roentgenol., *97:* 614, 1966.

Gould, H. R., and Titus, C. O.: Internal orbital fractures: the value of laminography in diagnosis. Am. J. Roentgenol., *97:* 618, 1966.

Hobeika, C. P.: Trauma involving the temporal bone. Otolaryngol. Clin. North Am., p. 433, June, 1969.

Hofman, W. B.: Injuries of the zygomatic and maxillary bones. Otolaryngol. Clin. North Am., p. 303, June, 1969.

LeForte, R.: Etude experimentale sur les fractures de la machoire superieure, Rev. Chir., *23:* 208, 1901.

Lewin, J. R., Rhodes, D. H., Jr. and Pavsek, E. J.: Roentgenologic manifestiations of fractures of orbital floor (blow-out fracture). Am. J. Roentgenol., *83:* 628, 1960.

Maller, R. W.: Fractures of the nasofrontal complex. Otolaryngol. Clin. North Am., p. 335, June, 1969.

Murray, R. S.: Orbital emphysema following fracture of nasal sinuses. J. Fac. Radiol. *1:* 121, 1949.

McGrigor, D. B. and Campbell, W.: The radiology of war injuries. Part IV, Wound of the face and jaw. Br. J. Radiol. *23:* 685, 1950.

McGrigor, D. B. and Samuel, E.: Radiology of war injuries. War wounds of orbit and eyeball. Br. J. Radiol., *18:* 284, 1945.

Rose, N. L. and Killey, H. C.: *Fractures of the Facial Skeleton.* E. & S. Livingstone Ltd., Edinburgh and London, 1968.

Schneider, M.: Aerosinusitis, a clinico-roentgenologic study. Am. J. Roentgenol., *53:* 563, 1945.

Schultz, R. C.: *Facial Injuries.* Year Book Medical Publishers, Inc., New York and Chicago, 1970.

Smith, B. and Regan, W. F.: Blow-out fracture of the orbit. Am. J. Ophthaolmol., *44:* 733, 1957.

Valvassori, G. E. and Hord, G. E.: Traumatic sinus disease. Semin. Roentgenol. *3:* 160, 1968.

Zatzkin, H. R.: *The Roentgen Diagnosis of Trauma.* Year Book Medical Publishers, Inc., New York and Chicago, 1965.

Zizmor, J., Smith, B., Fasano, C., and Converse, J. M.: Roentgen diagnosis of blow-out fractures of the orbit. Am. J. Roentgenol., *87:* 1009, 1962.

Chapter 17

Benign Tumors of the Paranasal Sinuses

Benign tumors of the paranasal sinuses are relatively infrequent. Primary benign tumors of the paranasal sinuses arise either from the lining mucosa or from the bony walls. Benign tumors of extrinsic origin may extend to and involve the paranasal sinuses. The following classification may be useful:

Primary Tumors
1. Epithelial Origin
   a. Surface epithelium
      (1) Squamous papilloma
      (2) Inverting papilloma
   b. Glandular
      (1) Mixed tumor (pleomorphic adenoma of accessory gland origin)
      (2) Adenoma
      (3) Others
2. Mesenchymal Origin
   a. Myoma
      (1) Leiomyoma
      (2) Rhabdomyoma
   b. Lipoma
   c. Fibroma
      (1) Myxofibroma
   d. Angiofibroma, Juvenile
   e. Angioma
      (1) Hemangioma
      (2) Lymphangioma
   f. Neurogenic
      (1) Meningioma
      (2) Neurofibroma
      (3) Neurolemmoma
   g. Osseous
      (1) Osteochondroma
      (2) Osteoblastoma
      (3) Osteoma
      (4) Giant cell tumor
      (5) Aneurysmal bone cyst
   h. Others

Tumor-like Lesions
1. Giant cell reparative granuloma
2. Fibroosseous lesions
   a. Ossifying fibroma
   b. Fibrous dysplasia
   c. Leontiasis ossea
3. Paget's disease

Secondary Tumors
1. Odontogenic origin
   a. Epithelial tumors
      (1) Ameloblastoma
      (2) Adenoameloblastoma
      (3) Calcifying epithelial odontogenic tumor
   b. Mesenchymal tumors
      (1) Odontoma
      (2) Odontogenic myxoma (myxofibroma)
      (3) Odontogenic fibroma
      (4) Benign cementoblastoma
      (5) Others
2. Pituitary origin
   a. Adenoma
3. Nasopharyngeal origin
   a. Chordoma
   b. Nasopharyngeal juvenile angiofibroma
4. Orbital origin
   a. Dermoid cyst

PRIMARY TUMORS OF THE PARANASAL SINUSES

Primary benign tumors of the paranasal sinuses are less common than malignant tumors by a ratio of approximately 1:3. They may affect any of the paranasal sinuses, but are found most commonly in the maxillary sinuses and, in decreasing frequency, the ethmoid, frontal, and sphenoid sinuses. The symptoms of these benign tumors are generally related to the size and location of the lesion. When the lesions are advanced and extend to surrounding structures, a multitude of clinical manifestations may appear.

The roentgen features of benign tumors are variable and nonspecific; they may appear as small well defined soft tissue masses or they may fill the sinus cavity and produce sufficient bony erosion and destruction to simulate malignant tumors. It is generally possible to diagnose the nature and extent of the lesions by roentgen examination. In perplexing cases tomograms are of value in assessing the disease process. Occasionally, even with meticulous roentgen studies, the type and extent of the tumor may be impossible to determine because of secondary inflammatory processes which opacify the affected sinus. A final diagnosis may be made only after microscopic examination of a tissue specimen.

Epithelial Papillomas-Inverting Papillomas

General Considerations

True epithelial papillomas arising in the mucous membrane of the paranasal sinuses and nasal cavity are rare. There are two types: hard (exophytic, squamous) and soft (inverting). The latter is clinically more significant than the former.

Inverting papilloma, a designation suggested by Ringertz (1938), is a firm, verrucous polypoid lesion arising in the ethmoid or maxillary sinuses. It arises nearly as often in the nasal cavity, but only rarely in the sphenoid and frontal chambers. It has a tendency to recur in 30 to 50% of patients and may transform into squamous cell carcinoma. It rarely, if ever, metastasizes, but may cause death from intracranial extension.

Pathologically, inverting papilloma is characterized by infoldings of the epithelium in a loose vascular connective stroma, resembling a genuine papilloma turned inside out. The epithelial cells may be of a well differentiated columnar or transitional type. Later, various degrees of anaplasia of the epithelium may be seen, including invasive carcinoma.

Inverting papilloma is most common in the male and the majority of cases occur in the middle aged group. Clinically there is nasal obstruction, discharge, and epistaxis. Displacement of the eyeball and deformity of the face in association with a discharging sinus tract may occur.

Roentgen Findings

The roentgen findings are nonspecific, and the chief manifestations are as follows (Figs. 17.1 to 17.4): (1) in the early stage an isolated, polyp-like lesion in the maxillary sinus, ethmoid sinus, or nasal cavity; (2) a soft tissue mass filling the maxillary sinus, ethmoid sinus, or nasal cavity; (3) in advanced cases erosion and destruction of the bony walls of the sinuses and of the nasal cavity.

Differential Diagnosis

Inverting papilloma should be differentiated from carcinoma when there is erosion or destruction of bone. The differential points are:

Parameter	Inverting Papilloma	Carcinoma
Character of growth	Slow	Rapid
Bony changes	Erosion and thinning by pressure	Destruction by invasion
Pain	Usually absent	Usually present
Clinical course	Protracted	Rapidly progressive

Maxillary Sinus Juvenile Angiofibroma

General Considerations

Juvenile angiofibroma is a benign, highly vascular, and noninfiltrative tumor. The involvement of the maxillary sinus is usually from anterior extension of a primary lesion in the nasopharynx. Primary juvenile angiofibroma of the maxillary sinus in the presence of a normal nasopharynx is extremely rare.

The disease usually occurs in young males and symptoms develop during the middle teens. There is nasal obstruction and discharge and episodes of epistaxis. Facial deformity may be present. Spontaneous regression of the lesion may occur during early adulthood.

Roentgen Findings

The roentgen findings are nonspecific (Fig. 17.5): (1) an expanding soft tissue mass in the maxillary sinus and opacification of the sinus; (2) pressure thinning, erosion, and destruction of the bony walls of the affected sinus; (3) involvement of the ethmoid sinus; (4) extension to the nasal cavity with bowing and displacement of the nasal septum to the opposite side.

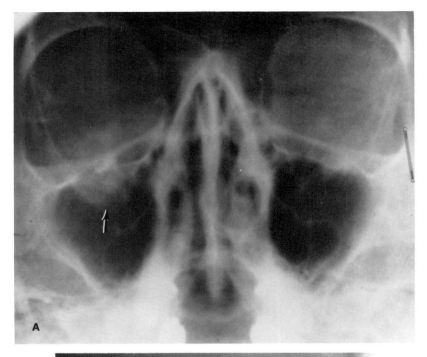

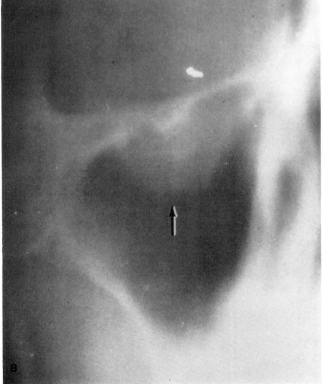

Figure 17.1 Inverting papilloma of the right maxillary sinus—early case. *A*, Waters view: a polyp-like lesion in the superior aspect of the right maxillary sinus. *B*, Posteroanterior tomogram: the polyp-like lesion is shown to good advantage. There is a possible calcific nidus in the central part of the lesion. No calcium was found in the specimen, and the cause of the hazy density has not been discovered. (Reprinted with permission from Radiology 91: 770, 1968.)

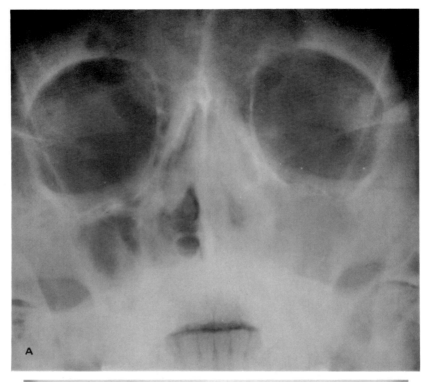

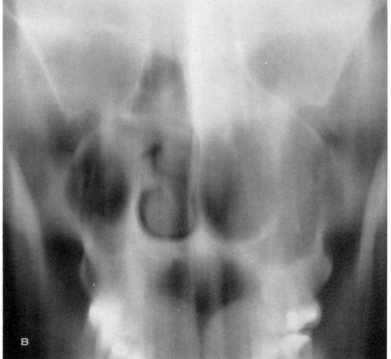

Figure 17.2. Inverting papilloma of the left maxillary sinus and nasal cavity — moderately advanced case. A, Waters view: complete opacity of the left maxillary sinus and left side of the nasal cavity. There is thickening of the lining mucosa of the right maxillary sinus. B, Posteroanterior tomogram: a large expanding soft tissue mass occupies the left maxillary sinus and the left side of the nasal cavity, with erosion of the walls of the sinus. There is also an irregular increase in the density in the right ethmoidomaxillary region. The cloudiness of the left ethmoid sinus is probably due to obstruction.

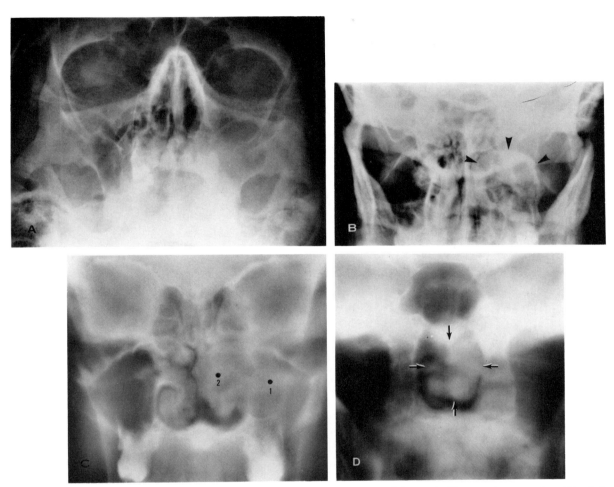

Figure 17.3. Inverting papilloma of the left maxillary sinus and left nasal fossa. *A*, Waters view: opacity of the left maxillary sinus. The medial wall is indistinct. *B*, Semiaxial view: soft tissue mass in the left maxillary sinus with thinning and expansion of the superoposterior wall of the sinus (*arrows*). *C*, Anteroposterior tomogram (mid cut): a large soft tissue mass occupying the entire left maxillary sinus (*1*) and extending to the left nasal fossa (*2*). There is destruction of the left middle and inferior turbinates. The medial wall of the left maxillary sinus is eroded and destroyed. *D*, Anteroposterior tomogram (posterior cut): polypoid mass in the nasal cavity extending to the anterior part of the nasopharynx.

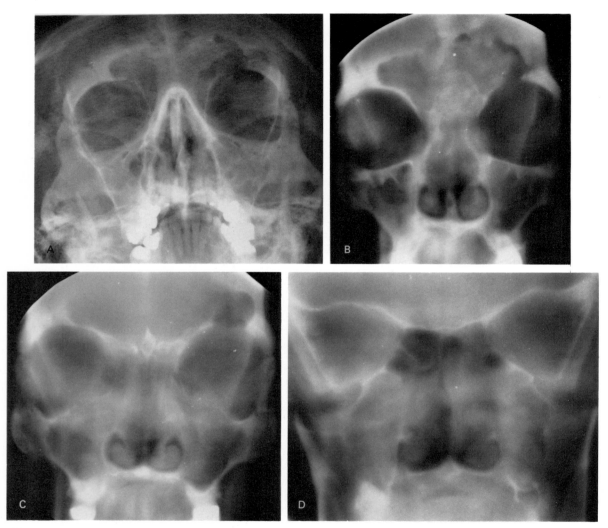

Figure 17.4. Inverting papilloma of the nasal cavity, maxillary, ethmoid and frontal sinuses—far advanced case. *A,* Waters view: complete opacity of both maxillary sinuses, more on the left side. There is cloudiness of the ethmoid and frontal sinuses. A bony defect is noted in the superior rim of the left orbit. *B,* Anteroposterior tomogram (anterior cut—19 cm): a huge soft tissue mass occupies the medial aspects of both maxillary sinuses, the upper two-thirds of the nasal cavity, and the anterior ethmoid and frontal sinuses. There is destruction of the middle turbinates. The lacrimal bone is eroded on each side. A bony defect resulting from tissue biopsy in the superior rim of the left orbit is shown to good advantage. *C,* Anteroposterior tomogram (middle cut—18 cm): the soft tissue mass occupies the superior and medial aspects of both maxillary sinuses, the upper two-thirds of the nasal cavity, and the ethmoid sinuses. There is erosion and destruction of the lamina papyracea and the floor of the orbit on each side. *D,* Anteroposterior tomogram (middle cut—17 cm): the soft tissue mass is seen in the upper two-thirds of the nasal cavity, both maxillary sinuses, and the left posterior ethmoid sinus.

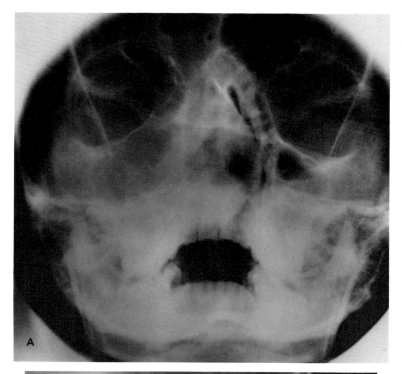

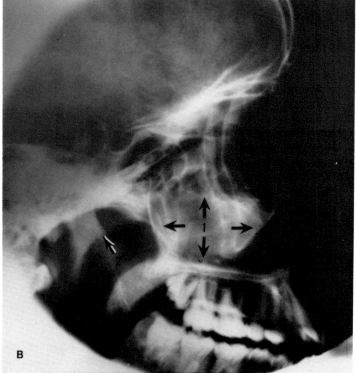

Figure 17.5. Juvenile angiofibroma of the right maxillary sinus. *A*, Waters view: a large homogeneous soft tissue mass occupies the entire right maxillary sinus with involvement of the right ethmoid sinus. There is thinning of the right inferior orbital rim and bowing of the nasal septum to the left. *B*, Lateral view: a soft tissue mass in the maxillary sinus (*1*), with a normal-appearing nasopharynx (*2*). (Reprinted with permission from Archives of Otolaryngology 89: 111, 1969.)

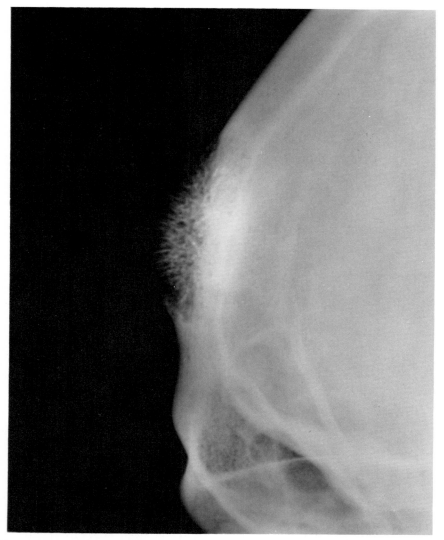

Figure 17.6. Hemangioma of the frontal bone. Lateral view of the skull: sharply defined zone of rarefaction with a characteristic sun-ray pattern in the frontal region of the skull.

Hemangiomas

General Considerations

The origin and nature of hemangiomas have been described in Part I, Chapter 6.

Hemangiomas of the paranasal sinuses are extremely rare and usually arise from the bony walls. They may occur in the maxilla or the frontal bone.

The symptoms are nonspecific and variable, depending upon the location and size of the tumor. Hemangiomas of the maxilla and frontal bone may be associated with palpable masses. The overlying skin may show telangiectatic areas.

Roentgen Findings

In hemangiomas of the maxilla or the frontal bone (Fig. 17.6) the following may be seen: (1) frontal projection—a sunburst appearance, produced by osseous spicules radiating outward from the area of involvement; (2) lateral projection—a sun-ray pattern, produced by radiopaque spicules running at right angles to the surface of the area of the lesion; (3) a honeycombed or soap bubble effect resulting from the formation of numerous thin trabeculae of new bone between the vascular sinuses.

In hemangioma of the maxilla or hard palate the lesion may extend to the maxillary sinus or the nasal cavity (Fig. 17.7). The roentgen findings in the maxillary sinus are: (1) thickening of the lining mucosa in the involved area; (2) soft tissue mass in the sinus; (3) erosion and destruction of walls of the sinus.

Meningioma of the Paranasal Sinuses

General Considerations

Meningioma is a slow growing, expansile intracranial tumor. Extracranial extension has been frequently reported, but extension into the nasal cavity or paranasal sinuses is infrequent.

Primary meningioma arising in the paranasal sinuses is extremely rare.

Pathogenetically, three theories have been advocated as to the origin of extracranial meningiomas: (1) the meningioma originates in the arachnoid cell nests of the dura which protrude through the dura to produce an extradural meningioma; (2) the meningioma arises from embryonal arachnoid nests which are pinched off and left behind during embryonic development. These arachnoid cell nests are situated outside the skull and vertebrae along the lines of fusion of embryonic bones; (3) blastomatous transfor-

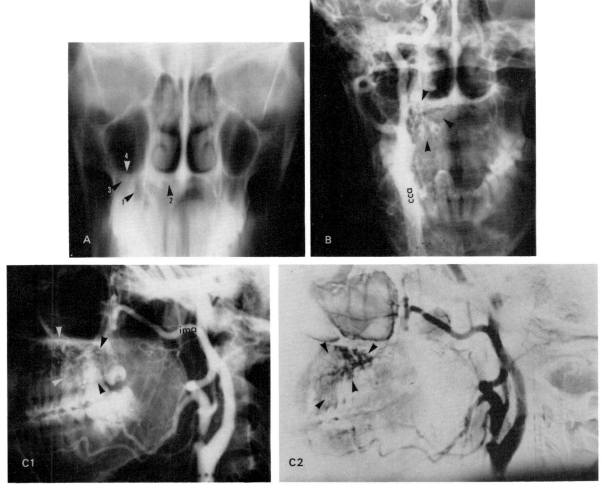

Figure. 17.7. Cavernous hemangioma of the maxilla and hard palate. *A*, Posteroanterior tomogram: soft tissue mass destroying the right maxilla (*1*) and the hard palate on the right side (*2*). The floor of the right maxillary sinus is thin and slightly irregular (*3*). There is some regional thickening of the lining mucosa of the sinus (*4*). *B*, Carotid angiogram (anteroposterior view): multiple anomalous vessels are seen within the soft tissue mass involving the maxilla and the hard palate (*arrows*). *C*, Carotid angiogram (lateral view): *1*, without subtraction: a large hemangioma is supplied by branches of the internal maxillary artery (*arrows*); *2*, with subtraction: the increased vascularity in the tumor is readily seen (*arrows*).

mation of heterotopic cellular elements of the dura.

Meningiomas of the paranasal sinuses occur most frequently in the frontal sinus, but can occur in the maxillary and ethmoid cells. They are found in young and middle aged adults.

Clinically, meningiomas of the frontal sinus may cause deformity of the forehead and displacement of the eyeball, with resulting proptosis and diplopia. When the tumor occurs in the maxillary sinus, there is swelling of the cheek and nasal obstruction. Displacement and protrusion of the eyeball may occur when there is upward expansion of the lesion.

Roentgen Findings

The roentgen findings are nonspecific (Fig. 17.8) and include: (1) soft tissue mass with homogenous density in the affected sinus; (2) expansion of the lesion with erosion and destruction of the bony walls of the affected sinus; (3) presence of psammoma bodies—these are caused by hyaline formation with sand-like calcification in the tumor mass.

Neurofibroma

General Considerations

Neurofibromas of the paranasal sinuses are extremely rare. They arise from the connective tissue sheath of nerves and may appear as solitary lesions or as part of a generalized neurofibromatosis. The symptoms are the same as those produced by any slow growing tumor. The most common symptoms are nasal obstruction and discharge. Bleeding or pain is seldom present.

Roentgen Findings

The roentgen changes are not characteristic and may be identical with other benign tumors (Fig. 17.9): (1) definable soft tissue mass within the sinus in the early case; (2) opacity of the affected sinus; (3) erosion and destruction of the bony walls of the affected sinus in the advanced case.

Osteochondroma

General Considerations

Osteochondroma is a very rare benign tumor arising from the bony walls of the paranasal sinuses. It is found most frequently in the sec-

ond or third decades of life. There is little sex predilection. The lesion is usually asymptomatic. It may cause enlargement and deformity of the face on the affected side.

Roentgen Findings

The roentgen findings are similar to those of osteochondroma elsewhere in the body: (1) irregular areas of increased density surrounded by areas of decreased density within the sinus; (2) expansion of the walls of the affected sinus with well defined borders.

Osteochondroma must be differentiated from osteomata, especially the cancellous type, and ossifying fibromas.

Osteoma

General Considerations

Osteoma is a slow growing benign bone tumor. It varies in size from a pea to a hen's egg and may fill the entire sinus and extend into neighboring structures. Osteoma is the most common tumor of the paranasal sinuses, with an incidence of approximately 0.25% in all routine examinations of the paranasal sinuses. The frontal sinus is the most common site of occurrence of osteoma and it is found less frequently in the ethmoid cells. Osteomas rarely occur in the maxillary or sphenoid sinuses.

Pathologically, osteomas may be divided into compact or hard and cancellous or soft varieties depending upon the constituents of the lesion. The cancellous osteoma probably represents a form of fibrous dysplasia and should not be considered a true bone neoplasm.

The disease is more common in the male than in the female. It most often occurs during puberty and in the second and third decades of life. Most of the patients are asymptomatic and the tumor is accidentally discovered during roentgen examination. When symptoms occur, they are usually the result of closure of the ostium of the affected sinus with retention of secretions or are secondary to expansion and erosion of the bony walls of the sinus. Osteoma of the frontal sinus may produce deformity of the forehead, involvement of the orbit, and extension to the anterior cranial fossa.

Roentgen Findings

The roentgen findings vary with the types of osteoma.

Compact Osteoma. It has the appearance of a

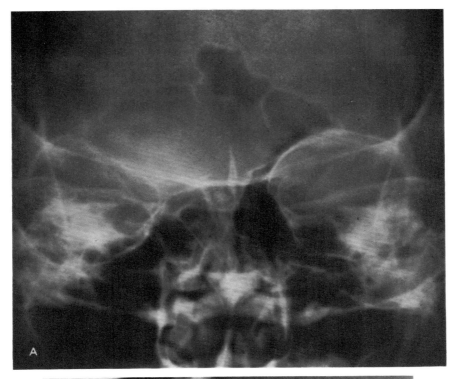

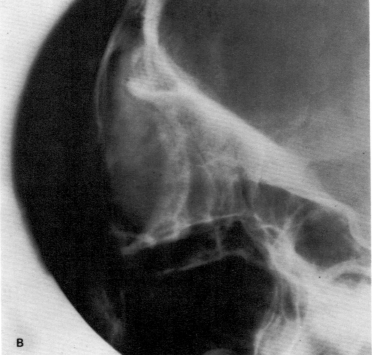

Figure. 17.8. Meningioma of the frontal sinus. *A*, Posteroanterior view: a large expanding soft tissue mass fills the right frontal sinus, with extension to the right ethmoid sinus. There is erosion and downward displacement of the roof and superior aspect of the medial wall of the right orbital cavity. *B*, Lateral view: the soft tissue mass is shown in the frontal sinus with extension to the anterior ethmoid sinus. (Reprinted with permission from Laryngoscope 80: 640, 1970.)

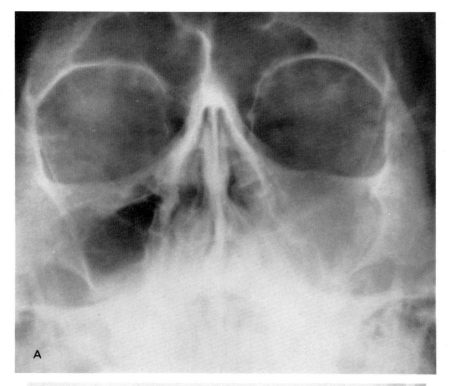

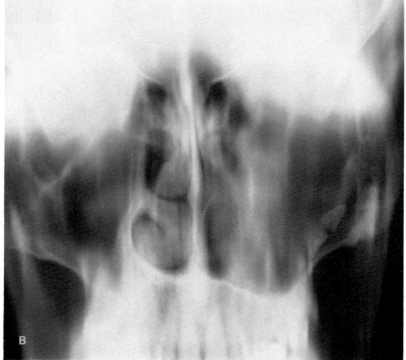

Figure 17.9. Neurofibroma of the left maxillary sinus. *A*, Waters view: complete opacity of the left maxillary sinus, with thinning of the left orbital floor. *B*, Posteroanterior tomogram: an ill defined soft tissue mass in the left maxillary sinus with erosion and destruction of the walls. There is involvement of the left nasal cavity with destruction of turbinates. A saucer-shaped erosion of the left alveolar ridge is also noted.

sharply defined, homogeneous rounded or lobulated, ivory-hard bony mass, either sessile or pedunculated (Fig. 17.10). This type of osteoma is common.

Cancellous Osteoma. It appears as an irregularly defined, rounded or lobulated mass of less density, and may be mistaken for a soft tissue mass. This type of osteoma is rare.

Giant Cell Tumor of the Jaws and Paranasal Sinuses

General Considerations

There is considerable diversity of opinion regarding bone lesions that contain a large number of multinucleated giant cells. According to Jaffe, true giant cell tumors do exist; however, a great majority of the giant cell lesions are reparative granulomas. Most authors concur in the classification of giant cell tumors as true neoplasms. Jaffe regards giant cell tumors as "a neoplasm of a definite kind, arising apparently from the undifferentiated supporting connective tissue of the marrow and clearly delimitable on the basis of its cytology."

Histologically, a true giant cell tumor forms by cellular proliferation. The picture is characterized by rather uniform and closely packed stromal cells showing nuclear preponderance. The giant cells are quite numerous and uniformly distributed. Hemorrhage is inconspicuous. Only rarely is osteoid or bone formation evident.

Recurrence after the surgical removal of a giant cell tumor is not uncommon, and sarcomatous transformation after radiation therapy or repeated curetting has frequently been reported.

The highest incidence of giant cell tumor occurs between the ages of 20 and 40, and it is rare in patients under 20 years of age. There is no sex predilection.

Giant cell tumor is uncommon in the mandible and maxilla. According to the Mayo Clinic study, the comparative incidence of giant cell tumor and giant cell reparative granuloma is 2:64 (Austin et al., 1959).

In the mandible the symphysis is the most common site. Giant cell tumors of the maxilla may involve the entire bone, including the sinus, but giant cell tumors arising in the paranasal sinuses per se are extremely rare. The maxillary sinus is the most frequently reported site, but it is often difficult to decide whether the primary growth originated in the maxilla or the sinus. Many authors believe that the lesion usually starts in the maxilla and then spreads to the maxillary sinus and/or the hard palate. The sphenoid, ethmoid, and frontal sinuses are rare sites of involvement.

The presenting symptoms of giant cell tumors of the paranasal sinuses are not characteristic and depend upon the location of the tumors. Proptosis occurs with ethmoid and sphenoid involvement. Swelling of the cheek and nasal obstruction are common complaints when the maxillary sinus is involved. Pain from nerve involvement has been reported.

Roentgen Findings

In the jaws, findings are as follows: (1) A well demarcated radiolucent area with a lobulated outline and coarse trabeculation (the "soap bubble" effect); (2) an area of radiolucency with ill defined margins; (3) Expansion and thinning of the cortex of the jaws; (4) Displacement of the teeth and resorption of the roots of the teeth; (5) Perforation of the alveolar process into the mouth or the maxillary sinus in advanced cases.

In the paranasal sinuses (Fig. 17.11), findings are as follows: (1) opacity of the affected sinus; (2) possible mass density within the affected sinus; (3) erosion and destruction of the bony wall of the sinus.

Other Rare Benign Tumors

Among other rare benign tumors of the paranasal sinuses are benign osteoblastoma (giant osteoid osteoma) and aneurysmal bone cyst. Two cases of benign osteoblastoma of the maxilla have been reported (Borella et al., 1967, and Kent et al., 1969). Radiographically, the lesion arises from the maxilla as a well circumscribed mass with irregular radiopaque densities. There is expansion and erosion of the cortex of the maxilla. A "nidus" is not seen as in cases of osteoid osteoma. There is no pronounced cortical new bone surrounding the lesion. Extension to the maxillary sinus may occur in the advanced case (Fig. 17.12).

Aneurysmal bone cyst is a cystic lesion of bone containing spaces filled with blood. Only six cases involving the facial bones have been reported in the literature. The radiographic appearance is not characteristic. It appears as an expanding tumor mass with erosion and destruction of the involved bone (Fig. 17.13).

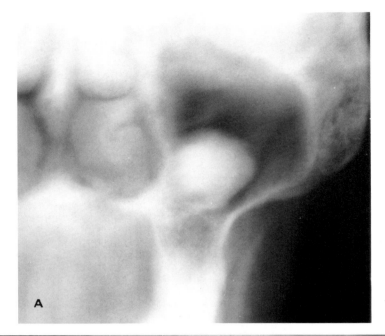

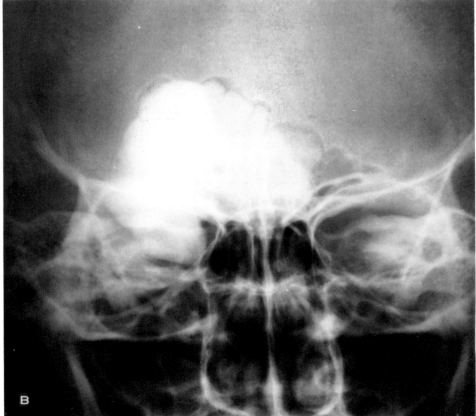

Figure 17.10. Osteomas of the maxillary and frontal sinuses. *A*, Anteroposterior tomogram: compact osteoma of the floor of the left maxillary sinus.

B, Posteroanterior view: a large, lobulated, compact osteoma of the frontal sinuses.

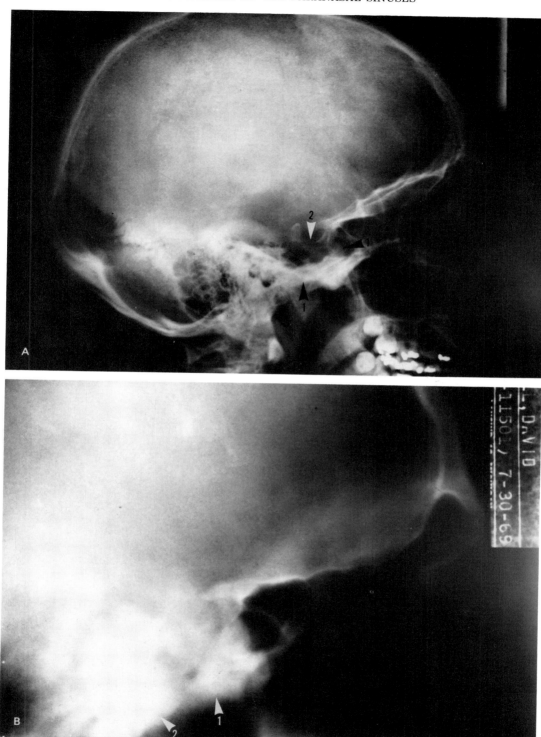

Figure 17.11. Giant cell tumor of the sphenoid sinus. *A*, Lateral view of the skull: a tumor mass is seen in the posterior part of the sphenoid sinus (*1*). There is demineralization and possible destruction of the floor of the sella turcica (*2*). *B*, Lateral tomogram of the sella turcica: a rounded tumor mass in the sphenoid sinus is apparent. There is destruction of the floor of the sphenoid (*1*) and clivus (*2*).

Tissue biopsy revealed a benign giant cell tumor of bone. (Reprinted with permission from Archives of Otolaryngology 94: 369, 1971.)

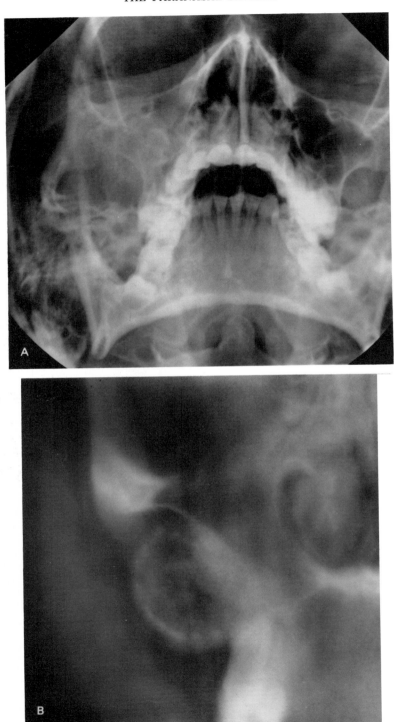

Figure 17.12. Benign osteoblastoma of the right maxilla with extension to the maxillary sinus. *A*, Waters view: opacity of the right maxillary sinus. There is a tumor mass with irregular calcification in the lateral aspect of the sinus. The lateral wall of the sinus is irregular and partially destroyed. *B*, Posteroanterior tomogram: a rounded, well corticated tumor mass with spotty calcifications is seen arising from the lateral wall of the right maxillary sinus. The lateral wall of the sinus is sclerotic and partially destroyed. There is extension of the lesion to the maxillary sinus with opacity of the floor. (Courtesy of Dr. Jess C. Galbreath.)

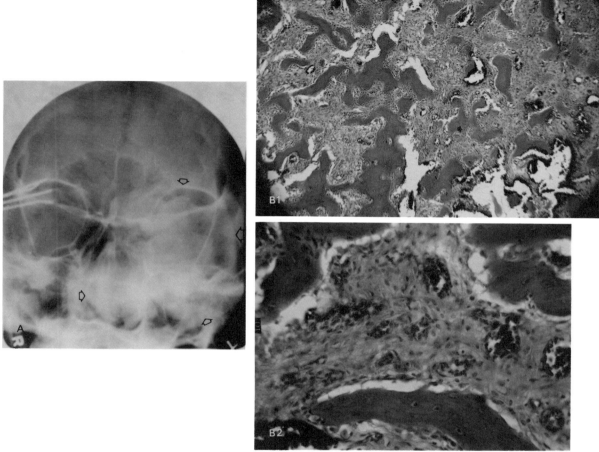

Figure 17.13. Aneurysmal bone cyst. *A*, Posteroanterior view of the paranasal sinuses: a large tumor mass of bony density involves the inferior aspect of the left frontal sinus, left orbit, left ethmoid, left maxillary sinus, and left nasal fossa, with deviation of the nasal septum to the right. The margin of the lesion is well defined, with a lobulated appearance (*arrows*). *B*, Histological appearance of the lesion: *1*, photomicrograph (H & E, × 60) showing fibrous tissue honey-combed with vascular spaces filled with blood; *2*, photomicrograph (H & E, × 200) showing vascular spaces lined by endothelial cells. There is no blood clot, indicating free venous circulation. (Reprinted with permission from Laryngoscope 77: 599, 1967.)

TUMOR-LIKE LESIONS OF THE PARANASAL SINUSES

Giant Cell Reparative Granuloma of the Jaws and Paranasal Sinuses

General Considerations

Giant cell reparative granuloma of the jaws is a local reparative reaction in response to an inflammatory process. It has been commonly confused with true giant cell tumor of bone.

Histologically, the lesion is characterized by the granulomatous appearance of its stroma. The stroma is composed of young connective tissue and fibroblasts, the cells of which show cystoplasmic preponderance. There is a rather irregular distribution of multinucleated giant cells. Hemorrhagic foci are almost always evident and there may be osteoid and immature bone formation.

Giant cell reparative granuloma is peculiar to the jaw bones, and more often occurs in the mandible than in the maxilla. It is most frequently found in the premolar region and symphysis of the mandible and in the canine region of the maxilla. In the maxilla, the maxillary sinus may be involved.

Giant cell reparative granuloma is also found on the gingiva or the alveolar borders of the jaws in the form of a sessile or pedunculated mass with erosion of the underlying bone. It is often called giant cell epulis or peripheral giant cell reparative granuloma. Epulis, often preceded by local mechanical injury to the site of origin, may also form in the socket of an extracted tooth. It may be the presenting sign of parathyroid adenoma with hyperparathyroidism.

Giant cell reparative granuloma of the paranasal sinuses is not a common disease. It can be found in any of the sinuses, but more often occurs in the maxillary and ethmoid chambers. When it is primary in the sinus there is often a severe chronic sinusitis with sclerosing osteitis; the giant cell reaction is indicative of a reparative process. If involvement of the sinus is secondary to primary maxillary disease, there is often bony destruction of the alveolar ridge.

The incidence is highest in young persons, most often between the age of 10 and 25, but can be observed in any age group from 5 to 70. Males and females are affected with about equal frequency.

Roentgen Findings

The roentgen features are nonspecific. The findings depend upon the location of the lesion.

Jaw Bone. In the jaw bone (Fig. 17.14), findings are as follows.

1. The lesion appears as a rounded or an oval area of radiolucency, which is sometimes faintly trabeculated. It thins and expands but does not perforate the cortex of the affected jawbone.

2. The lesions can appear as multiple radiolucencies with bony trabeculation.

3. Malposition of the teeth and resorption of the roots of the teeth upon which the lesion encroaches are not unusual findings. There is destruction of the lamina dura.

Alveolar Ridge. In the alveolar ridge (Fig. 17.15), findings are as follows.

1. A saucer-shaped area of pressure erosion is often seen on the surface of the alveolar ridge in the region of the lesion. The margin of the erosion may be dense and sclerotic.

2. When extensive, the lesion may cause destruction of the alveolar ridge of the maxilla and extend into the maxillary sinus.

Maxillary Sinus. In the maxillary sinus (Figs. 17.16 and 17.17), findings include: (1) erosion and destruction of the alveolar ridge when the lesion of the sinus is secondary to that of the maxilla; (2) opacity of the affected sinus; (3) thinning and expansion of the affected sinus; (4) destruction of the bony walls of the affected sinus.

Differential Diagnosis

True Giant Cell Tumor. The histological differentiation between true giant cell tumor and giant cell reparative granuloma is often possible, but it can also be extremely difficult. In some cases the final diagnosis may depend upon the clinical course of the disease.

"Brown Tumor" of Hyperparathyoidism. This usually occurs in advanced hyperparathyroidism. Histologically, it is similar to giant cell reparative granuloma. It can be excluded on the basis of blood chemistry (Fig. 17.18).

Ossifying Fibroma (Fibroosteoma)

General Considerations

Ossifying fibroma is essentially a cellular fibroma in which, according to the stage of differentiation, various amounts of calcified intercellular substances are produced.

Ossifying fibroma has been considered a variant of fibrous dysplasia. However, a number of investigators have attempted to differentiate the two. Sherman and Sternberg (1948) believe that unilocular lesions are caused by ossifying fibroma, whereas Pindborg (1951) states that the sharply circumscribed lesion may be considered an ossifying fibroma. Smith (1965) states that rimming of the bone trabeculae with osteoblasts is not a feature in fibrous dysplasia, but is frequently seen in ossifying fibroma. Osteoid is not often present in fibrous dysplasia, but is common in ossifying fibroma. Schmaman et al. (1970) demonstrated that immature woven bone with curved trabeculae is seen in fibrous dysplasia, whereas lamellar bone with parallel trabeculae occurs in ossifying fibroma.

The current theories of pathogenesis favor the hypothesis that ossifying fibroma is a benign tumor developing from cell nests in normal bone and is a disease entity separate from fibrous dysplasia. The tumor usually continues to grow without regard to skeletal maturity and may reach considerable size.

Ossifying fibroma generally occurs in the mandible and maxilla. In the maxilla, the lesion arises from the medullary or subperiosteal portion of the bone. It is often found in the

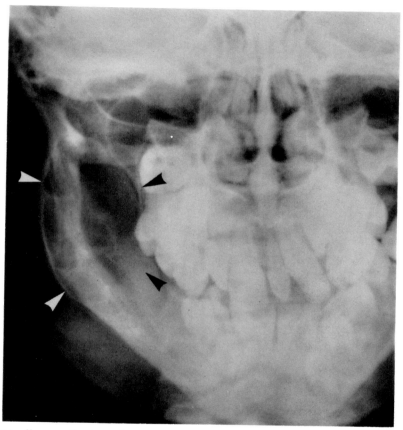

Figure 17.14. Giant cell reparative granuloma—typical giant cell reparative granuloma of the mandible. Anteroposterior view of the mandible: a large oval-shaped radiolucency with faint trabeculation involves the ramus of the right mandible. There is thinning and expansion of the cortex in the diseased area.

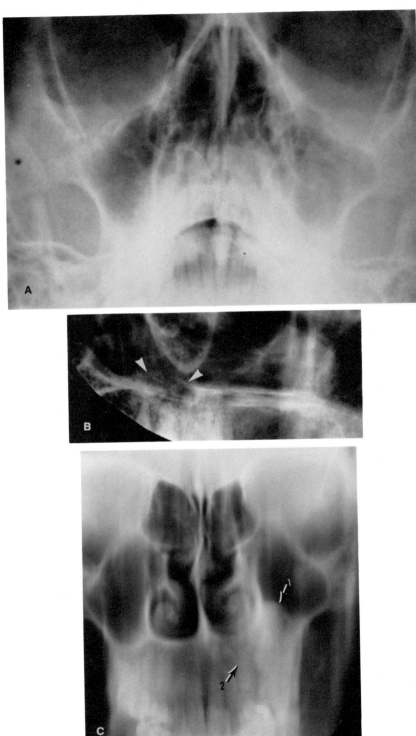

Figure 17.15. Giant cell epulis of the left upper gum with extension to the floor of the left maxillary sinus. *A*, Waters view: opacity of the left maxillary sinus, without evidence of bone destruction. There is soft tissue swelling in the left cheek. *B*, Lateral view: a small area of bony destruction and erosion in the floor of the left maxillary sinus in the region of the premolar teeth (*arrows*). *C*, Posteroanterior tomogram: erosion and destruction of the left alveolar ridge with a soft tissue density in the floor of the left maxillary sinus (*1*). There is a mottled destruction of the left side of the maxilla (*2*).

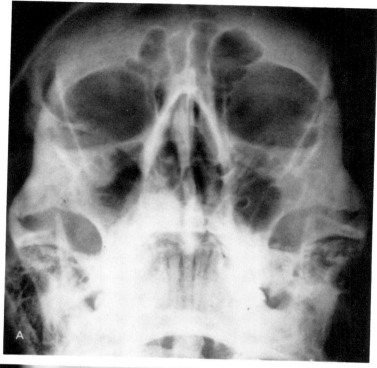

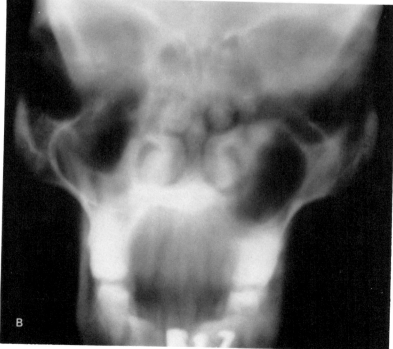

Figure 17.16. Early giant cell reparative granuloma of the left maxillary sinus. *A*, Waters view: slightly increased density in the floor of the left maxillary sinus. There is mucosal thickening in the right maxillary sinus. *B*, Anteroposterior tomogram: erosion of the left alveolar ridge, left side of the hard palate, and inferior part of the medial wall of the maxillary sinus.

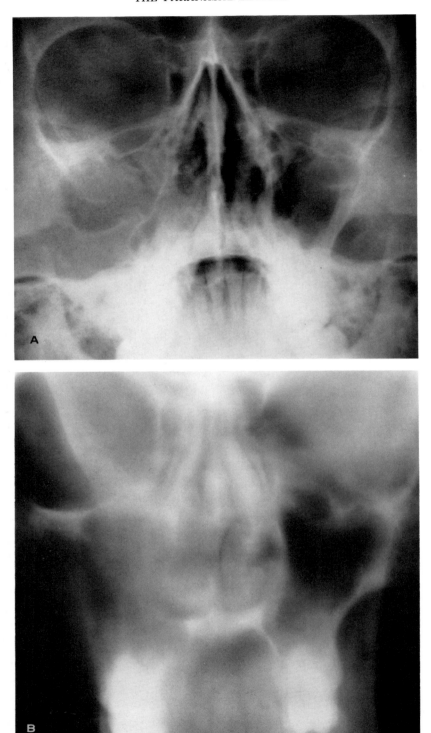

Figure 17.17. Advanced giant cell reparative granuloma of the right maxillary sinus. *A*, Waters view: opacity of the right maxillary sinus with indistinctness of the lateral wall. *B*, Anteroposterior tomogram: a large soft tissue mass occupies the entire right maxillary sinus, with erosion of the lateral and medial walls and the zygomatic process of the maxilla. There is erosion and destruction of the right alveolar ridge and right side of the hard palate.

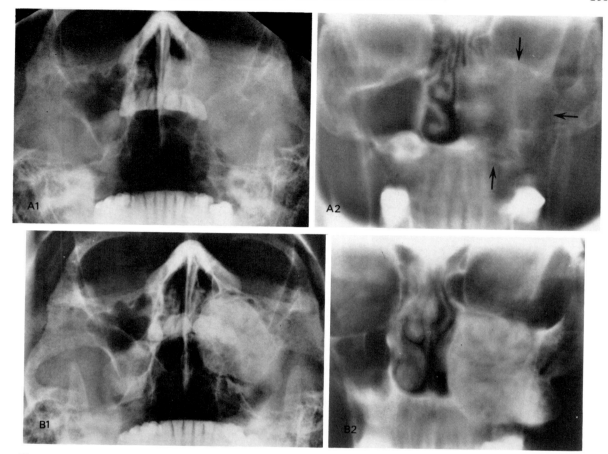

Figure 17.18. Left maxillary "brown tumor" of hyperparathyroidism *A*, preoperative. *1*, Waters view: complete opacity of the left maxillary sinus with erosion and destruction of the lateral wall and the floor of the sinus. *2*, Anteroposterior tomogram: an expanding soft tissue mass occupies the entire left maxillary sinus, with erosion of the lateral wall and the floor of the sinus (*arrows*). The upper alveolar ridge is partially eroded and destroyed (*arrow*). There is cloudiness of left ethmoid sinus and left nasal fossa, with enlargement of the turbinates. *B*, Postoperative (15 months after parathyroidectomy). *1*, Waters view: calcification and ossification of the soft tissue mass with restoration of the lateral wall and floor of the sinus. *2*, Anteroposterior tomogram: the calcified and ossified expanding soft tissue mass is clearly shown. (Courtesy of Department of Radiology, St. Joseph Hospital, Houston, Texas.)

second and third decades of life, but it is recognized only when the growth expands the bone, producing deformity and malocclusion. The disease is more common in the female.

Roentgen Findings

The roentgen appearance depends upon the relative amounts of fibrous tissue, newly formed osteoid, and calcified matrix. It varies from complete radiolucency containing little bone to uniform radiopacity containing masses of calcified matrix and well calcified bone. The chief roentgen features are as follows (Figs. 17.19 and 17.20): (1) a well defined osteolytic lesion in the

maxilla or mandible surrounded by a lamella of bone and no periosteal reaction in the early phase of the disease; (2) a well defined osteoblastic lesion producing thickened dense bone and containing confluent calcific densities in the advanced case; (3) expansion of the maxilla with obliteration of the maxillary sinus.

Differential Diagnosis

Ossifying fibroma of the maxilla should be differentiated from osteogenic sarcoma or chondrosarcoma which has not yet extended beyond the walls of the maxillary sinus. Ossifying fibroma of the maxilla has also to be differen-

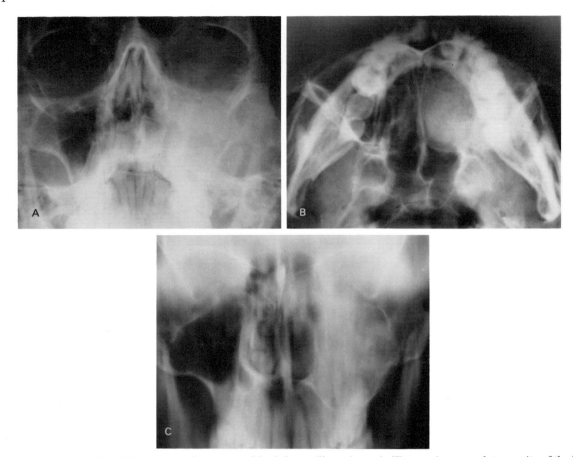

Figure 17.19. Ossifying fibroma, osteolytic type, of the left maxillary sinus. *A*, Waters view: complete opacity of the left maxillary sinus. There is slight thinning and expansion of the lateral wall of the sinus. *B*, Base view: a rather homogeneous density is seen in left maxillary sinus, with expansion and thinning of the posterolateral and middle walls. *C*, Posteroanterior tomogram: a soft tissue mass with rather faint ill defined confluent radiopaque densities occupies the entire left maxillary sinus. There is expansion and thinning of the lateral wall and the floor of the sinus. No definite destruction of the wall is noted. The left ethmoid sinus is slightly cloudy, probably due to obstructive change. There is slight thickening of the turbinates in the left nasal fossa.

tiated from monostotic fibrous dysplasia, but differentiation of these two lesions on roentgen grounds alone may not be possible.

Fibrous Dysplasia

General Considerations

Fibrous dysplasia is a disorder of bone characterized by the proliferation of fibroosseous tissue in the interior of the affected part or parts of the bone. It is not a tumor, but commonly produces a tumor-like lesion. The normal anatomic appearance of the bone is altered. The involved bone shows a disordered growth pattern with a slow expansion in various directions rather than a localized swelling.

The etiology of fibrous dysplasia is unknown. Some investigators attribute the initiation of the lesions to trauma, whereas other believe that the lesions represent nonspecific reparative responses to various injuries. However, the most widely accepted concept is that these lesions are attributable to developmental defects caused by faulty embryogenesis because they may be present in childhood, enlarge rapidly during the period of general body growth, and cease growth after puberty (Lichtenstein and Jaffe).

Fibrous dysplasia has been divided into three

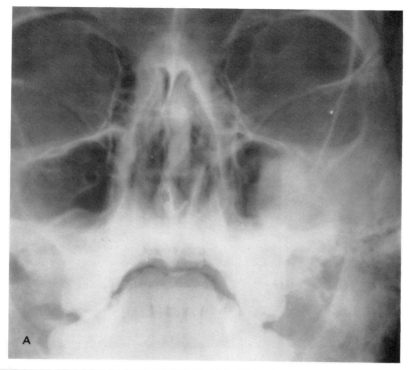

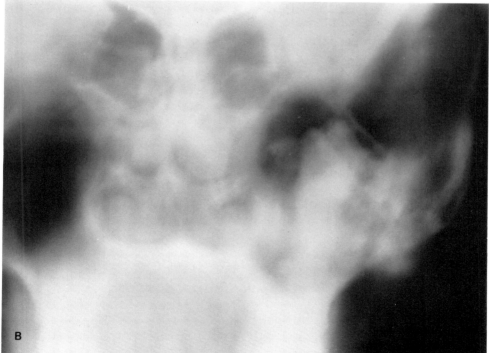

Figure 17.20. Ossifying fibroma, osteoblastic type, of the left maxillary sinus. A, Waters view: a large, lobulated, osteoblastic lesion occupies the lateral aspect of the left maxillary sinus. B, Anteroposterior tomogram: the lobulated contours and calcific densities of the mass are clearly shown. There is no destruction of bone.

types: monostotic, regional, and polyostotic. In the head and neck region the monostotic type may involve any of the facial bones, but the maxilla is the most common site. In regional fibrous dysplasia there is involvement of several adjacent bones of the face, calvarium, and anterior cranial fossa, resulting in leontiasis ossea. The polyostotic type may occur with or without systemic manifestations. The association of the bony changes with pigmentation of the skin and precocious puberty in the female is commonly called Albright's syndrome. The polyostotic type rarely involves the jaw bones. Although the majority of the lesions are found in young people in their first and second decades of life, the onset of the symptoms has been reported as late as 73 years of age. The polyostotic type of the disease predominates in the female; in the monostotic and regional types the sex distribution is almost equal.

In the monostotic form with involvement of the maxilla the most common symptom is swelling of the face and cheek. The lesions are usually unilateral and may produce facial deformity without pain, proptosis of the eye with visual disturbances, and nasal obstruction and epistaxis.

Roentgen Findings

The roentgen appearance depends upon the degree of mineralization of the abnormal bone. In many instances radiolucent changes predominate. These are irregular in appearance and often contain islands of well calcified bone. In other instances the degree of radioopacity results in a ground glass or orange peel appearance.

In monostotic fibrous dysplasia of the maxilla the roentgen manifestations are variable and the chief findings are: (1) predominantly osteoblastic changes with dense, ivory-like thickening of bone and partial obliteration of the maxillary sinus (Fig. 17.21); (2) mixed types with obliteration of the sinus; (3) a predominantly osteolytic variety with expansion and thinning of the cortex. The maxillary sinus is often filled with a soft tissue mass (Fig. 17.22); (4) involvement of the malor bone. Since in this disease no anatomic limits seem to be respected, there is often involvement of the malar bone with rather poor margination of the lesion.

Leontiasis Ossea

General Considerations

Leontiasis ossea is a rare condition of overgrowth of the bones of the skull and face, giving a lion-like facial appearance. Its etiology is unknown, but most observers believe it to be a regional form of fibrous dysplasia. The involvement of the skull and face may be unilateral or diffuse. It is a slow growing hyperostosis involving the frontal and temporal regions of the skull, the floor of the anterior cranial fossa, and the facial bones. The resultant deformity encroaches upon the orbital cavities and paranasal sinuses and obliterates the foramina of the floor of the anterior cranial fossa.

The disease may appear in early life, but because the lesion is asymptomatic in the early stages, it may not be appreciated until swelling and deformity of the face are obvious. As the disease progresses, involvement of the cranial nerves may occur, with various neurological manifestations. Sensory disturbances along the divisions of the fifth cranial nerve are often observed. Involvement of the first cranial nerve results in a loss of the sense of smell. Encroachment upon the optic foramen may cause loss of vision.

Roentgen Findings

The roentgen features are relatively characteristic and are as follows (Fig. 17.23): (1) ivory-like bony density in the affected portion of the skull, mainly in the frontal and temporal regions and face; (2) expansion and deformity of the skull and face; (3) encroachment upon, and deformity of, orbital cavities, paranasal sinuses, and nasal cavity; (4) thickening of the floor of the anterior cranial fossa with obliteration of the foramina.

Differential Diagnosis

Leontiasis ossea should be differentiated from hyperparathyroidism, Paget's disease, neurofibromatosis, osteoblastic metastasis, and meningioma.

Paget's Disease

General Considerations

Paget's disease, or osteitis deformans, is a disease of the skeleton of unknown etiology. In the early stages of the disease resorption of bone is prominent. Later, there is a mixture of destruction and repair of bone. In the final stage, the reparative process is prominent and osteosclerosis is marked. The pelvis, lumbar spine, skull, femur, and tibia are the common sites of involvement, although any bone, including those of the face, may be affected. The disease process may be limited to a part or all of a single

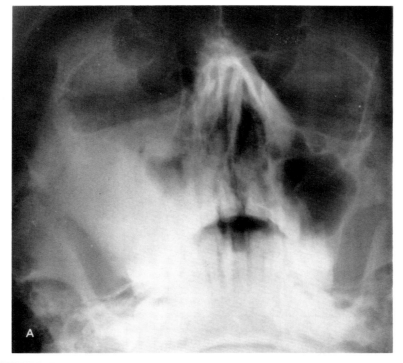

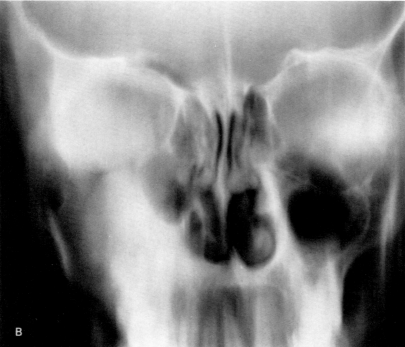

Figure 17.21. Fibrous dysplasia, osteoblastic type, of the right maxillary sinus. *A*, Waters view: the lateral wall of the right maxillary sinus is thickened, expanded, and dense, with almost complete obliteration of the sinus cavity. There is less sharply defined involvement of the zygoma. *B*, Posteroanterior tomogram: the lesion is shown to good advantage. The margin of the involved lateral wall is smooth and the density of the lesion is homogeneous.

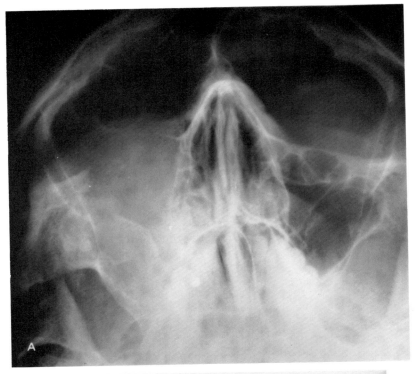

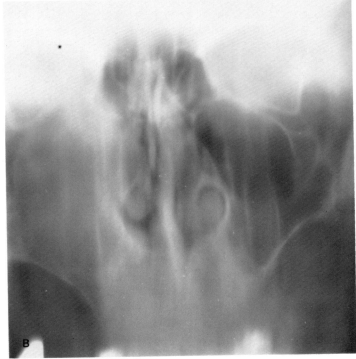

Figure 17.22. Fibrous dysplasia, osteolytic type, of the right maxillary sinus. *A*, Waters view: there is complete opacity of the right maxillary sinus with a soft tissue mass. The mass extends upwardly with expansion and erosion of the roof of the sinus. There is involvement of the zygoma. *B*, Posteroanterior tomogram: the sinus walls are thinned and expanded but otherwise intact.

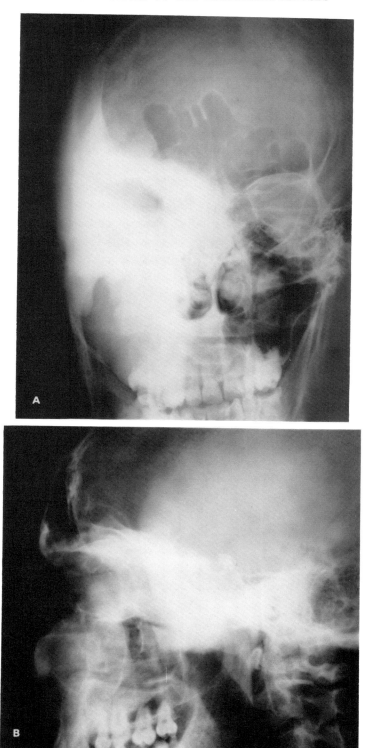

Figure 17.23. Leontiasis ossea. *A*, Posteroanterior view of the skull: ivory-like bone density involving the right side of face and skull; *B*, Lateral view of the skull: the lesion involves the maxilla, orbital cavity, temporal bone, and base of the skull.

bone, but in the disseminated form it may involve almost all bones. In the face, the thickened bone may encroach upon or obliterate a sinus cavity. When the ostium of a sinus becomes obstructed, sinusitis may result. The jaws are commonly involved and the maxilla alone is most often affected. Involvement of both maxilla and mandible is rare. The involved jaw may become abnormally large, resulting in malocclusion of the teeth.

Paget's disease occurs mainly in the middle and advanced years, although it has been reported in persons less than 20 years of age.

Roentgen Findings

The roentgen findings in the skull and facial bones depend upon the state of the disease.

In the early stage there are osteolytic changes in the affected bones which produce a ground glass appearance. These changes include: (1) osteoporosis circumscripta of the skull; (2) osteolytic lesions aroung the roots of the teeth, and resorption of the roots of the teeth; (3) absence of the lamina dura around the teeth in the involved area; (4) pronounced radiolucency of the jawbone.

In the advanced stage of the disease, osteoblastic changes become a prominent feature in the calvarium and facial bones. These changes include: (1) enlargement of the skull with "cotton-wool" appearance (Fig. 17.24A); (2) platybasia from softening of the base of the skull; (3) thickening of the maxillae with encroachment upon and reduction in the size of the maxillary sinuses (Fig. 17.24B); (4) thickening of the base of the skull with reduction in the size of the sphenoid sinuses (Fig. 17.24C); (5) hypercementosis of the roots of the teeth caused by excessive deposition of poorly differentiated cementum on the roots of the teeth. The presence of hypercementosis is of value in differentiating Paget's disease from other osseous lesions involving the jaws; (6) formation of new bone on the edentulous alveolar ridge and on the inferior border of the mandible.

Lesions to be differentiated from Paget's disease include fibrous dysplasia and osteoblastic metastasis from carcinoma of the prostate.

SECONDARY TUMORS OF THE PARANASAL SINUSES

Odontogenic Origin

Ameloblastoma (Adamantinoma, Multilocular Cyst)

General Considerations. Ameloblastomas are slowly growing but destructive tumors of epithelial odontogenic origin. They arise from the embryonal remnants of the dental lamina and the enamel organ, or the basilar layer of the oral epithelium. These tumors tend to recur and metastases have been reported.

From the pathological standpoint an ameloblastoma is characterized by a fusiform or cylindrical expansion of the bone. Perforation rarely occurs. Numerous cysts are characteristic of the tumors, but occasionally the tumor may appear solid. Microscopically the characteristic features of ameloblastoma are the columnar, ameloblast-like cells in the periphery and loosely arranged central cells which resemble the stellate reticulum in the normal enamel organ.

Ameloblastomas may occur anywhere in the jaws but are most commonly seen in the molar and premolar regions. The tumors may involve the maxillary sinus, with expansion and erosion of the walls.

Ameloblastomas may occur at any age and have been reported in infants. The majority of cases are seen after age 30. There is no sex predilection.

Clinically, the tumor is painless and slow growing. It may cause some degree of facial deformity and looseness of the teeth. Spontaneous fractures and infection may occasionally occur.

Roentgen Findings. The roentgen features are variable. The chief findings are as follows:

1. The cyst of the jaw may be unilocular, resembling a dentigerous, radicular, or primordial cyst. This is not frequent.

2. A more typical appearance is that of multilocular radiolucencies of the jaw, either honeycombed or bubble-like in appearance. The compartments are rounded and separated by distinct septa which are variable in size; the appearance may simulate a cluster of grapes. The cortex of the involved jaw is usually expanded. Destruction of the cortex and invasion of the soft tissues may occur (Fig. 17.25A).

3. The cysts may be so small and the bony trabeculation so dense that the whole mass appears solid. The solid form usually occurs in the body of the mandible.

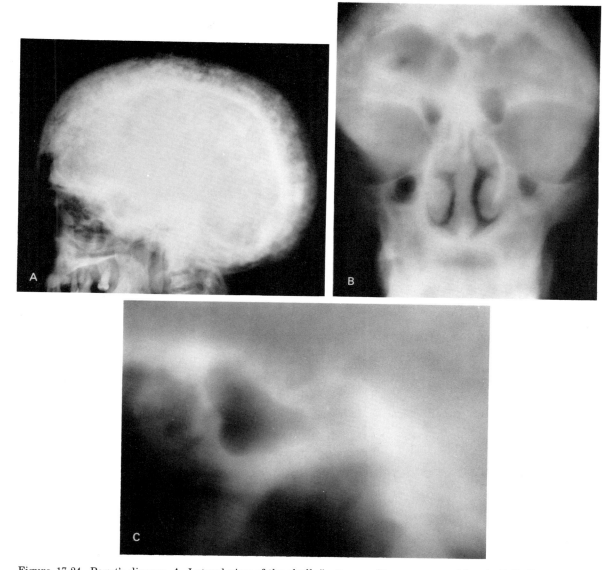

Figure 17.24. Paget's disease. *A*, Lateral view of the skull: "cotton-wool" appearance of the skull; *B*, Posteroanterior tomogram of the skull: extensive involvement of the facial bones with encroachment upon the maxillary, ethmoid, and frontal sinuses; *C*, Lateral tomogram of the sphenoid sinus: marked thickening of the base of the skull with partial obliteration of the sphenoid sinus.

4. Erosion of the roots of a tooth, appearing as a small excavation at the edge of a root, is considered a sign of ameloblastoma, but it is an uncommon finding.

5. When the maxillary sinus is involved, the cavity is filled and expanded. The bony walls are thin, decalcified, and eroded. In advanced cases adjacent structures may be invaded (Figs. 17.25*B* and 17.26). All or a portion of a tooth may be embodied in the tumor.

Calcifying Epithelial Odontogenic Tumor

General Considerations. Calcifying epithelial odontogenic tumors are rare. Not more than 50 cases have appeared in the literature since the first report by Pindborg in 1958. The tumor is probably derived from the reduced enamel organ of an embedded tooth. It may occur in the space from which an embedded tooth has been removed.

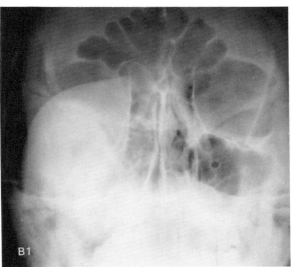

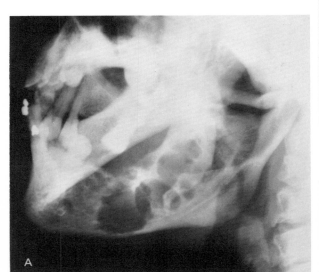

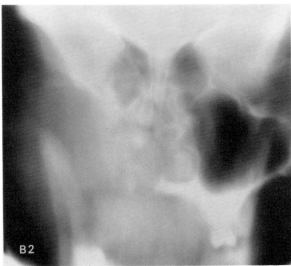

Figure 17.25. Ameloblastoma. *A*, Oblique view of the mandible: multilocular rounded radiolucencies, resembling a cluster of grapes, involving the ascending ramus and posterior part of the body of the right mandible. There is a small area of cortical erosion in the superior aspect of the lesion.

B, Ameloblastoma of the right maxillary sinus. *1*, waters view: opacity of the right maxillary sinus with a hugh soft tissue mass involving the right cheek. *2*, anteroposterior tomogram: a large soft tissue mass occupies the right maxillary sinus and the right side of the nasal cavity. There is erosion of the zygomatic process and the lateral wall of the sinus as well as erosion and destruction of the floor of the right orbit and the right alveolar ridge and hard palate.

Pathologically, the tumor consists of peculiar polyhedral epithelial cells which exhibit extensive intracellular degeneration. The degenerated cytoplasm has an affinity for mineral salts which are deposited in the form of Liesegang's rings. In some parts of the tumor the calcified cells coalesce to form a conglomerate. The older the tumor, the more pronounced is the calcification.

The tumors occur in the jawbones, most frequently in the mandible. The majority are located in the premolar and molar areas. In the maxilla, the tumor may expand and invade the maxillary sinus.

Calcifying epithelial odontogenic tumors occur in the same age group as does ameloblastoma. The recorded ages range from 13 to 78 years, with the bulk of cases in the middle aged

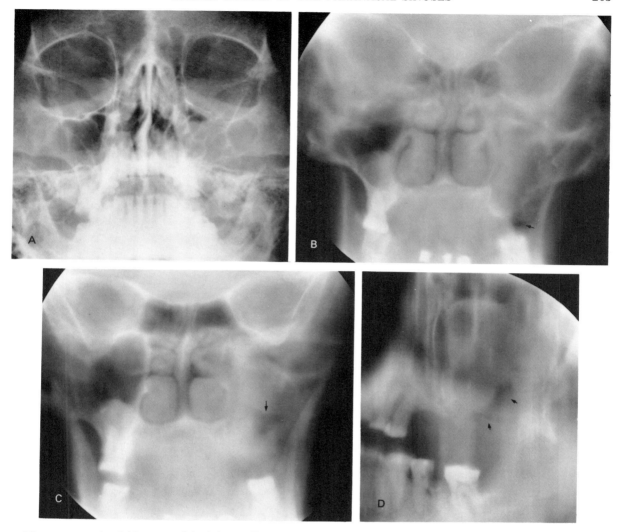

Figure 17.26. Ameloblastoma of the left maxillary sinus with antro-oral fistula. *A*, Waters view: soft tissue mass in the lower two-thirds of the left maxillary sinus with indistinctness of the lateral wall. *B*, Anteroposterior tomogram (middle cut): a soft tissue mass occupies most of the sinus with erosion of the lateral wall. There is a small area of radiolucency in the region of the missing third molar (*arrow*). A small serous cyst is noted in the floor of the right maxillary sinus. *C*, Anteroposterior tomogram (posterior cut): air (*arrow*) is seen in the lower part of the soft tissue mass leading to the region of the missing third molar. *D*, lateral tomogram of the left maxillary sinus: the communication between the sinus cavity and the oral cavity is seen posteriorly (*arrows*).

group. There appears to be no sex predilection. Clinically the tumors behave like ameloblastomas, are locally invasive, and tend to recur locally.

Roentgen Findings. The chief roentgen features are as follows:

1. The lesion shows a multiloculated, honeycombed appearance with irregular and ill defined borders. Areas of calcification or conglomerates of calcified densities are present within the lesion.

2. The tumor occurs in combination with an embedded tooth.

3. When the maxillary sinus is involved, the sinus cavity is filled with a calcified mass. The walls are thin, decalcified, and eroded, but are rarely destroyed (Fig. 17.27).

Differential Diagnosis. Calcifying epithelial odontogenic tumor should be differentiated from ameloblastoma. Pathologically, in calcifying epithelial odontogenic tumors, there are no ameloblast-like cells or central cells similar to stel-

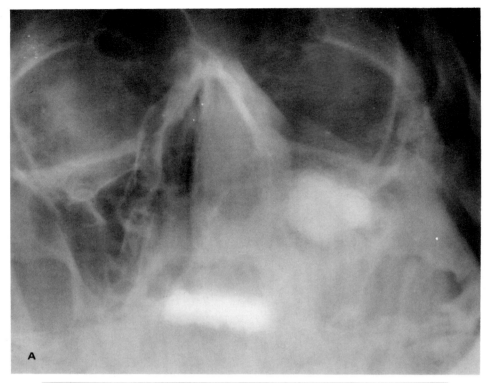

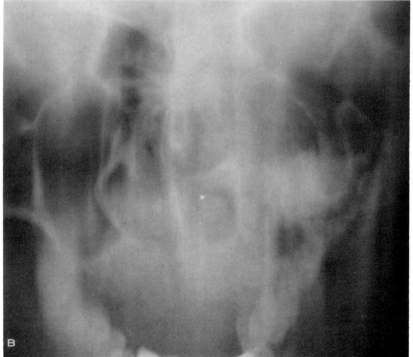

Figure 17.27. Calcifying epithelial odontogenic tumor of the left maxillary sinus. *A*, Waters view: the antrum is completely opacified and contains a lobulated, calcific mass. *B*, Posteroanterior tomogram: the calcified mass is associated with irregularity and spotty calcification of the left alveolar ridge. The cloudiness of the left ethmoid sinus and the loss of definition of the left middle turbinate are possibly due to direct extension.

late reticulum. Radiologically the tumor is, as a rule, more radiopaque than ameloblastoma because of the presence of mineral salts.

Odontoma

General Considerations. Odontomas are considered to be tumors formed by aggregation of masses of the various elements which form teeth. Usually they contain variable amounts of calcium. The tumor is often surrounded by a radiolucent area of varying width called the fibrous capsule.

Odontomas may develop in either jaw, but are found more frequently in the mandible. They occur in the area in which there has been failure of formation of the normal teeth. The most common sites are the upper central and lower incisors and the lower molar regions. They usually occur in young adults. There is no sex predilection.

Roentgen Findings. Odontomas are, as a group, called "composite odontomas" and are divided into complex and compound types.

COMPLEX ODONTOMA. It appears as a radiopaque mass of irregular configuration in which various components of teeth may be identified.

COMPOUND ODONTOMA. It is composed of masses of small, malformed teeth within the conglomerate mass. An individual tooth may be recognized.

Both types of odontomas may occur in the maxilla. These tumors tend to raise and displace the floor of the maxillary sinus and may extend into the sinus cavity. The walls of the sinus may be expanded and thinned. The changes may closely simulate a tumor arising from the sinus.

Odontogenic Myxoma (Myxofibroma)

General Considerations. Odontogenic myxoma or myxofibroma is a slow growing and nonmetastasizing tumor composed of mucous connective tissue. It may develop from retained islands of undifferentiated embryonic tissue or from connective tissue of the dental papillae. It is found almost exclusively in the tooth-bearing areas of the jawbones and is often associated with dental abnormalities such as unerupted or displaced teeth.

Pathologically, the tumor consists of myxomatous connective tissue resembling the stellate reticulum found in developing teeth. Occasionally fragments of odontogenic epithelium are found within the tumor.

Clinically the tumors are locally invasive and have a tendency to recur. Most of the cases are found between the second and third decades of life, but may be seen before the age of 10 and after the age of 50. There is no sex predilection. The mandible and maxilla are involved with about equal frequency. In the mandible the tumor often involves the ramus and the body; the symphysis is less often involved. The patient may exhibit facial deformity, but pain is uncommon. Exophthalmus may be present when the maxillary sinus is involved.

Roentgen Findings. The roentgen findings are nonspecific:

1. The lesions show fairly well defined radiolucent areas with a resultant honeycombed appearance. The compartments tend to be angular and may be separated by straight septa which form rectangular or triangular spaces.

2. Large lesions may produce thinning and expansion of the cortex; however, destruction of the cortex is seldom seen.

3. The lesions may occasionally appear unilocular.

4. The teeth are often displaced, occasionally with root resorption.

5. In the maxilla the tumors may extend to and invade the maxillary sinus. The tumor masses may fill the sinus cavity, but show distinct convex upper borders. Varying degrees of expansion and erosion of the bony walls of the sinus may occur, but destruction of the walls is rare (Fig. 17.28).

Differential Diagnosis. The unilocular type should be differentiated from dentigerous cyst. When the lesion is multilocular, giant cell reparative granuloma, fibrous dysplasia, odontogenic fibroma, and ameloblastoma should be ruled out.

Benign Cementoblastoma (True Cementoma)

General Considerations. Benign cemetoblastoma is a rare tumor of cementum which, although it develops during the period of tooth formation, may not be recognized until later in life. The tumors occur at the roots of the teeth and fuse with the roots. Histologically the benign cementoblastoma consists of numerous round bodies, often fused, with an appearance typical of cementum. The tumors tend to be encapsulated by fibrous tissue.

Cementoblastomas occur more often in the mandible than in the maxilla. A premolar or molar tooth is most frequently involved. There is probably no sex predilection. The tumors are

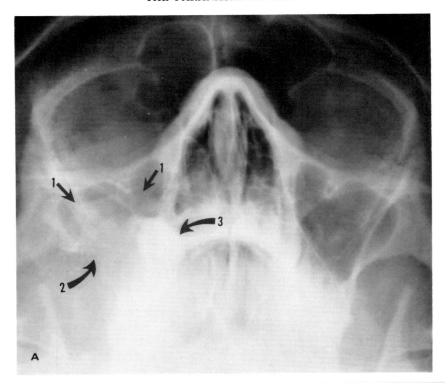

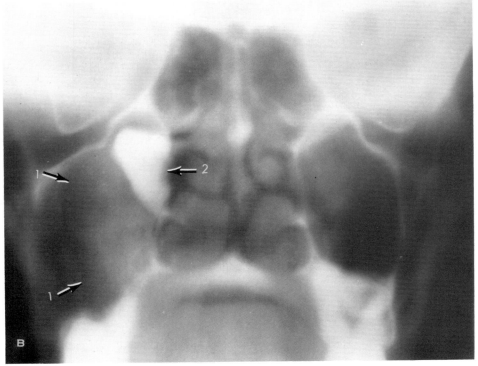

Figure 17.28. Odontogenic myxoma of the right maxillary sinus. *A*, Waters view: there is a soft tissue tumor mass with a well defined upper margin in the right maxillary sinus (*1*). Destruction of the lateral wall of the sinus is apparent (*2*). A tooth is faintly visualized in the medial aspect (*3*). *B*, Anteroposterior tomogram: a soft tissue mass occupies the majority of the sinus (*1*). The lesion appears to be attached to the alveolar ridge, which is partially destroyed. There is erosion and destruction of the lateral wall of the sinus. A molar tooth is displaced to the superomedial aspect of the sinus (*2*).

usually found in young adults under age 25. They grow slowly and are usually asymptomatic and without physical signs. Larger tumors may cause pain and swelling of the jaws.

Roentgen Findings (Fig. 17.29). These are as follows: (1) the lesion shows a well defined, radiopaque dense mass surrounded by a radiolucent zone representative of a fibrous capsule; (2) the lesion is attached to the root of the tooth; (3) the tooth related to the tumor is usually intact; (4) in the maxilla, the lesion may extend to the maxillary sinus. It may fill the sinus cavity with expansion and erosion of the walls of the sinus.

Pituitary Origin

Adenoma of the Pituitary Gland

General Considerations. Because of its anatomic proximity, the sphenoid sinus is prone to involvement by tumors of the pituitary gland. The most common benign neoplasms of the pituitary gland are adenomas. Of these, the chromophobe variety is the most frequent cause of enlargement of the sella turcica with concomitant involvement of the ethmoid and/or sphenoid sinuses.

Roentgen Findings. The roentgen manifestations in well established cases are (Figs. 17.30 to 17.32): (1) enlargement of the sella turcica; (2) depression of the floor of the sella turcica in whole or in part; a double contour effect may be produced by asymmetric enlargement; (3) elongation, thinning, and posterior bent of the dorsum sella; (4) erosion of the posterior clinoid

processes; (5) undercut appearance of the anterior clinoid processes; (6) extension of the lesion into the sphenoid and ethmoid sinuses, nasopharynx, and nasal cavity.

Nasopharyngeal Origin

Nasopharyngeal Juvenile Angiofibroma

Nasopharyngeal juvenile angiofibroma is an essentially benign but highly vascular tumor. The tumor is locally invasive and may erode adjacent paranasal sinuses and penetrate into normal fissures and foramina. In an advanced stage the lesion often extends into the maxillary, ethmoid, and sphenoid sinuses and the nasal cavity (Fig. 17.33). These are described in detail in Part III, Chapter 24.

Chordoma

Chordoma is a locally invasive, benign tumor occurring in the region of the clivus or of the upper portion of the cervical spine. The predominant feature is bony destruction. In advanced cases there is extension to the nasopharynx and involvement of the sphenoid sinus and sella turcica. This is discussed in Part III, Chapter 24.

Orbital Origin

Tumors of the orbital cavity may involve the paranasal sinuses by direct extension. Dermoid cysts and retinal anlage tumors may extend to the maxillary and ethmoid sinuses.

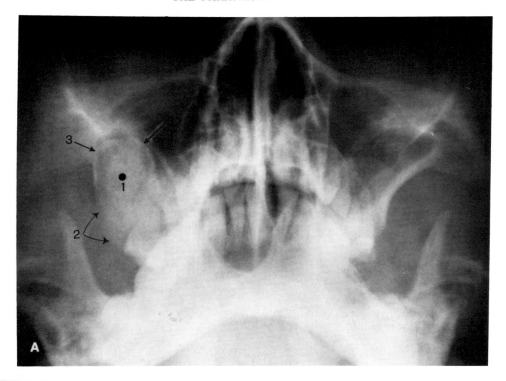

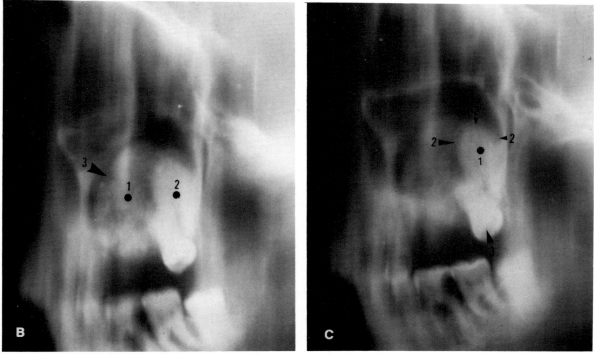

Figure 17.29. Benign cementoblastoma. *A*, Waters view: an oval-shaped radiopaque mass involves the lateral aspect of the right maxillary sinus (*1*). There is expansion and erosion of the lateral wall of the sinus (*2*). A radiolucent zone is noted along the periphery of the tumor mass (*3*). *B*, Lateral tomogram: two well defined radiopaque masses occupy the lower part of the maxillary sinus (*1 & 2*). The fibrous capsule of the anterior mass is partially visualized (*3*). The second molar was recently extracted. *C*, Lateral tomogram: the posterior tumor mass is well demonstrated (*1*) and the fibrous capsule is clearly shown (*2*). The tumor mass is attached to the root of the third molar (*3*).

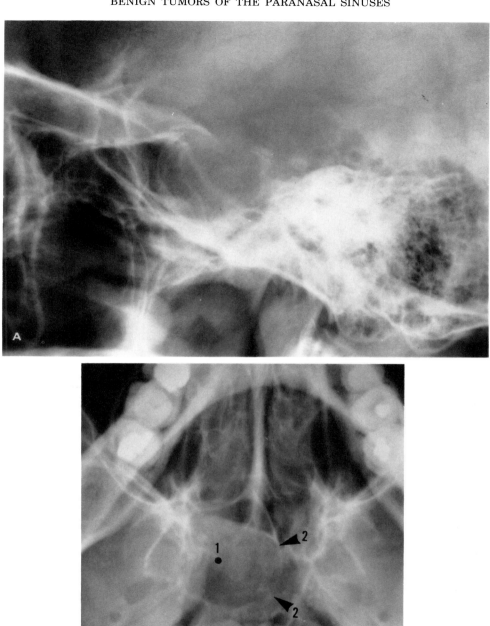

Figure 17.30. Pituitary adenoma with encroachment upon the sphenoid sinus. *A*, Lateral soft tissue film of the skull: enlargement of the sella turcica with double contour of the floor. There is encroachment upon the sphenoid sinus. *B*, Base view of the skull: mass lesion is in the region of the right side of the sphenoid sinus (*1*) with thinning and bowing of the bony septum of the sinus to the left side (*2*).

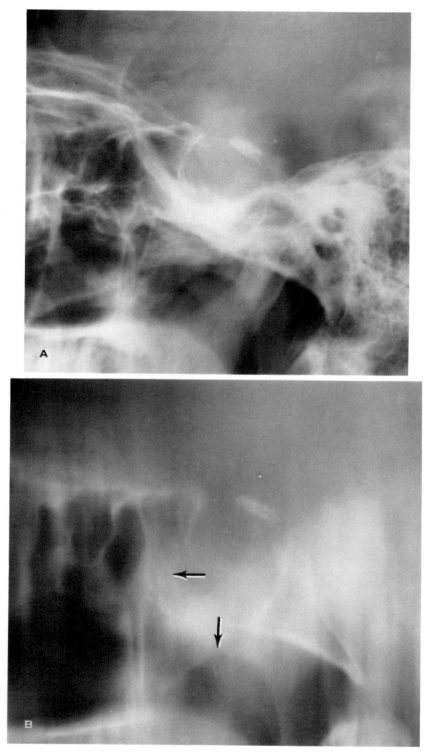

Figure 17.31. Advanced pituitary adenoma with extension to the sphenoid sinus and the nasopharynx. *A*, Lateral view of the skull: enlargement of the sella turcica with ballooning of the posteroinferior aspect; *B*, Lateral tomogram of the sella turcica: mass lesion in the sella turcica extending to the sphenoid sinus and the nasopharynx.

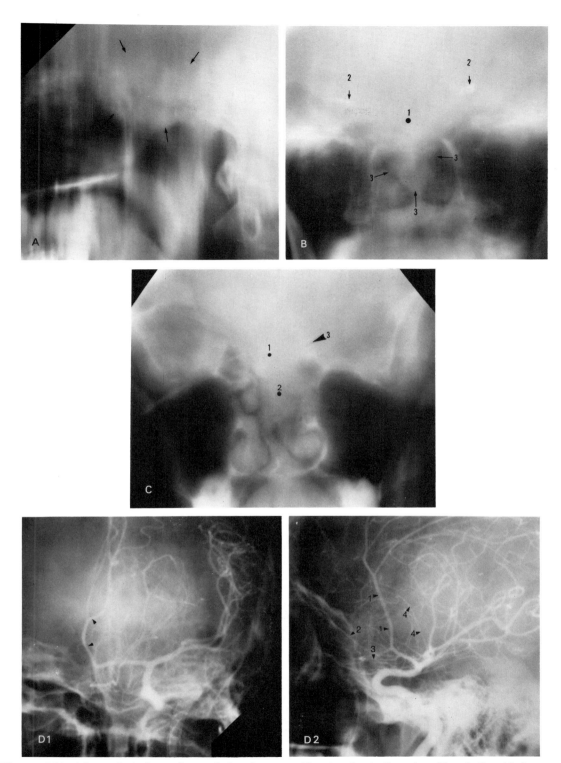

Figure 17.32. Advanced pituitary chromophobe adenoma with extension to the sphenoid and ethmoid sinuses, nasal cavity, and nasopharynx. *A*, Lateral tomogram of the sella turcica: a large destructive lesion involves the sella turcica and sphenoid sinus (*arrows*). *B*, Anteroposterior tomogram (posterior cut): a large soft tissue mass with complete destruction of the sella turcica and sphenoid sinuses (*1*). Remnants of the anterior clinoid processes are seen (*2*). There is extension of the lesion to the nasopharynx (*3*). *C*, Anteroposterior tomogram (middle cut): destruction of the ethmoid sinuses, particularly on the left (*1*). There is an extension into the left nasal fossa with destruction of the middle turbinate (*2*). The left lamina papyracea is destroyed (*3*). *D*, Carotid angiogram: *1*, anteroposterior view: rounded displacement of the left anterior cerebral artery to the right (*arrows*). *2*, lateral view: stretching and the posterior displacement of the anterior cerebral artery (*1*), and the inferior displacement of the frontopolar artery (*2*). There is some depression of the ophthalmic artery (*3*). Posterior displacement and straightening of the frontoascending artery (*4*) are also noted.

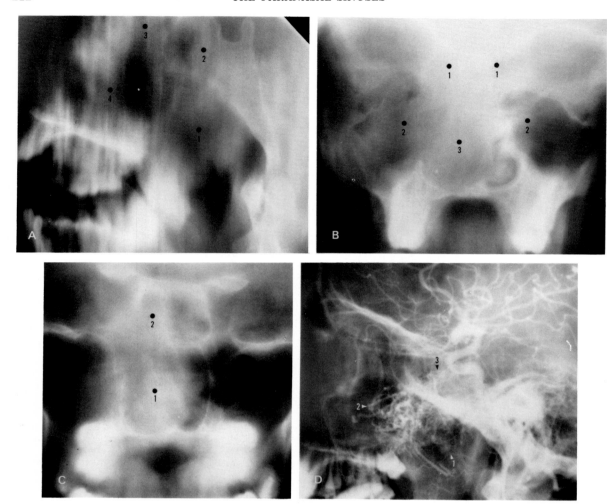

Figure 17.33. Advanced nasopharyngeal juvenile angiofibroma. *A*, Lateral tomogram of the nasopharynx: a huge tumor mass in the nasopharynx (*1*) extending into the sphenoid (*2*), ethmoid (*3*), and maxillary sinuses (*4*). *B*, Anteroposterior tomogram (middle cut): the mass extends to the ethmoid (*1*) and maxillary sinuses (*2*) and nasal cavity (*3*). *C*, Anteroposterior tomogram (posterior cut): tumor mass in the nasopharynx (*1*) extending to the right sphenoid sinus (*2*). *D*, Right lateral carotid angiogram: tumor vessels in the nasopharynx (*1*) extending to the maxillary and sphenoid sinuses (*2* & *3*).

REFERENCES

Aaron, I.: Ossifying fibroma of the maxillary sinus. Report of a case. Am. J. Cancer, *9:* 551, 1937.

Alford, T. C., Winship, T.: Epithelial papillomas of the nose and paranasal sinuses. Am. J. Surg., *106:* 764, 1963.

Austin, L. T. et al: Giant cell reparative granuloma and related conditions affecting the jaw bones. Oral Surg. Oral Med. Oral Pathol., *12:* 1285, 1959.

Ballenger, J. J.: *Diseases of the Nose, Throat and Ear*, 11th ed. p. 178–186. Lea and Febiger, Philadelphia, 1969.

Borello, E. D., Sedano, H. D.: Giant osteoid osteoma of the maxilla. Oral Surg. Oral Med. Oral Pathol., *23:* 563, 1967.

Brunner, H.: Fibrous dysplasia of facial bones and para—nasal sinuses. Arch. Otolaryngol., *55:* 43, 1952.

Brunner, H., Spiesman, I. C.: Osteoma of the frontal and ethmoid sinuses. Ann. Otol. Rhinol. Laryngol., *57:* 714, 1948.

Bucy, P. C., Capp, C. S.: Primary hemangioma of bone—with special reference to roentgenologic diagnosis. Am. J. Roentgenol., *23:* 1, 1930.

Choukas, N. C., Toto, P. D., Valaitis, T.: Sclerosing cavernous hemangioma of the maxilla. Oral Surg. Oral Med. Oral Pathol., *16:* 17, 1963.

Clark, H. B., Holte, N. O.: Paget's disease simulating osteomyelitis of the mandible. Case report. Northwest

Dentistry, *29:* 247, 1950.

Cody, C. C.: Inverting papillomata of the nose and sinuses. Laryngoscope, *77:* 584, 1967.

Cohen, A., Rosenwasser, H.: Fibrous dysplasia of the temporal bone. Arch. Otolaryngol., *89:* 31, 1969.

Conley, J., Healey, W. V., Blaugrund, S. M., Perzin, K. H.: Nasopharyngeal angiofibroma of the juvenile. Surg. Gynecol. Obstet. *126:* 825, 1968.

Cummings, C. W., Goodman, M. L.: Inverted papillomas of the nose and paranasal sinuses. Arch. Otolaryngol., *92:* 445, 1970.

Dahlin, D. C.: *Bone Tumors,* 2nd ed., p. 18–115. Charles C Thomas, Springfield, Illinois, 1967.

Dutta, A.: Ameloblastic odontoma. Oral Surg., *20:* 827, 1970.

Edeiken, J., Hodes, P. J. *Roentgen Diagnosis of Diseases of Bone,* p. 6.185–6.202. Williams & Wilkins, Baltimore, 1969.

Eggston, A. A., Wolef, D.: *Histopathology of the Ear, Nose and Throat,* p. 489–844. Williams & Wilkins, Baltimore, 1947.

Fetissof, A. G.: Pathogenesis of osteomas of nasal accessory sinuses. Ann. Otol., *38:* 404, 1929.

Finerman, W. B.: Juvenile nasopharyngeal angiofibroma in female. Arch. Otolaryngol., *54:* 620, 1951.

Friedberg, S. A., Eisenstein, R., Wallner, L. J.: Giant cell lesions involving the nasal acccessory sinuses. Laryngoscope, *79:* 763, 1969.

Friedman, W. H., Pervez, N., Schwartz, A. E.: Brown tumor of the maxilla in secondary hyperparathyroidism. Arch. Otolaryngol., *100:* 157, 1974.

Gay, I., Viskoper, J. R., Chowers, I.: Maxillary tumor as a presenting sign of secondary hyperparathyroidism due to renal insufficiency. J. Laryngol. Otol., *85:* 737, 1971.

Georgiade, N., Masters, F., Horton, C., Pickrell, K.: Ossifying fibromas (fibrous dysplasia) of the facial bones in children and adolescents. J. Pediat., *46:* 36, 1955.

Gorlin, R. J., Goldman, H. M.: *Thomas' Oral Pathology,* 6th ed., Vol. 1, p. 481–507, 560–574. C. V. Mosby, St. Louis, 1970.

Gorlin, R. J., Pindborg, J. J., Redman, R. S., Williamson, J. J., Hansen, L. S.: The calcifying odontogenic cyst: a new entity and possible analogue of the cutaneous calcifying epithelioma of Malherbe. Cancer, *17:* 723, 1964.

Gross, C. W., Montgomery, W. W.: Fibrous dysplasia and malignant degeneration. Arch. Otolaryngol., *85:* 653, 1967.

Halper, H.: Chordomata. Br. J. Radiol., *22:* 88, 1949.

Hamlin, W. B., Lund, P. K.: Giant cell tumors of the mandible and facial bones. Arch. Otolaryngol., *86:* 658, 1967.

Harma, R.: Nasopharyngeal angiofibroma. Acta Otol. Suppl., *146:* 1, 1958–1959.

Hera, J. F., Brown, A. K.: Paranasal juvenile angiofibroma. Arch. Otolaryngol., *76:* 457, 1962.

Hermia, M., Koskinen, O.: Metastasizing nasopharyngeal angiofibroma. Arch. Otolaryngol., *89:* 107, 1969.

Hill, C. L.: Meningioma of the maxillary sinus. Arch. Otolaryngol., *76:* 547, 1962.

Holman, C. B., Miller, W. E.: Juvenile nasopharyngeal fibroma; roentgenologic characteristic. Am. J. Roentgenol., *94:* 292, 1965.

Hora, J. F., Brown, A. R.: Paranasal juvenile angiofibroma. Arch. Otolaryngol., *76:* 547, 1962.

Jaffe, H. L.: Fibrous dysplasia of the bone. Bull. N. Y. Med., *22:* 588, 1946.

Jaffe, H. L.: *Tumors and Tumorous Conditions of the*

Bones and Joints, p. 425–450. Lea & Febiger, Philadelphia, 1958.

Jaffe, H. L., Lichtenstein, L., Portis, R. B.: Giant cell tumor of bone — its pathologic appearance, grading, supposed variants and treatment. Arch. Pathol., *30:* 993, 1940.

Jones, W. A., Gerrie, J., Pritchard, J.: Cherubism. A familia fibrous dysplasia of the jaws. J. Bone Joint Surg., *32:* B:334, 1950.

Kent, J. N., Castro, H. F., Girotti, W. R.: Benign osteoblastoma — case report and review of the literature. Oral Pathol., *27:* 209, 1969.

Kelemen, G., Holmes, E. M.: Cavernous hemangioma of the frontal bone. J. Laryngol. Otol., *62:* 557, 1968.

Krolls, S. O., Pindborg, J. J.: Calcifying epithelial odontogenic tumor — a survey of 23 cases and discussion of histomorphologic variations. Arch. Pathol., *98:* 206, 1974.

LaDow, C. S., Henefer, E. P., McFall, T. A.: Central hemangioma of the maxilla with Von Hippel's disease. J. Oral Surg., *22:* 252, 1964.

Lampertico, P., Russell, W. O., MacComb, W. S.: Squamous papilloma of upper respiratory epithelioma. Arch. Pathol., *75:* 293, 1963.

Lee, F. M. S.: Ameloblastoma of the maxilla with probable origin in a residual cyst. Oral Surg., *20:* 799, 1970.

Lichtenstein, L.: Polyostotic fibrous dysplasia. Arch. Surg., *36:* 874, 1938.

Lichtenstein, L., Jaffe, H. L.: Fibrous dysplasia of bone. Arch. Pathol., *33:* 777, 1942.

Lindstrom, C. G., Lindstrom, D. W.: On extracranial meningioma. Acta Otolaryngol., *68:* 451, 1969.

Maybery, T. E., Devine, K. D., Harrison, E. G., Jr.: The problems of malignant transformation in the nasal papilloma. Otolaryngology, *82:* 296, 1965.

Mainzer, F., Stargardter, F. L., Connolly, E. S., Eyster, E. F.: Inverting papilloma of the nose and paranasal sinuses. Radiology, *92:* 964, 1969.

Majoros, M.: Meningioma of the paranasal sinuses. Laryngoscope, *80:* 640, 1970.

Malcolmson, K. C.: Ossifying fibroma of the sphenoid. J. Laryngol., *81:* 87, 1967.

Males, J. L., Howard, W. J., Mask, D. R., Townsend, J. L., Snow, J. B., McPherson, H. T.: Primary hyperparathyroidism presenting as giant cell tumor of the maxilla. Arch. Intern. Med., *132:* 107, 1973.

Maniglia, A. J., Mazzarella, L. A., Minkowitz, S., Moskowitz, H.: Maxillary sinus angiofibroma treated with cryosurgery. Arch. Otolaryngol., *89:* 111, 1969.

Martin, H., Ehrick, H. E., Abels, J. C.: Juvenile nasopharyngeal angiofibroma. Ann. Surg., *127:* 513, 1948.

Maxwell, M. M., Blackstone, C. H.: Odontoma of the maxillary sinus. J. Oral Surg., *10:* 22, 1952.

Michael, L. A.: Giant cell tumor of the sinuses. Laryngoscope, *69:* 320, 1959.

New, G. B., Devine, N. D.: Neurogenic tumor of nose and throat. Arch. Otolaryngol., *46:* 163, 1947.

Norris, H. J.: Papillary lesion of the nasal cavity and paranasal sinuses. Part II — inverting papillomas, a study of 29 cases. Laryngoscope, *73:* 1, 1963.

Peimer, R.: Benign giant cell tumors the skull and nasal sinuses. Arch. Otolaryngol., *60:* 186, 1954.

Pendergrass, E. P., Schaeffer, J. P., Hodge, P. J.: *The Head and Neck in Roentgen Diagnosis,* 2nd ed., vol. 1, p. 503–523. Charles C Thomas, Springfield, Illinois, 1956.

Pantazopoulos, P. E., Nomikos, N.: Adamantinoma of the maxillary sinus. Ann. Otol., *75:* 1160, 1966.

Pernier, J. L., Bhaskar, S. N.: Aneurysmal bone cysts of

the mandible. Oral Surg. Oral Med. Oral Pathol., *11:* 1018, 1958.

Pindborg, J. J.: Calcifying epithelial odontogenic tumor. Cancer *11:* 838, 1958.

Pugh, D. O.: Fibrous dysplasia of the skull, probably explanation for leontiasis ossea. Radiology, *44:* 548, 1945.

Ringertz, N.: pathology of malignant tumors arising in nasal and paranasal cavities and maxilla. Acta Otolaryngol. Suppl., *27:* 31, 1938.

Schlumberger, H. C.: Fibrous dysplasia of single bone (monostotic fibrous dysplasia). Milit. Surgeon, *99:* 504, 1946.

Schmaman, A., Smith, I., Ackerman, L. V.: Benign fibro-osseous lesions of the mandible and maxilla. Cancer, *26:* 303, 1970.

Shahen, H. B.: Psammoma in the maxillary antrum. J. Laryngol., *46:* 117, 1931.

Sherman, R. S., Glauser, O. J.: Radiological identification of fibrous dysplasia of the jaws. Radiology, *71:* 553, 1958.

Sherman, R. S., Sternberg, W.: The roentgen appearance of ossifying fibroma of bone. Radiology, *50:* 595, 1948.

Shklar, G., Meyer, I.: Vascular tumors of the mouth and jaws. Oral Surg., *19:* 335, 1965.

Skolnik, E., Fornatto, E.: Ossifying fibroma of the paranasal sinuses. Ann. Otol. Rhinol. Laryngol., *64:* 689, 1955.

Skolnik, E. M., Leewy, A., Friedman, J. E.: Inverted papilloma of the nasal cavity. Arch. Otolaryngol., *84:* 61, 1938.

Smith, J. F.: Fibrous dysplasia of the jaws. Arch. Otol., *81:* 592, 1965.

Sonesson, A.: Odontogenic cysts and cystic tumors of the jaws. A roentgen-diagnostic and patho-anatomic study. Acta Radiol. Suppl., *81:* 104, 1950.

Stafne, E. C.: *Oral Roentgenographic Diagnosis,* 3rd ed., p. 168–217. W. B. Saunders Company, Philadelphia, 1969.

Stimson, P. G., Luna, M. A., Butler, J. J.: Seventeen-year history of a calcifying epithelial odontogenic (Pindborg) tumor. Oral Surg. Oral Med. Oral Pathol., *25:* 204, 1968.

Thoma, K. H., Goldman, H. M.: Central myxoma of the jaw. Am. J. Orthodontics (Oral Surg. Sect.), *33:* 532, 1947.

Thomas, G. K., Kasper, K. A.: Ossifying fibroma of the frontal bone. Arch. Otolaryngol., *83:* 43, 1966.

Vianna, M. R., Horizonte, M. G.: Aneurysmal bone cyst in the maxilla: report of case. J. Oral Surg., *20:* 432, 1962.

Wang, S. Y.: An aneurysmal bone cyst in the maxilla. Plast. Reconst. Surg., *25:* 62, 1960.

Ward, G. E., Hendrick, J. W.: *Diagnosis and Treatment of Tumors of the Head and Neck,* p. 306–426. Williams & Wilkins, Baltimore, 1950.

Ward, P. H., Alley, C., Owen, R.: Monostotic fibrous dysplasia of the maxilla. Laryngscope, *79:* 1295, 1969.

Windholz, F.: Cranial manifestations of fibrous dysplasia of bone: their relation to leontiasis ossea and to simple bone cysts of the vault. Am. J. Roentgenol., *58:* 51, 1947.

Wolfowitz, B. L.: Ameloblastoma of the maxilla. J. Laryngol. Otol., *86:* 1085, 1972.

Young, F., Putney, F. J.: Ossifying fibroma of the sinuses. Ann. Otol. Rhinol. Laryngol., *77:* 425, 1968.

Chapter 18

Malignant Tumors of the Paranasal Sinuses

Malignant tumors of the paranasal sinuses, either primary or secondary are relatively uncommon. They may be classified as follows:

Primary Tumors
1. Carcinoma
   a. Squamous cell carcinoma
      (1) Epidermoid carcinoma
      (2) Clear cell variant
      (3) Spindle cell variant
      (4) Lymphoepithelioma
      (5) Transitional cell carcinoma
   b. Adenocarcinoma of respiratory mucosa
   c. Carcinomas of accessory gland origin
      (1) Adenoid cystic carcinoma
      (2) Mucoepidermoid carcinoma
      (3) Medullary adenocarcinoma
      (4) Malignant mixed tumor
      (5) Other
   d. Unclassified carcinoma
2. Soft Tissue Sarcoma
   a. Rhbdomyosarcoma
      (1) Alveolar type
      (2) Embryonal type
   b. Leiomyosarcoma
   c. Fibrosarcoma
   d. Unclassified sarcoma
3. Bone Sarcoma
   a. Osteosarcoma
   b. Chondrosarcoma
   c. Ewing's sarcoma
4. Malignant Melanoma, Primary
5. Plasmacytoma, Primary
6. Malignant Lymphoma
   a. Histiocytic lymphoma
   b. Lymphocytic lymphoma
   c. Burkitt's lymphoma
7. Others

Secondary Tumors
1. Direct Extension
   a. Nasopharynx
   b. Nasal cavity
   c. Oral cavity
   d. Orbital cavity
   e. Pituitary fossa
   f. Infratemporal fossa
2. Hematogenous Spread
   a. Breast
   b. Lung
   c. Kidney
   d. Prostate
   e. Others

Perineural Metastasis
1. Trigeminal Nerve
2. Facial Nerve

PRIMARY MALIGNANT TUMORS

General Considerations

Malignant tumors of the paranasal sinuses are relatively rare. The community level incidence is 0.26 to 0.31% of the cancer population. In cancer centers the incidence is much higher. At the University of Texas M. D. Anderson Hospital and Tumor Institute at Houston, of 51,296 patients seen between 1948 and 1968, the percentage distribution of malignant disease of the paranasal sinuses was 0.7%.

Malignant tumors of the paranasal sinuses have a large variety of histological types. In general, squamous cell carcinomas appear to be dominant. Other carcinomas, soft tissue sarcomas, and malignant lymphomas are less common. Bone sarcoma, plasmacytoma, and malignant melanoma are rare. Among the unusual neoplasms reported have been leiomyomas and malignant teratomas.

215

Carcinoma of the paranasal sinuses tends to occur in middle aged patients. Patients with sarcomas and malignant lymphomas comprise a younger age group. Rhabdomyosarcoma usually occurs in children and young adults.

The maxillary sinus is the most common site of malignant disease, with the ethmoid sinus less frequently involved. Primary malignant disease of the frontal and sphenoid sinuses is rare.

The symptomatology of malignant tumors of the paranasal sinuses varies and depends upon the location of the lesion, the rate of growth, the size of the tumor, and the presence or absence of bone destruction. Early symptoms and signs are often not characteristic and are generally confused with inflammatory conditions. Detection of malignant disease in its early stage in these areas is chiefly made by roentgen examination. In the advanced lesion, identification of the site of origin and determination of the extent of the lesion also depend upon the roentgen findings.

Systems of Classification

Many systems of classification have been advocated for malignant tumors of the paranasal sinuses, but none has been generally accepted. The TNM system developed by the American Joint Committee on Cancer Staging and End Results Reporting is too indefinite and difficult to apply to tumors of the paranasal sinuses. Clinically, Ohngren's line is a simple and useful guide to operability and prognosis. This line is drawn from the inner canthus of the eye to the mandibular angle and divides the facial bones, as seen in profile, in two parts: anteroinferior and posterosuperior (Fig. 18.1). The anterior, lateral, and a part of the medial wall of the maxillary sinus lie in front of the line, and the remainder of the medial wall and the posterior and superior walls of the maxillary sinus, the ethmoid, and the sphenoid sinuses lie posterior to the line. All tumors lying posterosuperior to the line are regarded as having a much poorer prognosis than those which lie anteroinferior. From the radiological viewpoint, Baclesse's classification, dividing cancers of the paranasal sinuses as to the site of origin and route of spread, is superior to systems related to anatomic or clinical findings. This classification, based on tomographic findings, allows a better understanding of the natural history of the disease and contributes definite indications as to the choice of treatment. The following classification is derived from that of Baclesse.

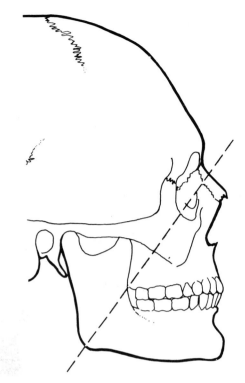

Figure 18.1. Ohngren's line. In general, the prognosis for tumors arising above the line is worse than for those below. This is primarily related to the structures involved.

Malignant Tumors of the Maxillary Sinus

Tumors of the Infrastructure. These tumors arise from the alveolar recess or inferolateral aspect of the sinus. The disease may spread inferiorly into the alveolar ridge, laterally into the soft tissues of the cheek below the zygoma, medially into the nasal cavity and hard palate, and posteriorly into the pterygoid plate (Fig. 18.2A).

Tumors of the Suprastructure. These tumors arise from the superior wall (sinus aspect of the orbital floor) and superolateral wall of the sinus. The disease may spread anterolaterally into the zygoma, superiorly into the floor of the orbit, posterolaterally into the infratemporal fossa, superomedially into the ethmoid sinus, medially into nasal cavity, and posteriorly into the pterygoid fossa and the floor of the middle cranial fossa (Fig. 18.2B).

Tumors of the Endosinus. These tumors involve the major part of the lining mucosa of the sinus. The disease may spread outwardly in all directions, with expansion and subsequent destruction of the walls at various points (Fig. 18.2C).

Tumors of the Medial Wall. These tumors arise from the medial wall of the sinus (nasoantral septum) and may stimulate primary tumors of the nasal cavity. The tumors are uncommon and often of mesenchymal origin. The disease may spread medially into the nasal cavity, superiorly into the ethmoid sinus, and laterally into other parts of the ipsilateral sinus (Fig. 18.2D).

Tumors of the Ethmoidomaxillary Region. These tumors arise in the superomedial portion of the maxillary sinus in the region of the ethmoidomaxillary plate. They involve the maxillary and ethmoid sinuses from the onset and spread rapidly into the medial wall and the medial portion of the floor of the orbit and nasal cavity (Fig. 18.2E).

Malignant Tumors of the Ethmoid Sinus

Tumors arising from the ethmoid sinus may be divided into three groups: anterior, middle, and posterior. The disease may spread medially into the contralateral ethmoid sinus, inferolaterally into the maxillary sinus, laterally into the orbit, inferiorly into the nasal cavity, posteriorly into the sphenoid sinus, the nasopharynx, and the base of the skull, superiorly into the frontal sinus, the cribriform plate, and the anterior cranial fossa, and anteriorly into the frontonasal region (Fig. 18.3).

Malignant Tumors of the Sphenoid Sinus

Primary malignant tumors of the sphenoid sinus may spread inferiorly into the nasopharynx, superiorly and posteriorly into the sella turcica and middle cranial fossa, and anteriorly into the ethmoid sinus and nasal cavity (Fig. 18.4).

Malignant Tumors of the Frontal Sinus

Tumors of the frontal sinus may spread anteriorly into the frontal bone and soft tissue of the forehead, posteriorly into the anterior cranial fossa, and inferiorly into the ethmoid sinus and orbit (Fig. 18.5).

Roentgen Findings

The roentgen manifestations of a malignant tumor of a paranasal sinus are opacity of the sinus, soft tissue mass arising in the sinus, sclerosis, erosion, or destruction of the bony

walls of the sinus (Figs. 18.6 to 18.19), and involvement of surrounding structures by tumor which has extended beyond the confines of the sinus.

Opacity of the Sinus

Obstruction of the normal drainage is the natural consequence of both inflammatory and malignant disease of the paranasal sinuses. Opacification of a sinus is, therefore, of limited value in differential diagnosis between these two conditions. Persistent opacity of one sinus, or unilateral opacity of several sinuses, should suggest the possibility of a malignant process.

Soft Tissue Mass

A soft tissue mass arising from the sinus and extending beyond the bony confines is highly suggestive of a malignant process; however, it can be simulated by mucocele, benign tumors, or granulomatous lesions.

Sclerosis of the Bony Wall of the Sinus

Bony sclerosis may result from osteoblastic reaction to tumor invasion and/or from an inflammatory process within the sinus. In the maxillary antra bony sclerosis is often the result of an osteitis caused by infection rather than osteoblastic reaction to tumor cells. This is best seen on the posterolateral wall of the sinus and in the adjacent pterygoid plate. In the absence of specific evidence of an inflammatory process, localized sclerosis of the wall of the maxillary sinus should indicate the possibility of a malignant lesion (Fig. 11.3).

Erosion of the Bony Wall of the Sinus

Bony erosion is usually caused by pressure from the expansion of a bulky soft tissue mass. It occurs in the advanced stage of cysts or benign tumors. However, it is occasionally found in adenocarcinomas and sarcomas of considerable size.

Destruction of the Bony Wall of the Sinus

Actual destruction of the bony wall indicates the invasive nature of the lesion and is reasonably reliable evidence of a malignant tumor (Fig. 18.6). Early destruction may be recognized by a fading or speckling of the bony wall. In this regard it should be borne in mind that the medial wall of the maxillary sinus sometimes

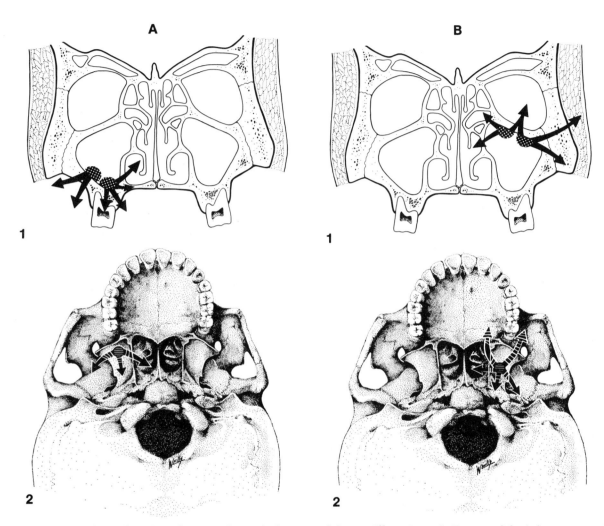

Figure 18.2. Sites of origin and routes of spread of tumors of the maxillary sinus. *A*, Tumors of the infrastructure: *1*, frontal view; *2*, base view. *B*, Tumors of the suprastructure: *1*, frontal view; *2*, base view. *C*, Tumors of the endosinus: frontal view. *D*, Tumors of the medial wall: frontal view. *E*, tumors of the ethmoidomaxillary region: *1*, frontal view; *2*, lateral view.

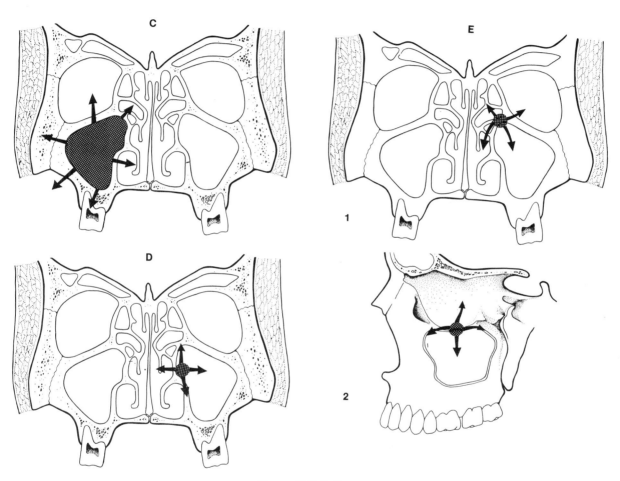

Figure 18.2 *C–E.*

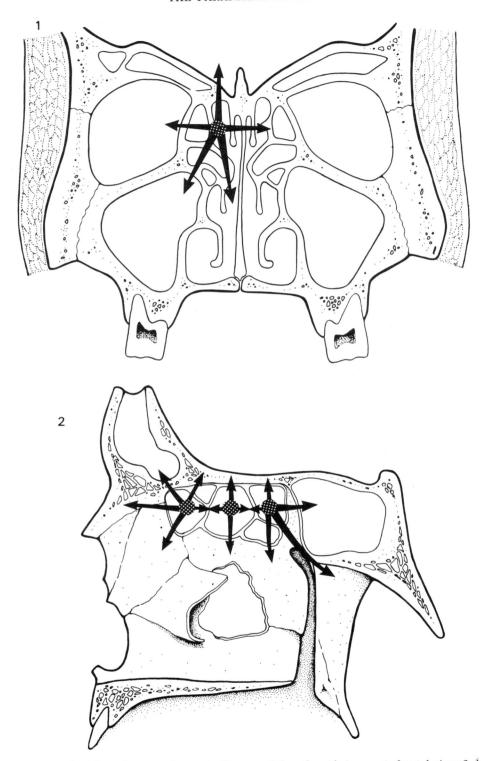

Figure 18.3. Sites of origin and routes of spread of tumors of the ethmoid sinuses: *1*, frontal view; *2*, lateral view.

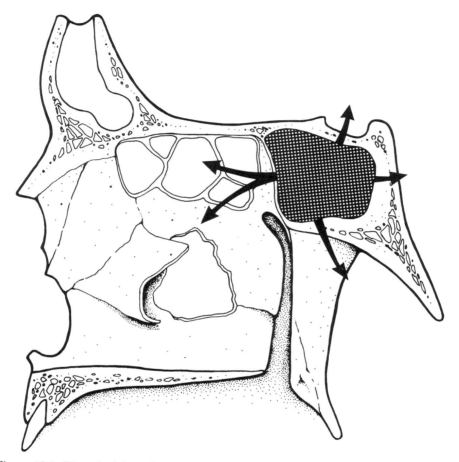

Figure 18.4. Sites of origin and routes of spread of tumors of sphenoid sinuses; lateral view.

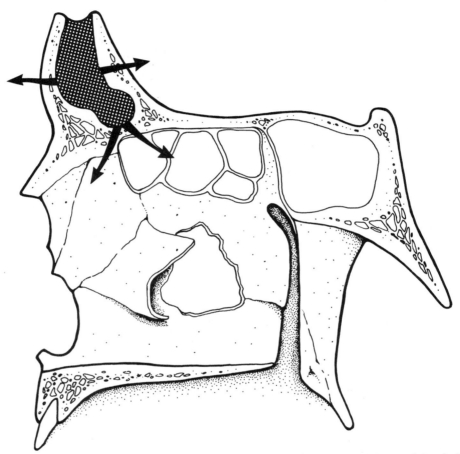

Figure 18.5. Sites of origin and routes of spread of tumors of the frontal sinuses; lateral view.

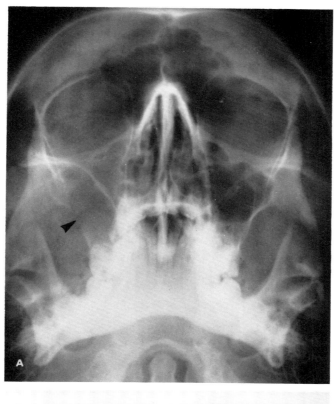

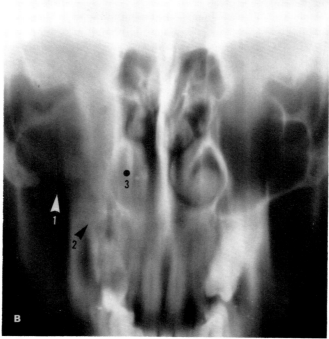

Figure 18.6. Squamous cell carcinoma of the infrastructure of the right maxillary sinus. *A*, Waters view: complete opacity of the right maxillary sinus with an area of destruction of the lateral wall (*arrow*). *B*, Posteroanterior tomogram: absence of the lateral wall of the right maxillary sinus due to destruction (*1*). There is also irregular destruction of the floor of the right maxillary sinus and of the alveolar ridge (*2*). The right inferior turbinate is enlarged, probably to extension of the lesion into the right nasal fossa (*3*). (Reprinted by permission from: G. H. Fletcher and B. S. Jing: The Head and Neck. Chicago: Year Book Medical Publishers, 1968.)

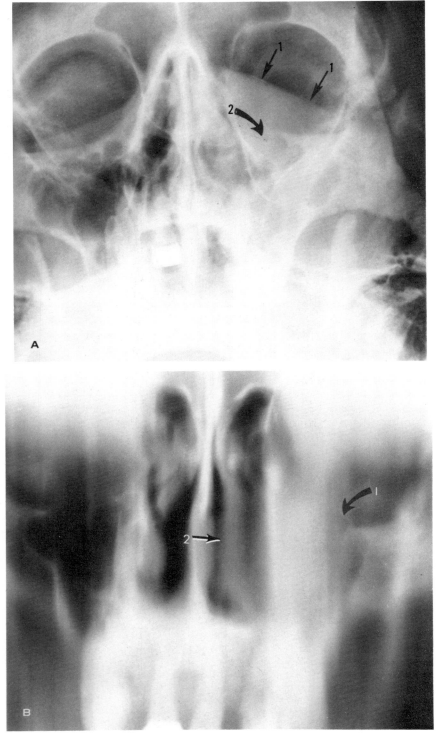

Figure 18.7. Fibrosarcoma of the suprastructure of the left maxillary sinus. *A*, Waters view: a large soft tissue tumor mass extends superiorly beyond the confines of the left maxillary sinus (*1*). There is destruction of the inferior rim of the left orbit (*2*). *B*, Posteroanterior tomogram (anterior cut): a large soft tissue tumor mass, mainly in the medial aspect of the left maxillary sinus, with destruction of the floor of the left orbit (*1*) and extension to the left side of the nasal cavity (*2*).

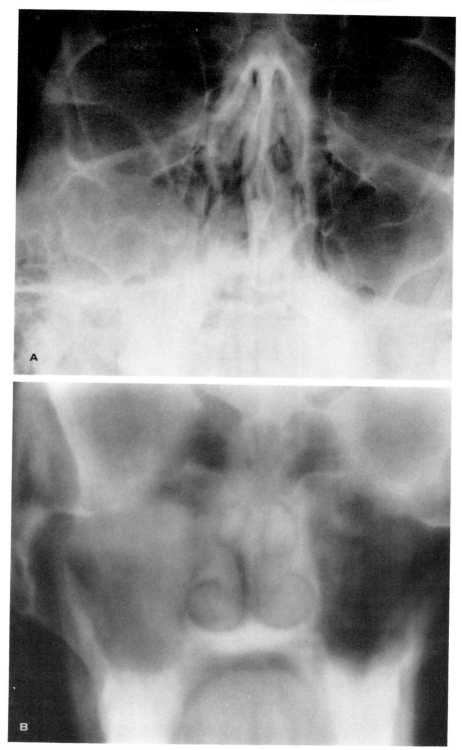

Figure 18.8. Embryonal rhabdomyosarcoma of the medial wall of the right maxillary sinus. *A*, Waters view: a large soft tissue mass occupies most of the right maxillary sinus; *B*, Anteroposterior tomogram: the soft tissue mass is clearly shown; although it fills the entire sinus, the destruction of the medial wall indicates the site of origin.

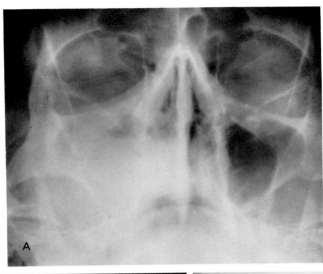

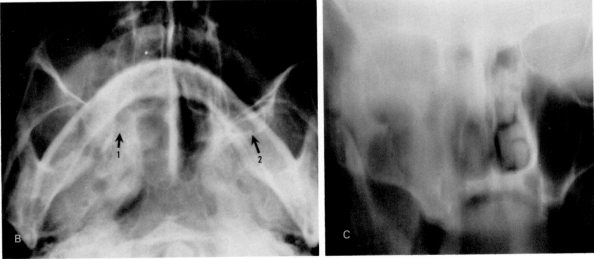

Figure 18.9. Hystiocytic lymphoma of the right maxillary sinus, endosinus type. *A*, Waters view: complete opacity of the right maxillary sinus with a fluid level in the superomedial aspect of the sinus. *B*, Base view: destruction of the posterior wall of the right maxillary sinus (*1*). The posterior wall of the left maxillary sinus is intact (*2*). *C*, Posteroanterior tomogram: a large soft tissue mass in the right maxillary sinus, destroying the medial wall and extending to the right nasal fossa. There is destruction of the floor of the right orbit and ethmoidomaxillary plate with involvement of the right ethmoid sinus.

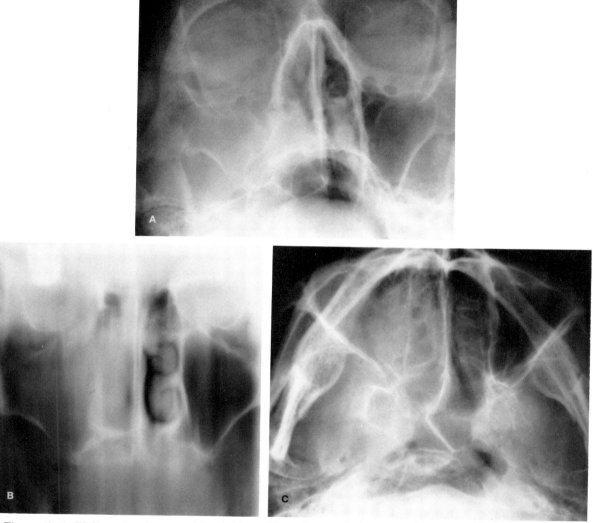

Figure 18.10. Malignant melanoma of the right maxillary sinus, endosinus type. *A*, Waters view: complete opacity of the right maxillary sinus with probable destruction of the inferolateral wall. *B*, Posteroanterior tomogram (middle cut): destruction of the inferolateral wall of the right maxillary sinus. There is extension of the lesion to the right ethmoid sinus and right nasal cavity. *C*, Base view: destruction of the posterior wall of the right maxillary sinus. There is cloudiness of the right maxillary and ethmoid sinuses.

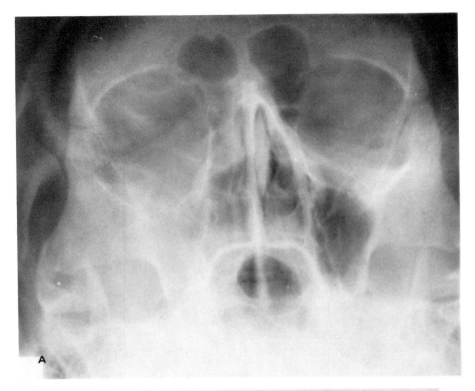

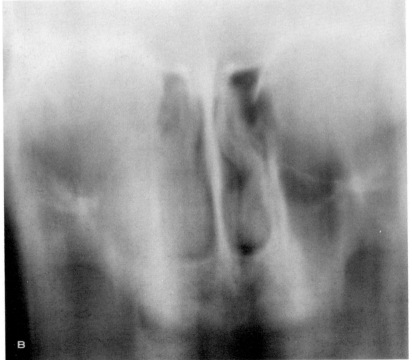

Figure 18.11. Plasmacytoma of the right maxillary sinus. *A*, Waters view: complete opacity of the right maxillary sinus, with a soft tissue mass extending superiorly to the orbital cavity. The floor of the right orbit is indistinct. *B*, Posteroanterior tomogram: a large soft tissue mass occupies most of the right maxillary sinus. There is erosion and destruction of the zygomatic process and the floor of the right orbit.

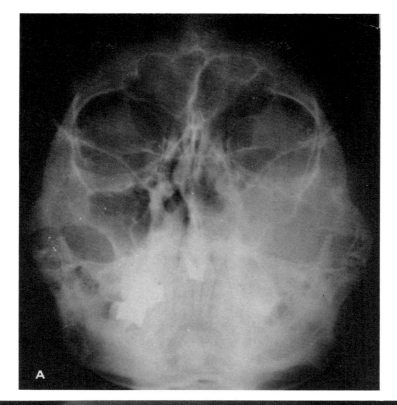

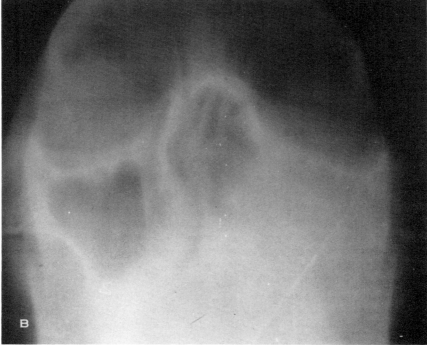

Figure 18.12. Leiomyosarcoma of the left maxillary sinus. A, Waters view: diffuse opacity of the left maxillary sinus. There is destruction of the inferolateral wall. The left nasal fossa is partially obliterated. B, Posteroanterior tomogram: a large soft tissue mass occupies the majority of the entire left maxillary sinus, destroying the medial wall and bulging into the nasal cavity. There is also destruction of the inferolateral wall and the floor of the sinus. (Reprinted with permission from Archives of Otolaryngology 90:114, 1969).

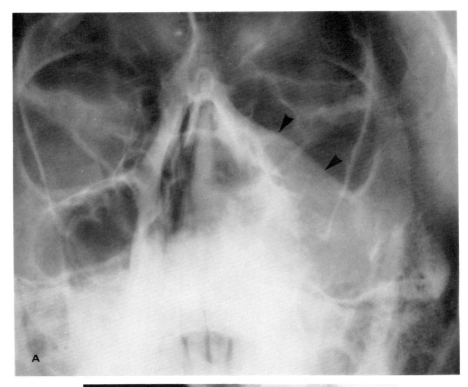

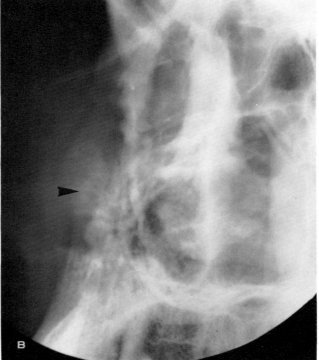

Figure 18.13. Osteogenic sarcoma of the right maxillary sinus. *A*, Waters view: complete opacity of the left maxillary sinus, with soft tissue mass extending beyond the confines superiorly (*arrows*). There is destruction of the walls of the sinus. *B*, Lateral view: bony destruction of the walls of the sinus, with sunburst periosteal thickening anteriorly (*arrows*). Osteogenic sarcoma should be differentiated from Ewing's sarcoma, because although they are similar radiographically, the clinical course and treatment differ.

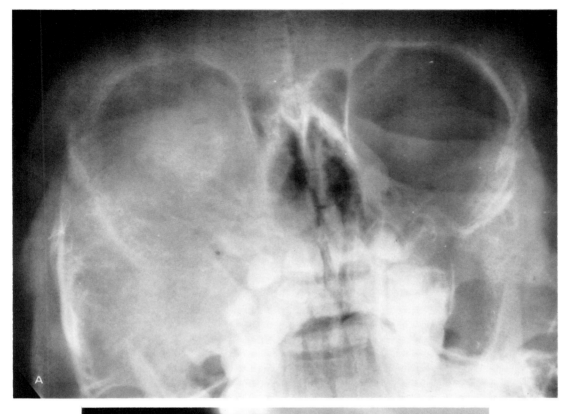

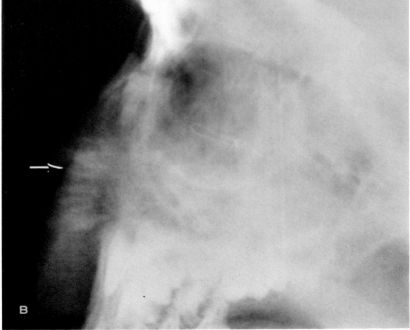

Figure 18.14. Ewing's sarcoma of the right maxillary sinus. *A*, Waters view: a large soft tissue mass over the right cheek, with extensive destruction of the walls of the right maxillary sinus; *B*, Lateral view: radiating spicules with sun-ray appearance arising from the anterior wall of the maxillary sinus and extending into the soft tissue component of the tumor. (Reprinted with permission from: G. H. Fletcher and B. S. Jing: The Head and Neck. Chicago: Year Book Medical Publishers, 1968.)

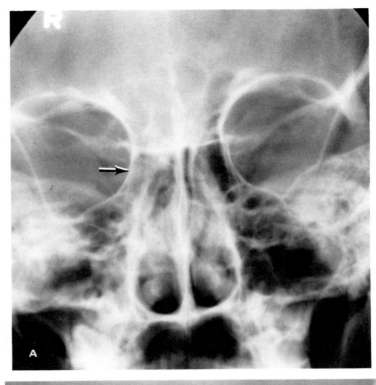

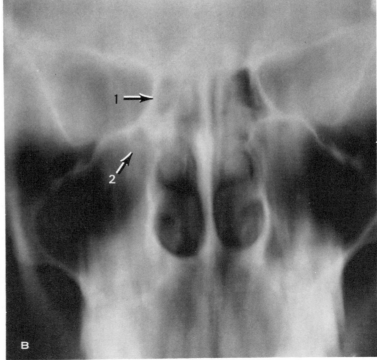

Figure 18.15. Squamous cell carcinoma of the right ethmoid sinus. *A*, Posteroanterior view: opacity of the right ethmoid sinus (*arrow*). There is indistinctness of the right ethmoidomaxillary plate. *B*, Posteroanterior tomogram: opacity with destruction of all of the ethmoid cells (*1*). There is absence of the right ethmoidomaxillary plate, with extension of the lesion to the superomedial aspect of the right maxillary sinus (*2*).

seems absent in normal subjects because of the thinness and obliquity of the bony plate. In estimating destruction of the medial wall great care must be taken not to confuse this normal variation with actual destruction. It is worth noting, however, that the absence of destruction from the radiological standpoint does not excluse a malignant process or early invasion of the bony wall.

Invasion of Surrounding Structures

In advanced malignant tumors of the paranasal sinuses, spread to the surrounding structures has often occurred when the patient is first seen.

Maxillary Sinus. In advanced lesions of the suprastructure of the maxillary sinus, destruction of the posterior wall with invasion of the pterygopalatine fossa is often observed. From this fossa the tumor may spread to the base of the skull, to the nasopharynx, and to the orbital cavity. Sclerosis of the pterygoid plate is often attributable to an inflammatory process within the maxillary sinus or nasopharynx rather than a reaction to invading tumor. However, when a coexistent osteolytic lesion is present in the vicinity of the sclerosis, the change in the pterygoid plate is assumed to be caused by invasion. Destruction of the posterolateral wall of the maxillary sinus can be followed by invasion of the infratemporal fossa, causing painful trismus owing to infiltration of the muscles of mastication. Destruction of the floor of the orbit is a frequent finding in suprastructure cancers of the maxillary antrum and the ethmoidomaxillary region. Actual invasion of the orbital cavity may lead to exophthalmos, deviation of the globe, and external ophthalmoplegia (Fig. 18.20). Cancer arising in the infrastructure of the maxillary sinus may spread to the hard palate and alveolar ridge, causing ulceration of the alveolus and dental pain; however, this type of lesion must be differentiated from primary malignancy of the hard palate or alveolar ridge (Fig. 18.21).

Ethmoidomaxillary Region and Ethmoid Sinus. In malignant disease of the ethmoidomaxillary region and the ethmoid sinus, bony destruction of the medial wall of the orbit with extension to the orbital cavity is often observed (Fig. 18.22). In its early stage, the bony change is often difficult to detect. Occasionally, destruction of the medial wall can be detected on conventional films, but tomographic studies are often necessary. In advanced lesions of the ethmoid sinus, there may be an upward spread with destruction of the roof of the sinus cribriform plate and eventual extension to the anterior cranial fossa.

Sphenoid Sinus. Primary malignant tumors of the sphenoid sinus are rare and are often advanced when the patient is first seen. The spread is principally in three directions; anteriorly into the ethmoid sinus, inferiorly into the nasopharynx, and superiorly and posteriorly into the sella turcica and the floor of the middle cranial fossa. Involvement of the cranial nerves is common with the latter (Fig. 18.23).

Frontal Sinus. Destruction of the upper medial margin of the orbit is considered a characteristic feature of primary malignancy of the frontal sinus. Destruction of the posterior wall of the sinus with extension to the anterior cranial cavity may occur. The disease can spread into the forehead with secondary infection (Fig. 18.24).

Differential Diagnosis

In the differential diagnosis of primary malignant tumors of the paranasal sinuses the principal point is the presence or absence of bone destruction. However, in early malignant tumors, bone involvement may not be demonstrated. A soft tissue mass which extends beyond the bony confines of the sinus is only suggestive of a malignant process. Sclerosis of the sinus wall and opacity of the sinus are not specific and cannot be relied upon. In making a differential diagnosis, the following conditions should be considered.

Chronic Infectious Sinusitis

In chronic sinusitis the involvement of several sinuses is likely to occur. In long standing chronic sinusitis uneven thickening of the lining mucosa and sclerosis of the bony walls are often observed and the diagnosis may be readily made. However, there is frequently a diffuse opacification of the affected sinuses from thickening of the lining mucosa. This may be associated with erosion and/or destruction of the bony walls (Figs. 18.25 and 18.26). Under these circumstances, biopsy is required to establish the diagnosis.

Granulomatous Disease of the Paranasal Sinus

Granulomatous disease may cause bone erosion and destruction that is suggestive of a malignant process. Biopsy is mandatory (see Chapter 15).

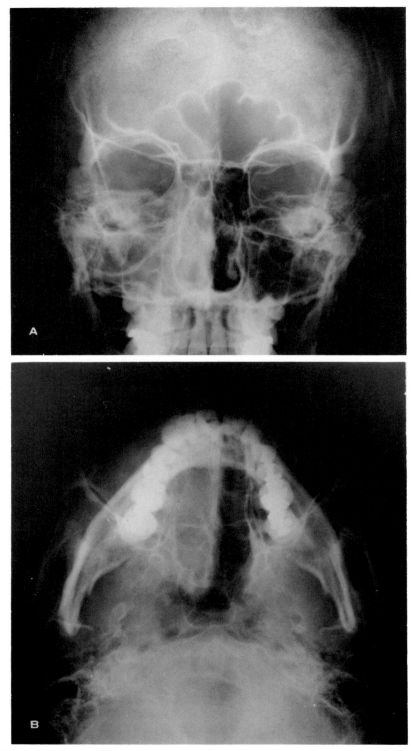

Figure 18.16. Malignant teratoma of the right ethmoid sinus. *A*, Posteroanterior view: opacity of the right maxillary, ethmoid, and frontal sinuses and right nasal fossa. There is destruction of the ethmoid cells on the right side. *B*, Submentovertical view: opacity and destruction of air cells of the right ethmoid sinus. The nasopharynx is normal in appearance.

A 39-year-old male was seen with complaints of right nasal obstruction, bloody discharge from the right nostril, and a vague sensation of discomfort in the right maxillary teeth and forehead. On physical examination, the right nasal fossa was occupied by a large reddish-purple tumor mass which was attached to the area of the middle turbinate and above. At surgery, a tumor mass weighing 27 g was removed from the region of the right ethmoid sinus. The maxillary sinus was filled with pus but not involved by tumor. Histological diagnosis was Schneiderian papilloma with focal areas of adenocarcinoma. The patient expired 2 months after surgery. At autopsy, the tumor was found to have grown in a dumbell

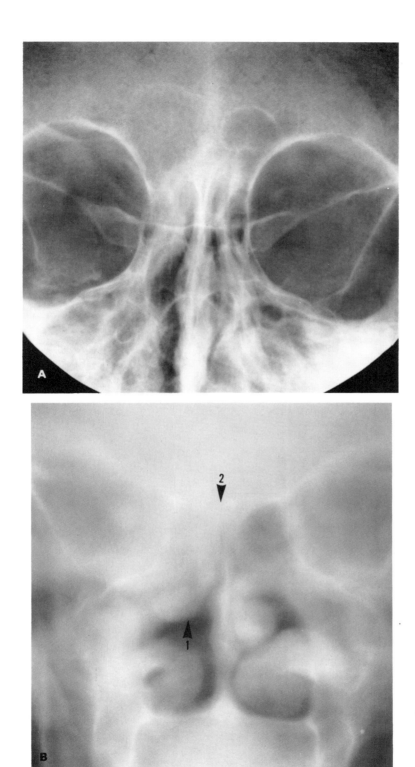

Figure 18.17. Adenocarcinoma of accessory gland of right ethmoid sinus. *A*, Posteroanterior view: opacity of the right ethmoid sinus and, to a lesser degree, of the left ethmoid sinus. *B*, Anteroposterior tomogram: a large soft tissue mass in the right ethmoid sinus, with destruction of the right middle turbinate (*1*). There is extension of the lesion to the olfactory region of the right nasal fossa. The right cribriform plate is destroyed (*2*). There is marked thickening of the lining mucosa of the maxillary sinuses.

fashion, completely filling and destroying the right ethmoid sinus and extending both intracranially and into the right maxillary sinus, hard palate, nasal fossa, and nasopharynx. The final pathological diagnosis was malignant teratoma of the ethmoid sinus. (Reprinted with permission from Cancer 21:714, 1968.)

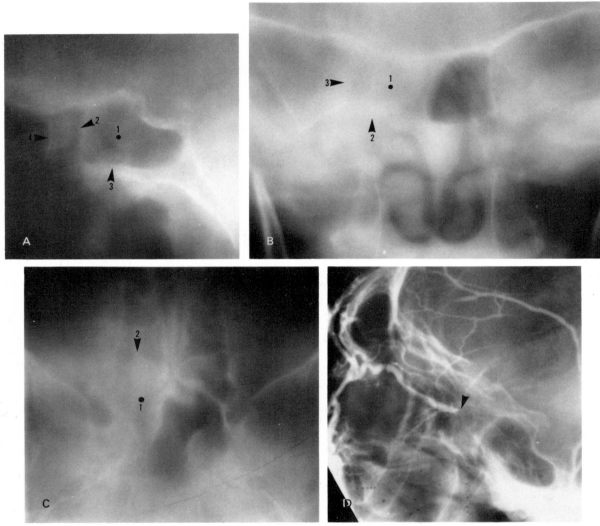

Figure 18.18. Squamous cell carcinoma of sphenoid sinus with extension to the ethmoid sinus. *A*, Lateral tomogram of sphenoid sinus: soft tissue density in the anterior part of the sphenoid sinus (*1*) with destruction of the anterior wall (*2*) and anterior part of the floor of the sinus (*3*). There is extension of the lesion into posterior ethmoid cells (*4*). *B*, Anteroposterior tomogram of sphenoid sinus: complete opacity of the right sphenoid sinus (*1*) with destruction of right lateral wall (*3*) and floor of the sinus (*2*). *C*, Tomogram of the base of the skull: mass is apparent in the right sphenoid sinus (*1*) with anterior extension to the posterior ethmoid cells (*2*). *D*, Orbital venogram (lateral projection): complete obstruction of the right superior ophthalmic vein in the region of posterior ethmoid cells (*arrow*). The vein is dilated and tortuous.

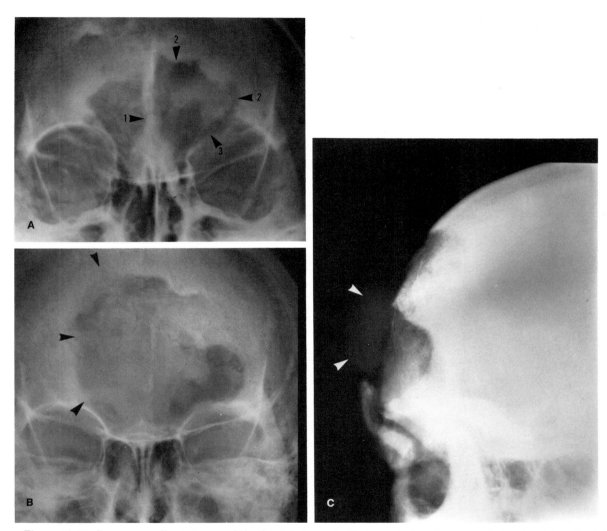

Figure 18.19. Squamous cell carcinoma of the frontal sinus. *A*, Posteroanterior view: mass lesion involving the left frontal sinus with partial destruction and sclerosis of the bony septum (*1*). There is absence of the mucoperiosteal line, with an irregular margin of the bony wall of the sinus (*2*). The right frontal sinus is cloudy, with sclerosis of the bony margin. There is extension of the lesion to the superomedial wall of the left orbit (*3*). *B*, Posteroanterior view (8 months after surgery): recurrence of the tumor with predominant involvement on the right. There is destruction of the bony wall and anterior part of the frontal bone (*arrows*). Postoperative changes are seen in the lateral aspect of the left frontal sinus. There is sclerosis of the superomedial wall of the left orbit. *C*, Lateral view: destruction of the frontal bone and frontal sinuses. In addition, there is a rounded soft tissue mass protruding into the forehead (*arrows*).

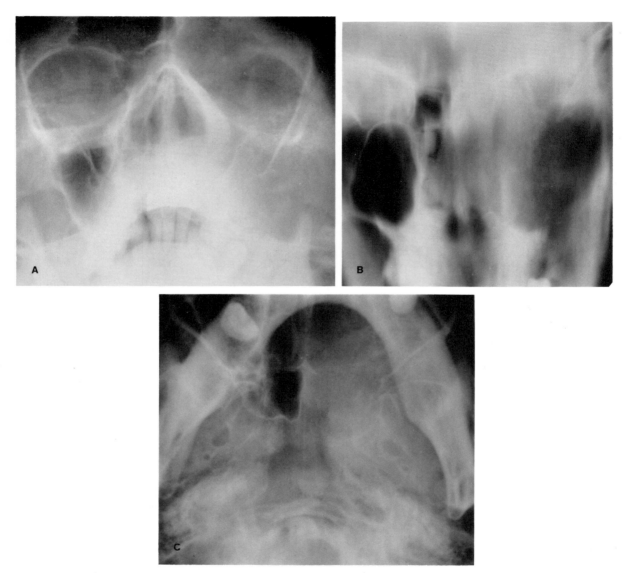

Figure 18.20. Advanced squamous cell carcinoma of the suprastructure of the left maxillary sinus. *A*, Waters view: complete opacity of the left maxillary sinus, with a soft tissue mass in the superior aspect of the sinus. There is destruction of the floor of the orbit, lateral wall, and floor of the sinus. *B*, Anteroposterior tomogram: additional destruction is shown in the left ethmoid sinus, medial wall of the left orbit, and the left alveolar ridge. There is also partial destruction of the nasal septum, with deviation to the right. *C*, Base view: destruction of the posterolateral wall of the sinus and left medial pterygoid plate. (Reprinted with permission from: G. H. Fletcher and B. S. Jing: The Head and Neck. Chicago: Year Book Medical Publishers, 1968.)

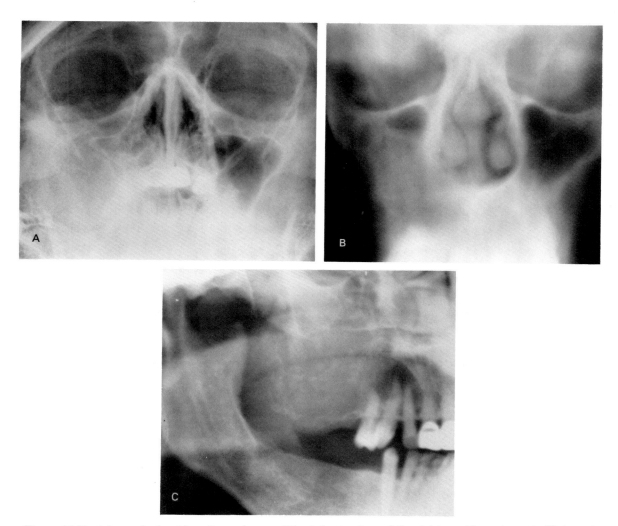

Figure 18.21. Advanced adenoid cystic carcinoma of the infrastructure of the right maxillary sinus. *A*, Waters view: complete opacity of the right maxillary sinus, with indistinctness of the lateral wall. *B*, Anteroposterior tomogram: a large expanding soft tissue mass occupies the lower two-thirds of the sinus and extends beyond the confines of the lateral aspect of the sinus. There is erosion and partial destruction of the zygomatic process, lateral wall, alveolar ridge, and hard palate. Minimal erosion and destruction of the lower part of the medial wall of the sinus is also noted. *C*, Panorex view of the maxilla: the involvement of the alveolar ridge is shown to good advantage. There is destruction of the ridge, with an expansile appearance.

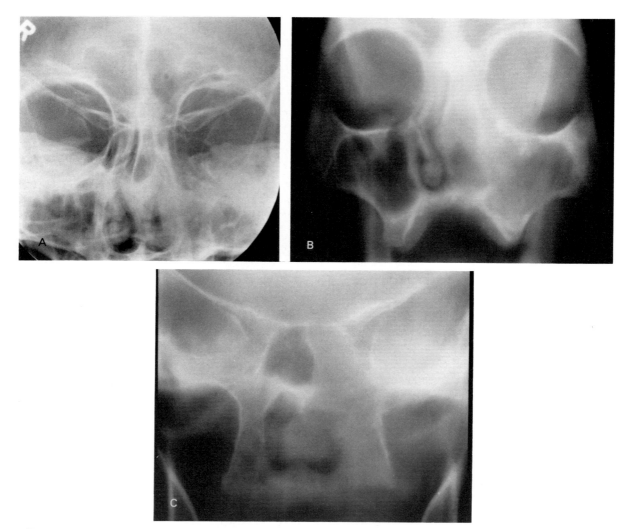

Figure 18.22. Advanced squamous cell carcinoma of the ethmoidomaxillary region. *A*, Posteroanterior view: opacity of both ethmoid sinuses, particularly the left, and the left nasal fossa and left maxillary sinus. There is destruction of the inferomedial wall of the left orbit. *B*, Anteroposterior tomogram (anterior cut): the tumor mass involving the ethmoid sinuses, left nasal fossa, and left maxillary sinus is well shown. There is erosion and destruction of the medial aspect of the inferior walls of the left orbit and medial wall of the left maxillary sinus. The nasal septum is also eroded and deviated to the right. *C*, Anteroposterior tomogram (posterior cut): the tumor mass extends to the left sphenoid sinus, with destruction of the floor. There is also partial destruction of the left pterygoid plates. A portion of the tumor mass is noted in the posterior part of the nasal cavity.

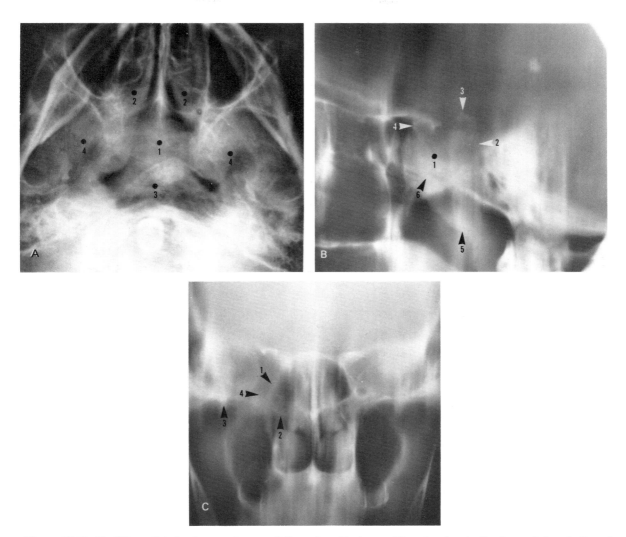

Figure 18.23. Undifferentiated adenocarcinoma of the sphenoid sinus with extension to the base of the skull and nasopharynx. *A*, Base view of the skull: destruction of the sphenoid sinus (*1*), posterior ethmoid sinuses (*2*), clivus (*3*), and greater wings of the sphenoid (*4*). *B*, Lateral tomogram of the sphenoid sinus: tumor mass in the sphenoid sinus (*1*) extending upward to the sella turcica with destruction of the dorsum sellae (*2*), posterior clinoid processes (*3*), and inferior aspect of the anterior clinoid processes (*4*). The tumor also extends downward to the nasopharynx (*5*), with destruction and sclerosis of the floor of the sphenoid sinus (*6*). *C*, Posteroanterior tomogram: destruction of the right lateral wall (*1*) and floor of the right sphenoid sinus (*2*). There is destruction of the floor of the right middle fossa (*3*). The right foramen rotundum is not visualized (*4*).

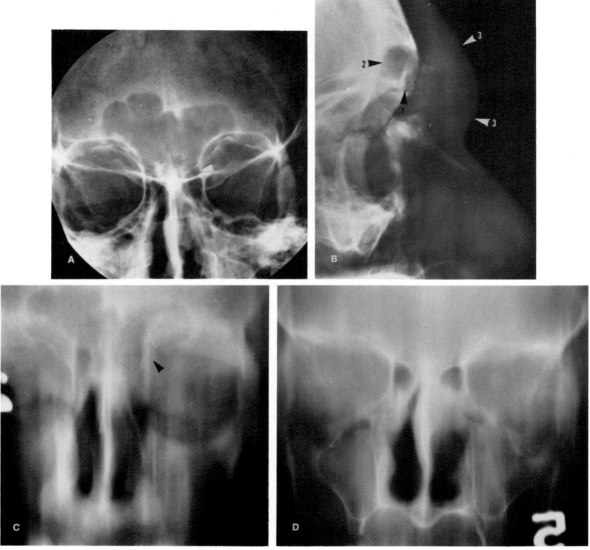

Figure 18.24. Squamous cell carcinoma of the frontal sinus with extension to the left orbit, left nasal fossa, and ethmoid sinuses. *A*, Caldwell view: poor definition of the mucoperisoteal line and disappearance of the bony septum of the frontal sinuses. There is destruction of the superior aspect of the medial wall of the left orbit (*arrow*). *B*, Lateral view: destruction of the anterior wall of the frontal sinus, with bony fragments displaced anteriorly (*1*). The posterior wall is also partially destroyed (*2*). There is anterior extension of the lesion to the soft tissues of the forehead (*3*). *C*, Anteroposterior tomogram (anterior cut): destruction of superomedial wall of the left orbit (*arrow*). There is opacity of the ethmoid sinuses, more on the left side. *D*, Anteroposterior tomogram (middle cut): extension of the lesion to the ethmoid sinuses is clearly shown. There is also involvement of the superior part of the left nasal fossa and superomedial aspect of the left maxillary sinus. There is rather diffuse chronic sinusitis of the maxillary sinuses.

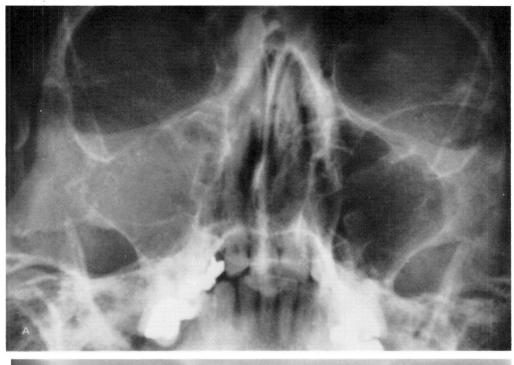

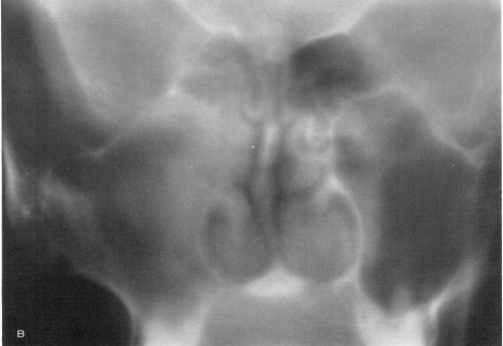

Figure 18.25. Acute and chronic inflammation with extensive necrosis of the right maxillary sinus. *A*, Waters view: complete opacification of the right maxillary sinus. There is no evidence of bony erosion or destruction. *B*, Anteroposterior tomogram: soft tissue density in the medial aspect of the right maxillary sinus. There is erosion of the superior part of the middle wall. The right middle turbinate is enlarged, with poorly defined contours. The right ethmoidomaxillary plate is rather indistinct.

Exclusion of a carcinoma is not possible without tissue biopsy.

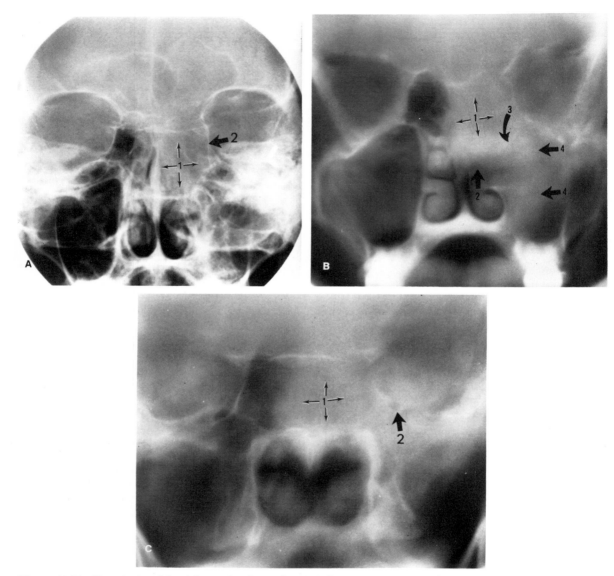

Figure 18.26. Chronic sinusitis with pseudopolyps (chronic inflammation with inspissated mucoid material). *A*, Caldwell view: there is a diffuse opacity of left ethmoid (*1*), with slight lateral displacement of the lamina papyrecea (*2*). *B*, Anteroposterior tomogram (anterior cut): a large mass lesion occupies most of the left ethmoid and superior part of the left nasal fossa (*1*). There is downward extension with involvement of the left middle turbinate (*2*). The left ethmoidomaxillary plate is partly destroyed (*3*) and a soft tissue mass is noted in the superior and medial aspect of the left maxillary sinus (*4*). *C*, Anteroposterior tomogram (posterior cut): there is opacity of left sphenoid sinus, with poor definition of the left lateral wall and the floor (*1*). The foramen rotundum is well preserved (*2*).

The lesion is indistinguishable from a malignant tumor radiographically.

Benign Tumors of the Paranasal Sinus

Benign tumors are usually well defined except at the site of attachment. In certain cases there are stipplings or calcific densities, the outlines of dental structures, etc. In advanced lesions bone erosion or destruction may result from pressure of the expanding tumor and stimulate malignant disease. The final diagnosis usually depends upon tissue biopsy (see Chapter 17).

Primary Versus Secondary Malignant Disease

Secondary invasion of the paranasal sinus from malignant tumors of adjacent structures is not uncommon. When the lesion is of limited size, the site of origin may be determined from relative positions of the soft tissue mass and bone involvement, but this is subject to interpretative error and cannot be completely relied upon.

Hematogenous metastasis to paranasal sinuses from a distant malignancy is rather infrequent. When present and associated with bone destruction the lesion is difficult to distinguish from a primary cancer. History and tissue biopsy offer the only methods of precise diagnosis.

SECONDARY INVOLVEMENT OF THE PARANASAL SINUSES

Secondary involvement of the paranasal sinuses can be attributable either to direct extension from tumors of adjacent structures or to hematogenous spread from tumors of the distant organs.

Direct Extension

Direct invasion of the paranasal sinuses from a malignant tumor of an adjacent structure is not infrequent. If the lesion is extensive when first seen it may be impossible to distinguish from a primary neoplasm of the sinus. Lesions which commonly cause confusion are carcinomas of the skin of the nose and of the cheek with inward penetration, carcinoma of the nasal cavity with secondary invasion of the ethmoids and maxillary sinuses, carcinomas of the nasopharynx with spread to ethmoid, maxillary, and sphenoid sinuses, and carcinomas of the oral cavity with involvement of the maxillary sinus.

Malignant Tumors of the Nasopharynx

In advanced malignant tumors of the nasopharynx, involvement of the surrounding structures is often present when the patient is first seen. Anteriorly, the lesion can spread to the nasal cavity with destruction of the nasal septum and hard palate. Further extension may involve the pterygoid plate and the maxillary and ethmoid sinuses. Superiorly, the lesion, through destruction of the floor of the sphenoid sinus, can extend to the sphenoid sinus, the pituitary fossa, and parasellar structures (Fig. 18.27).

Malignant Tumors of the Oral Cavity

Secondary invasion of the maxillary sinus may result from the direct extension of a carcinoma of the alveolar ridge or hard palate. When an antrooral fistula develops, differentiation between a carcinoma of the oral cavity involving the antrum and an antral carcinoma invading the mouth may be impossible (Fig. 18. 28).

Malignant Tumors of the Nasal Cavity

Malignant tumors of the nasal cavity are divided into superior and inferior groups. Tumors of the superior group arise in the olfactory region of the nasal cavity above the lower border of the superior turbinate. The disease may spread laterally into the ethmoid sinus, posteriorly into the sphenoid sinus, and inferiolaterally into the maxillary sinus (Fig. 18.29).

The lesions of the inferior group arise in the lateral wall, including the middle and inferior turbinates and the nasal septum, and are usually carcinomas or sarcomas. The tumors may spread laterally into the maxillary sinus (Fig. 18.30).

Malignant Tumors of the Pituitary Fossa

Malignant tumors of the pituitary fossa are rare. Even in the early stages, however, the lesions may extend downward to the sphenoid sinus. Radiologically there is tumor mass in the pituitary fossa with destruction of the floor of the fossa. The sphenoid sinus is usually opaque and shows evidence of the mass lesion. When the lesion is extensive it is often difficult to differentiate between primary and secondary tumors of the sphenoid sinus (Fig. 18.31).

Malignant Tumors of the Orbital Cavity

Malignant tumors of the orbital cavity may involve the maxillary and ethmoid sinuses by direct extension (Figs. 18.32 and 18.33).

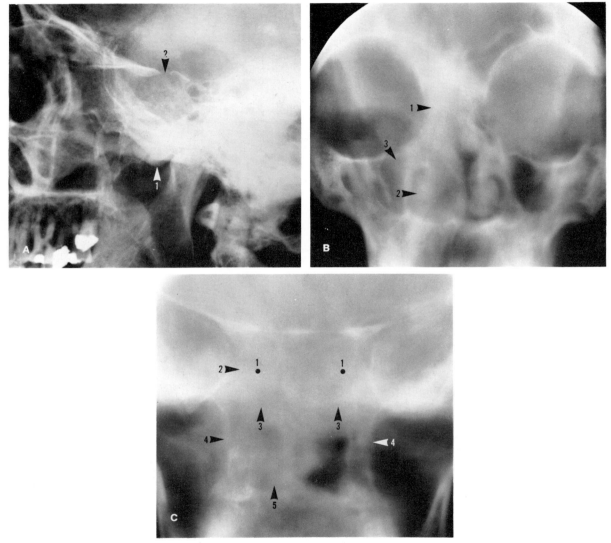

Figure 18.27. Recurrent squamous cell carcinoma of the nasopharynx with involvement of the paransal sinuses. *A*, Lateral view of the nasopharynx: recurrent tumor mass in the nasopharynx (*1*) with extension to the sphenoid sinus and sella turcica (*2*). There is destruction of the dorsum sellae. *B*, Posteroanterior tomogram (anterior cut): the tumor mass extends to the ethmoid sinuses (*1*), right nasal fossa (*2*), and superomedial aspect of the right maxillary sinus (*3*). *C*, Posteroanterior tomogram (posterior cut): mass lesion in the sphenoid sinuses (*1*) with destruction of the right lateral wall (*2*) and floor of the sinuses (*3*). There is sclerosis and destruction of pterygoid plates, more on the right side (*4*). A tumor mass is seen in the anterior part of the nasopharynx, especially on the right (*5*).

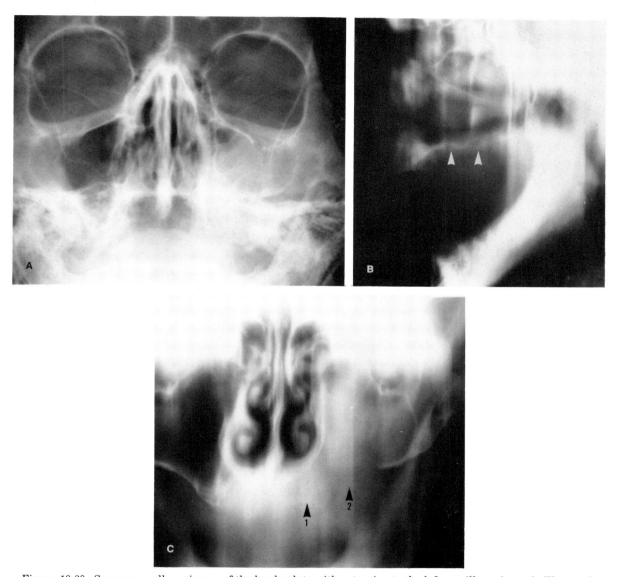

Figure 18.28. Squamous cell carcinoma of the hard palate with extension to the left maxillary sinus. *A*, Waters view: opacification of the left maxillary sinus; *B*, Lateral tomogram of the hard palate: destructive lesion of the hard palate (*arrows*); *C*, Posteroanterior tomogram of the paranasal sinuses: destruction of the hard palate (*1*) and the floor of the left maxillary sinus (*2*).

The differentiation of palatine and alveolar ridge tumors from tumors of the infrastructure of the maxillary antra may be difficult both clinically and radiographically.

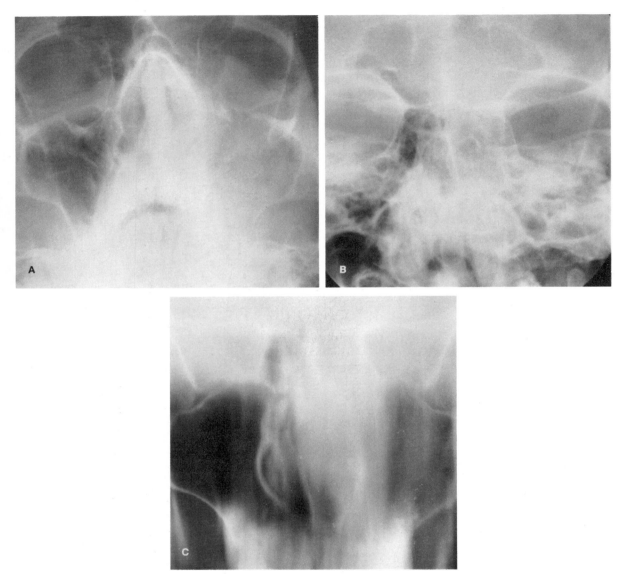

Figure 18.29. Advanced olfactory neuroblastoma of the nasal cavity. *A*, Waters view; a soft tissue mass is seen in the inferomedial aspect of the left orbit. There is opacity of the left maxillary, ethmoid, and frontal sinuses. The left lamina papyrecea is slightly displaced laterally. The left nasal fossa is clouded and there is thickening of the nasal septum. *B*, Posteroanterior view: opacity of the left ethmoid sinus with slight lateral displacement of the lamina papyrecea. Cloudiness of the left maxillary and frontal sinuses is again noted. The left ethmoidomaxillary plate is indistinct. *C*, Anteroposterior tomogram: a large soft tissue mass fills the left ethmoid sinus and left nasal fossa. There is considerable lateral displacement of the left lamina papyrecea. The tumor mass bulges into the right nasal fossa, with destruction of the nasal septum. There is destruction of the left ethmoidomaxillary plate and the medial wall of the left maxillary sinus, with tumor mass extending to the superomedial portion of the sinus. (Reprinted with permission from: G. H. Fletcher and B. S. Jing: The Head and Neck. Chicago: Year Book Medical Publishers, 1968.)

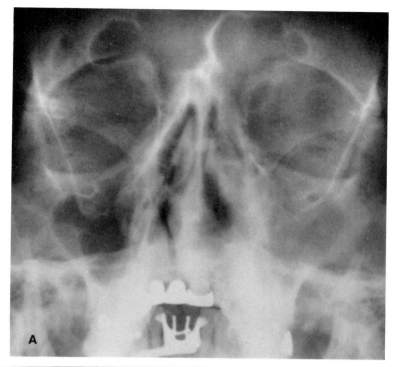

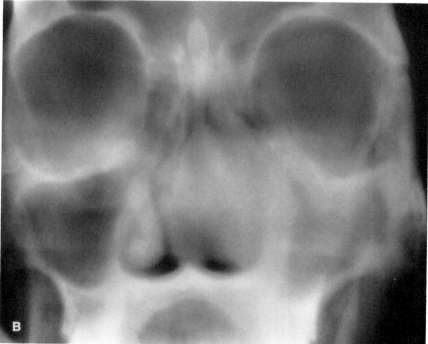

Figure 18.30. Plasmacytoma of nasal cavity with extension to the left maxillary sinus. *A*, Waters view: opacity of the left maxillary sinus and soft tissue density in the left nasal fossa. *B*, Anteroposterior tomogram; there is a large soft tissue mass in the left nasal fossa, with displacement of the nasal septum to the right. There is erosion and destruction of the left lateral wall of the nasal cavity, with extension of the lesion to the left maxillary sinus.

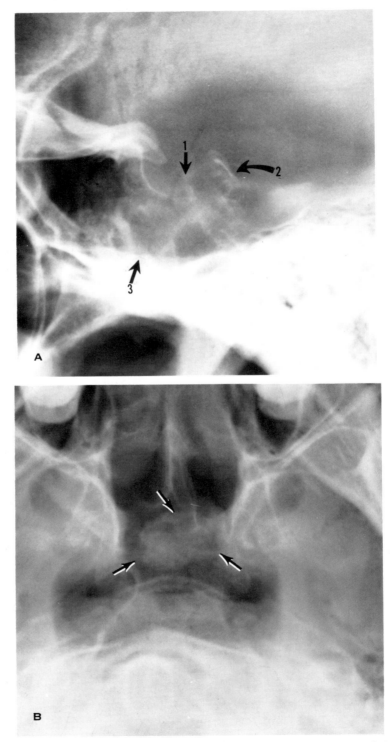

Figure 18.31. Undifferentiated carcinoma of the pituitary fossa with extension to the sphenoid sinus. *A*, Lateral view of the sella turcia: irregular densities in posteroinferior aspect of the pituitary fossa (*1*) with destruction of the floor of the fossa and dorsum sellae (*2*). An extension of the lesion to the posterior two-thirds of the sphenoid sinus is clearly shown (*3*). *B*, Base view of the skull: a well defined tumor mass is noted in the sphenoid sinuses, mainly on the left side.

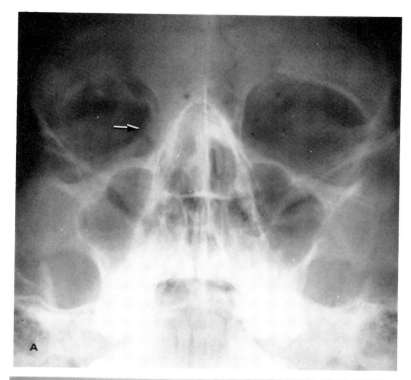

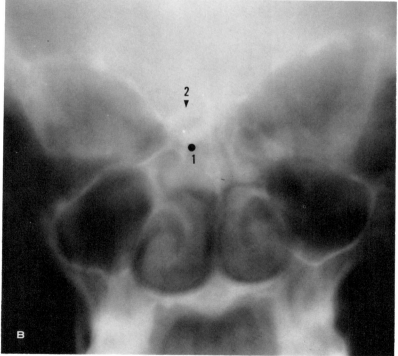

Figure 18.32. Recurrent retinoblastoma of the right eye with metastasis to right ethmoid sinus and right nasal fossa. *A*, Waters view: opacity of the right ethmoid sinus and a soft tissue density in the medial aspect of the right eye (*arrow*). A soft tissue mass is noted in the right nasal fossa. There is hypoplasia of the frontal sinuses, and the right orbital cavity is underdeveloped. *B*, Anteroposterior tomogram: a soft tissue mass is seen in the superior aspect of the right nasal fossa and the right ethmoid region (*1*). The right cribriform plate is indistinct (*2*). The left ethmoid sinus is slightly increased in density.

A 23-year-old female had enucleation of the left eye for retinoblastoma at the age of 7 months. At the age of 3 her right eye was enucleated because of similar disease and postoperative external radiation was given from two ports. Swelling of the right lateral aspect of the nose was noticed for 2 months before roentgen examination. Biopsy diagnosis vᵒ undifferentiated malignant neoplasm, small cell type, consistent with recurrent retinoblastoma.

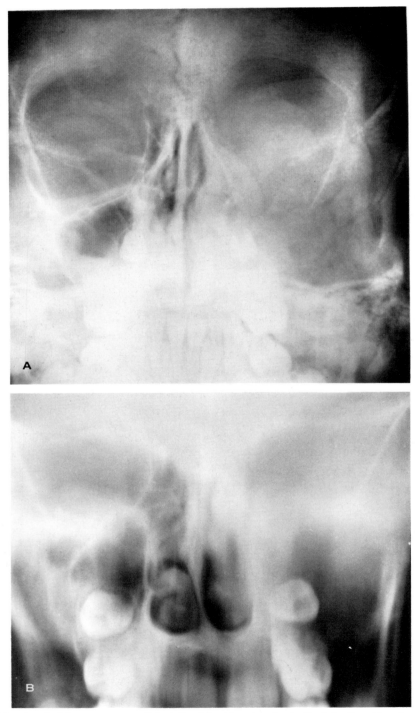

Figure 18.33. Alveolar rhabdomyosarcoma of the left orbit with extension to the left ethmoid and maxillary sinuses. *A*, Waters view: there is a large soft tissue mass in the left orbital region, with destruction of the floor of the left orbit. The left maxillary and ethmoid sinuses are opacified. The left ethmoidomaxillary plate is not clearly shown. *B*, Posteroanterior tomogram: the left orbital outline is lost. There is destruction of the left ethmoidomaxillary plate and the walls of the left maxillary sinus, with soft tissue mass in the maxillary region. There is also destruction of the left zygoma. The left ethmoid sinus is dense due to secondary tumor invasion.

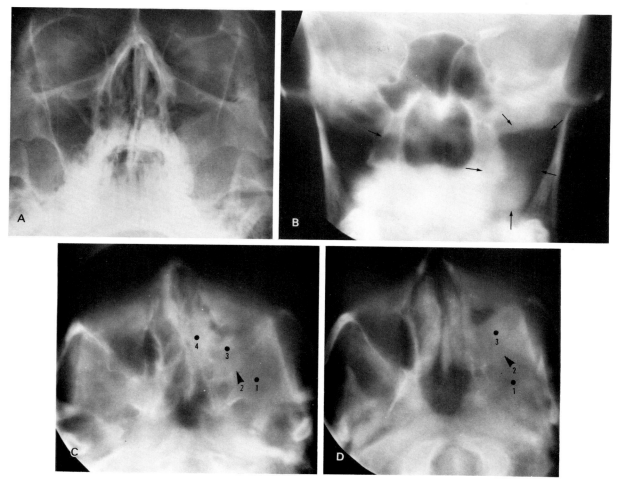

Figure 18.34. Embryonal rhabdomyosarcoma of the left infratemporal fossa with involvment of the left maxillary sinus and orbital cavity. *A*, Waters view: soft tissue density in the lower and lateral aspects of the left maxillary sinus. The lateral wall of the left maxillary sinus is intact. *B*, Anteroposterior tomogram (posterior cut): a huge soft tissue mass is present in the left infratemporal fossa (*arrows*). There is destruction of the lateral lamina of the left pterygoid plate. The lateral lamina of the right pterygoid plate is normal (*arrow*). *C*, Base tomogram (at the level of the orbit): there is a large soft tissue mass in the left infratemporal fossa (*1*), with destruction of the posterior wall of the orbit (*2*) and extension to the orbital cavity (*3*). There is involvement of the left ethmoid sinus (*4*). The destruction of the floor of the left orbital cavity is shown on the anteroposterior tomogram. *D*, Base tomogram (at the level of the maxillary sinus): the large infratemporal tumor mass (*1*) has destroyed the posterior wall of the maxillary sinus (*2*) and extended to the sinus cavity (*3*).

Malignant Tumors of the Infratemporal Fossa

The infratemporal fossa is an irregularly shaped space, situated below and medial to the zygomatic arch. It contains the lower part of the temporalis and the pterygoidei internus and externus muscles, the internal maxillary vessels, and the mandibular and maxillary nerves. Malignant tumors of the fossa predominantly arise from the muscle and are sarcomatous in nature. The lesions may grow quite large and extend anteriorly to the maxillary sinus and the orbit, medially to the pterygoid plate, laterally to the mandible and the zygoma, superiorly to the floor of the middle cranial fossa, posteriorly to the ear and the mastoid process, and inferiorly to the alveolar border of the maxilla (Fig. 18.34).

Hematogenous Metastasis

Hematogenous spread to the paranasal sinuses from malignant tumors of distant organs is rather uncommon. As noted, the lesions are difficult to distinguish from primary tumors. The most common sites of origin are kidney, lung, prostate, and breast (Figs. 18.35 and 18.36). Other primary sites, such as the larynx and retroperitoneal space, are rare.

PERINEURAL METASTASIS

General Considerations

Perineural metastasis is an important but neglected avenue of spread of tumors of the head and neck. Recognition of this form of metastatic disease is essential for adequate therapy.

Certain cranial nerves, both centrally and peripherally, are confined by bony foramina and canals; enlargement of the nerves by malignant permeation of the perineural spaces may be reflected by erosion of the surrounding bony envelope. The cranial nerves most amenable to the radiographic demonstration of malignant permeation are the trigeminal and facial. The remaining 10 cranial nerves, although often involved, traverse relatively large foramina and fissures which are not affected by expansion of the neural trunks.

In malignant tumors of the maxillary sinus and skin of the cheek, perineural metastasis usually occurs along the infraorbital nerve of the maxillary division of the trigeminal nerve. In advanced lesions, extension centrally to the gasserian ganglion may occur (Fig. 18.37).

Spread from the ganglion to other divisions of the trigeminal nerve may follow.

Roentgen Findings

The roentgen diagnosis of perineural metastasis involving the infraorbital nerve is dependent upon enlargement of the infraorbital foramen and canal. In more advanced lesions sequential enlargement of the foramen rotundum and foramen ovale may occur. Occasionally, plain films will suggest these changes, but tomography is usually essential for purposes of confirmation and to determine the extent of involvement. The pertinent findings are as follows (Figs. 18.38 to 18.40): (1) enlargement and erosion of the infraorbital foramen and canal; (2) destruction of the floor of the orbit in the vicinity of the infraorbital canal; (3) enlargement and erosion of the foramen rotundum; (4) enlargement of the homolateral foramen ovale. The presumption of tumor in the gasserian ganglion implies an extremely poor prognosis.

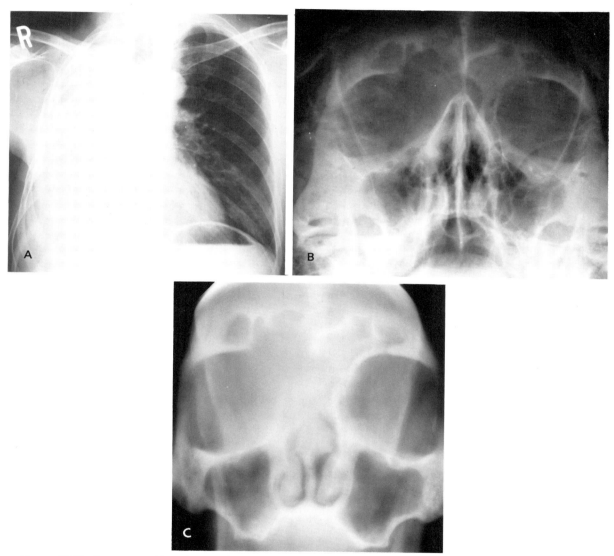

Figure 18.35. Squamous cell carcinoma of the lung with metastasis to the paranasal sinuses. *A*, Posteroanterior chest film: A right pneumonectomy has been performed for carcinoma of the lung. *B*, Waters view: there is destruction of the right ethmoid and frontal sinuses. The medial wall of the right orbit is also destroyed. *C*, Posteroanterior tomogram: the extent of the destructive lesion is more readily appreciated.

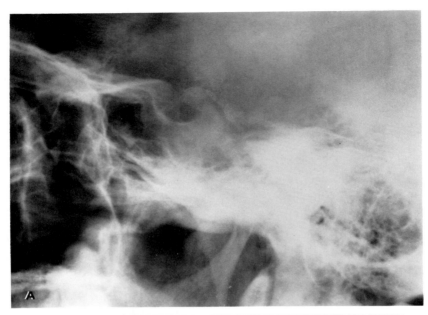

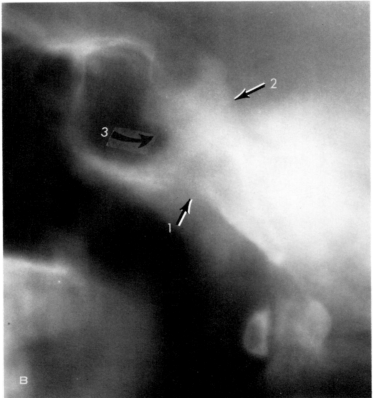

Figure 18.36. Carcinoma of the prostate with metastasis to the clivus, sphenoid sinus, and sella turcica. *A*, Lateral view of the skull: an osteoblastic process involves the clivus and dorsum sella. *B*, Lateral tomogram: the lesion is shown to be of the mixed variety, but predominately osteoblastic. The clivus (*1*), dorsum sella (*2*), and posterior aspect of the sphenoid sinus (*3*) are involved.

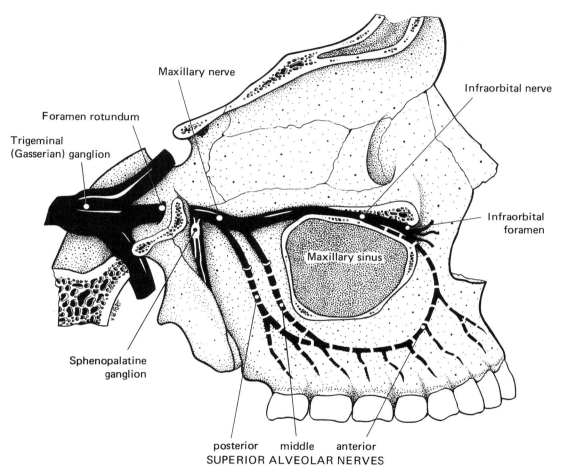

Figure 18.37. Course and branches of the maxillary division of the trigeminal nerve.

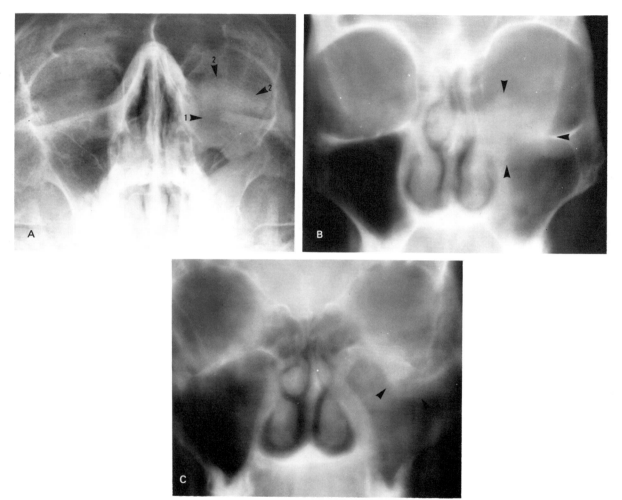

Figure 18.38. Squamous cell carcinoma of the suprastructure of the left maxillary sinus with involvement of the left infraorbital nerve. *A*, Waters view: destruction of the medial aspect of the floor of the left orbit (*1*) with a soft tissue mass in the inferior aspect of the orbit (*2*). There is marked thickening of the lining mucosa of the left maxillary sinus. *B*, Anteroposterior tomogram (anterior cut): destruction of the medial two-thirds, including infraorbital canal, of the floor of the left orbit. There is an associated soft tissue mass (*arrows*). The inferior part of the medial wall of the left orbit is partially destroyed. *C*, Anteroposterior tomogram (middle cut): irregular destruction of the floor of the left orbit associated with an ill defined soft tissue mass. There is thickening of the mucosa of the medial wall of the left maxillary sinus and the left ethmoidomaxillary plate. The left infraorbital canal is destroyed and surrounded by a soft tissue mass (*arrows*).

At surgery the tumor invasion had not reached the foramen rotundum.

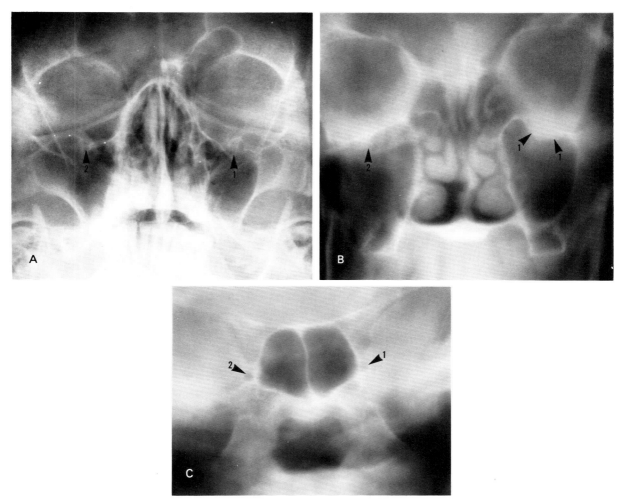

Figure 18.39. Basal cell carcinoma of the left cheek with perineural metastasis along infraorbital nerve of the mandibular division of the trigeminal nerve. *A*, Waters view: enlargement of the left infraorbital foramen (*1*) with thickening of lining mucosa in the superior aspect of the maxillary sinus. The right infraorbital foramen is normal (*2*). *B*, Posteroanterior tomogram (middle cut): enlargement of the left infraorbital canal with erosion of the roof of the maxillary sinus (*1*). There is a normal-appearing infraorbital canal on the right side (*2*). *C*, Posteroanterior tomogram (posterior cut): loss of definition of the left foramen rotundum (*1*) with regional bony destruction. The right foramem rotundum is normal in appearance (*2*).

The presence of tumor centrally to the gasserian ganglion was confirmed at surgery.

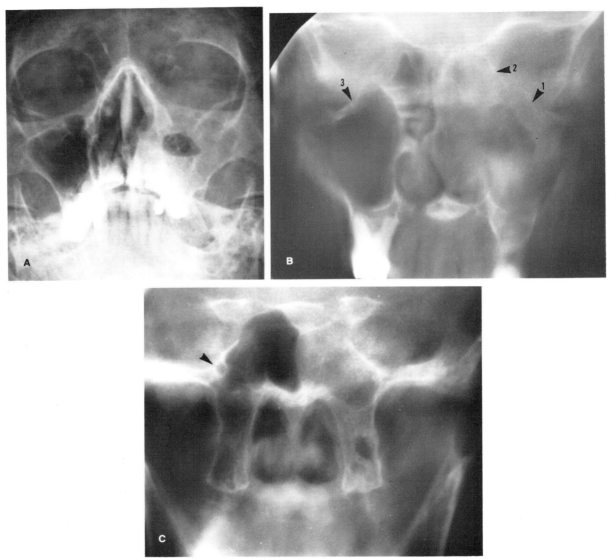

Figure 18.40. Adenoid cystic carcinoma of the left maxillary sinus with perineural metastasis involving the left infraorbital canal and foramen rotundum. *A*, Waters view: there is opacity of the left maxillary sinus with a fluid level. *B*, Anteroposterior tomogram (middle cut): an extensive soft tissue mass involves the left nasal fossa and the left ethmoid and maxillary sinuses. There is thickening and sclerosis of the left orbital floor (*1*). The left infraorbital canal is not visualized. The left lamina papyracea is partially destroyed (*2*). The right infraorbital canal is normal (*3*). *C*, Anteroposterior tomogram (posterior cut): the left foramen rotundum is poorly defined with associated regional soft tissue density and bony destruction. There is opacity of the left sphenoid sinus. The right rotundum is normal (*arrow*). (Reprinted with permission from Annals of Otology, Rhinology, and Laryngology, 81: 591, 1972.)

REFERENCES

Abt, A. B., and Toker, C. T.: Malignant teratoma of the paranasal sinuses. Arch. Pathol., *90:* 176, 1970.

Alexander, I. W.: Primary tumors of the sphenoid sinus. Laryngoscope, *73:* 537, 1963.

Baclesse, F.: Les cancer du sinus maxillaire de l'ethmoide et des fosses nasales. Ann. Otolaryngol., *69:* 465, 1953.

Batsakis, J. G.: Mucous gland tumors of the nose and paranasal sinuses. Ann. Otol. Rhinol. Laryngol., *79:* 557, 1970.

Boone, M., Harle, T. S., and Higholt, H. W.: Paranasal sinus and nasal cavity. In *Atlas of Tumor Radiology, The Head and Neck*, edited by G. H. Fletcher and B. S. Jing, Year Book Medicine Publishers, Chicago, 1968.

Boone, M., and Harle, T. S.: Malignant tumors of the paranasal sinuses. Semin. Roentgenol. *3:* 202, 1968.

Brownson, R. J., and Ogura, J. H.: Primary carinoma of the frontal sinus. Laryngoscope, *81:* 71, 1971.

Bunting, J. S.: The anatomical influence in the megavoltage radiotherapy of carcinoma of the maxillary antrum. Br. J. Radiol., *38:* 235, 1965.

Castro, E. B., Lewis, J. S., and Strong, E. W.: Plasmacytoma of paranasal sinuses and nasal cavity. Arch. Otolaryngol., *97:* 326, 1973.

Choukas, N. C.: Lymphosarcoma of the maxilla. Oral Surg. Oral Med. Oral Pathol., *23:* 567, 1967.

Crone, R. P.: Malignant amelanotic melanomas of the nasal septum and maxillary sinus. Laryngoscope, *76:* 1826, 1966.

Dodd, G. D., Collins, L. C., Egan, R. L., and Herrera, J. R.: The systematic use of tomography in the diagnosis of carcinoma of the paranasal sinuses. Radiology, *72:* 379, 1959.

Dodd, G. D., Dolan, P. A., Ballantyne, A. J., Ibanez, M. L., and Chau, P.: The dissemination of tumors of the head and neck via the cranial nerves. Radiol. Clin. North Am., *8:* 445, 1970.

Dodd, G. D., and Jing, B. S.: Radiographic findings in adenoid cystic carcinoma of the head and neck. Ann. Otol. Rhinol. Laryngol., *81:* 8, 1972.

Finkelstein, J. B.: Osteosarcoma of the jaw bones. Radiol. Clin. North Am. *8:* 425, 1970.

Frazell, E. L., and Lewis, J. S.: Cancer of the nasal cavity and accessory sinuses. Cancer *16:* 1293, 1963.

Frazell, E. L., and Strong, E. W.: Osteogenic Sarcoma of the Maxilla with Cervical Lymph Node Metastasis. Texas Rep. Biol. Med., (Suppl.), *24:* 303, 1966.

Freedman, H. M., DeSanto, L. W., Devine, K. D., and weiland, L. H.: Malignant melanoma of the nasal cavity and paranasal sinuses. Arch. Otolaryngol., *97:* 322, 1937.

Gallagher, J. C.: Upper respiratory melanoma pathology and growth rate. Ann. Otol. Rhinol. Laryngol., *79:* 551, 1970.

Helmus, C.: Extramedullary plasmacytoma of the head and neck. Laryngoscope, *74:* 553, 1964.

Holdcraft, J., and Gallagher, J. C.: Malignant melanomas of the nasal and paranasal sinus mucosa. Ann. Otol. Rhinol. Laryngol., *78:* 1, 1969.

Jing, B. S.: Roentgen diagnosis of malignant disease of paranasal sinuses and nasal cavity. Ann. Otol. Rhinol. Laryngol., *79:* 584, 1970.

Larsson, L. G., and Martensson, G.: Carcinoma of the paranasal sinuses and nasal cavities. Acta Radiol., *42:* 149, 1954.

Lederman, M.: Cancer of the upper jaw and nasal chambers. Proc. R. Soc. Med., *62:* 65, 1969.

Lederman, M.: Tumors of the upper jaw: natural history and treatment. J. Laryngol. Otol., *84:* 369, 1970.

Lewis, J. S.: Sarcoma of the nasal cavity and paranasal sinuses. Ann. Otol. Rhinol. Laryngol., *78:* 778, 1969.

McDonald, E. J.: The present incidence and survival pictures in cancer and the promise of improved prognosis. Bull. Am. Coll. Surg., *33:* 75, 1948.

McDonald, E. J.: *The Survery of Cancer in Texas, 1944–1966. Present Status and Results*. The University of Texas M. D. Anderson Hospital and Tumor Institute, Houston, 1968.

MacComb, W. S., and Fletcher, G. H.: *Cancer of the Head and Neck*. p. 329–356. Williams & Wilkins, Baltimore, 1967.

Mainzer, F., Stargardeter, F. L., Connolly, E. S., and Eyster, E. F.: Inverting papilloma of the nose and paranasal sinuses. Radiology, *92:* 964, 1969.

Medellin, H., and Wallace, S.: Angiography in neoplasms of the head and neck. Radiol. Clin. North. Am., *8:* 307, 1970.

Morgenstein, K. M.: Bronchongenic carcinoma metastatic to the maxillary sinus. Laryngoscope, *78:* 262, 1968.

Ohngren, L. G.: Malignant tumors of the maxillo-ethmoid region. Acta Otolaryngol. Suppl., *19:* 1, 1933.

Osborn, D. A., and Wallace, M.: Carcinoma of the frontal sinus associated with epidermoid cholesteatoma. J. Laryngol. Otol., *81:* 1021, 1967.

Patchefsky, A., Sundmaker, W., and Marden, P. A.: Malignant teratoma of the ethmoid sinus. Cancer *21:* 714, 1968.

Paulus, D. D., Jr., and Dodd, G. D.: The roentgen diagnosis of tumor of the nasal cavity and accessory paranasal sinuses. Radiol. Clin. North Am., *8:* 434, 1970.

Rocca, A. N., Smith, J. L., and Jing, B. S.: Osteosarcoma and parosteal osteogenic sarcoma of the maxilla and mandible. Am. J. Clin. Pathol., *54:* 625, 1970.

Rocca, A. N., Smith, J. L., MacComb, W. S., and Jing, B. S.: Ewing's sarcoma of the maxilla and mandible. Oral Surg. Oral Med. Oral pathol., *25:* 194, 1968.

Silver, W. E., Daly, J. E., and Friedman, M.: Reticulum cell sarcoma of the nose and paranasal sinuses. Arch. Otolaryngol., *87:* 532, 1968.

Sofferman, R. A., and Cummings, C. W.: Malignant lymphoma of the paranasal sinuses. Arch. Otolaryngol., *101:* 287, 1975.

Tiwari, R. M.: Hodgkin's disease of the maxilla. J. Laryngol. Otol. *87:* 85, 1973.

Touma, Y. B.: Extramedullary plasmacytoma of the head and neck. J. Laryngol. Otol., *85:* 125, 1971.

Walike, J. W., and Bailey, B. J.: Head and neck hemangiopericytoma. Arch. Otolaryngol., *93:* 345, 1971.

Wong, D. S., Fuller, L. M., Buller, J. J., and Shullenberger, C. C.: Extranodal non-Hodgkin's lymphomas of the head and neck. Am. J. Roentgenol., *123:* 471, 1975.

Wright, J. L. W., Phelps, P. D., and Maclver, A. G.: Sarcoma of the ethmoid in children. J. Laryngol. Otol., *85:* 57, 1971.

Chapter 19

Postirradiation and Postoperative Changes in the Paranasal Sinuses

Treatment of tumors of the paranasal sinuses includes radiation therapy, surgery, or both, depending upon the type and stage of the disease. In any case, radiological evaluation of the disease after treatment is essential. An immediate postirradiation or postoperative roentgen examination should be made in all cases as a base line study for the evaluation of subsequent radiological changes. A thorough understanding of the various types of surgery performed and of the changes resulting from radiation therapy is essential in the interpretation of posttreatment roentgen findings.

POSTIRRADIATION CHANGES IN THE PARANASAL SINUSES

Changes in the paranasal sinuses after radiation therapy vary from simple sinusitis to life-threatening sarcomatous degeneration. In general, they may be grouped as follows: sinusitis, recurrence, radiation necrosis, and postirradiation sarcoma.

Postirradiation Sinusitis

During and after radiation therapy for tumors of the paranasal sinuses, there is usually a nonspecific sinusitis caused by tissue reaction and edema. Radiographically, there is thickening of the mucosa and cloudiness of the sinus. Fluid levels may be present (Fig. 19.1). Occasionally these changes may be difficult to differentiate from those caused by residual tumor and a definite radiological diagnosis may be possible only on serial examinations.

Postirradiation Recurrence of Tumor in the Paranasal Sinus

Tumor recurrence in a paranasal sinus after radiation therapy is not infrequently encountered. Radiological examination, when compared with the postirradiation base line study, can determine with accuracy the location and extent of the recurrence in the majority of cases. As in the pretreatment phase, this information is essential for proper management (Fig. 19.2).

Postirradiation Necrosis of Jawbone and Wall of Maxillary Sinus

General Considerations

Necrosis of bone may be a complication of radiation therapy in the treatment of tumors of various parts of the body. In 1922 Ewing defined the pathological changes in bone after external irradiation as a radiaton osteitis. Radiation necrosis of bone is, to a considerable extent, the sequela of an ischemic infarction resulting from a radiation-induced thrombosis of the regional blood vessels. The changes in the bone depend upon the dose of radiation and the presence or absence of associated infection. After irradiation the devitalized tissue is particularly susceptible to secondary infection and osteomyelitis may occur. The osteomyelitis may be accompanied by pathological fractures. The tissue dose of external radiation which may bring about bony changes is estimated to be in the vicinity of 3,000 R. Bone death can be expected after about 5,000 R (Woodard and Coley, 1947). The changes in bone may occur from several months to 5 years after radiation therapy. This complication is less common with supervoltage than with orthovoltage therapy, but it is still liable to occur with an excessive total dose, repeated treatment courses, and/or overlapping fields.

Microscopically, radiation necrosis shows var-

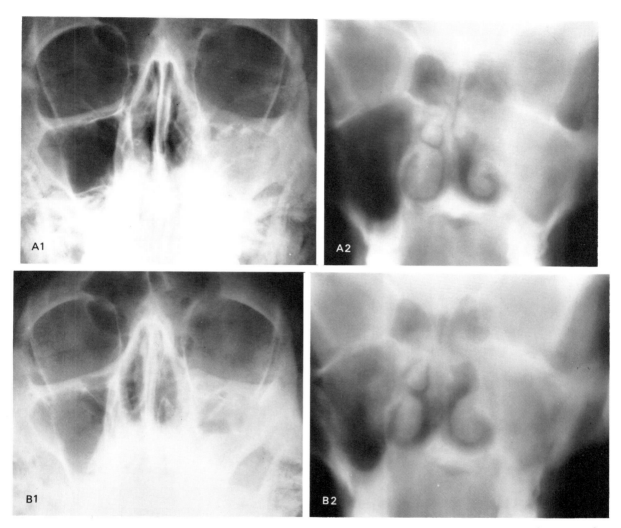

Figure 19.1. Postirradiation sinusitis of the left maxillary sinus. *A*, Preirradiation. *1*, Waters view: almost complete opacity of the left maxillary sinus. There is a small fluid level superiorly and medially. *2*, Anteroposterior tomogram: tumor mass in the left maxillary sinus with extension to the left ethmoid sinus. The tumor probably originated in the ethmoidomaxillary region. *B*, Postirradiation, (10 months after radiation therapy). *1*, Waters view: (the base line roentgen examination showed clearing of the left maxillary sinus). There is recurrent opacification of the sinus with a fluid level. *2*, Anteroposterior tomogram: there is thickening of the lining mucosa of the left maxillary sinus and of the medial aspect of the right maxillary sinus. The sinus walls are sclerotic and no interim destruction is seen. The findings favor inflammatory disease.

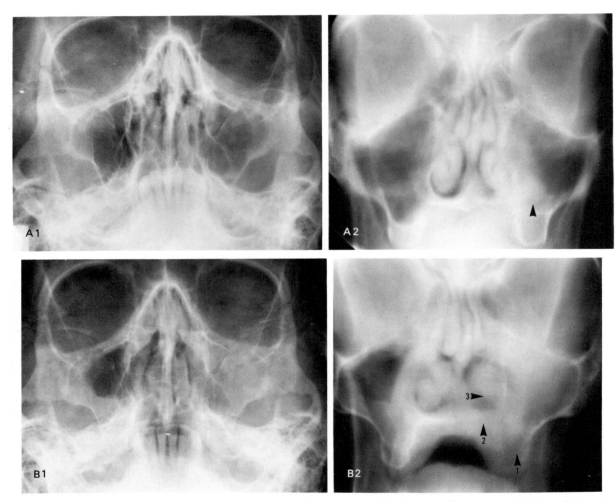

Figure 19.2. Postirradiation recurrence of squamous cell carcinoma of the left maxillary sinus. *A*, Preirradiation: *1*, Waters view: slight cloudiness of the left maxillary sinus; *2*, Anteroposterior tomogram: tumor mass, infrastructure, of the left maxillary sinus (*arrow*). *B*, Postirradiation (8 months after radiation therapy): *1*, Waters view: opacification of the left maxillary sinus and thickening of the lining mucosa of the right maxillary sinus; *2*, Anteroposterior tomogram: recurrence of the lesion with destruction of the left alveolar ridge (*1*) and left side of the hard palate (*2*), and involvement of the left nasal fossa (*3*).

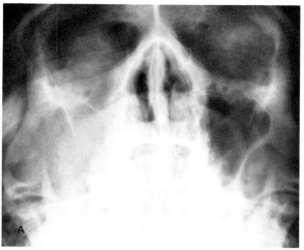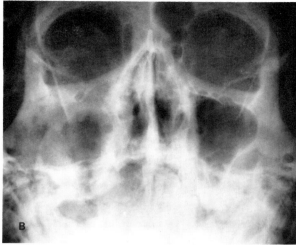

Figure 19.3. Postirradiation necrosis of the right maxilla. *A*, Waters view before external irradiation: opacification of the right maxillary sinus with destruction of the lateral wall and zygomatic process. *B*, Waters view 21 months after external irradiation: regression of the lesion involving the right maxillary sinus. The right maxilla, the right zygoma, and the floor of the right orbit appear more dense, with areas of rarefaction consistent with radiation necrosis.

This 45-year-old male was treated with external radiation for a squamous cell carcinoma of the right maxillary sinus. A tumor dose of 6017 rads was given in 6 weeks with ^{60}Co. Clinically, radiation necrosis was suspected 1½ years later. The patient had a partial resection of the maxilla and curettage of the maxillary sinus after postirradiation roentgen examination. Postirradiation necrosis of the maxilla was confirmed.

ious combinations of aseptic necrosis, disintegration of necrotic bone, and compensating reparative reaction. The changes may be confined to a limited area or may be widespread.

In tumors of the face and oral cavity the possibility of significant radiation necrosis is often augmented by the presence of infection. When radiation damage to bone occurs, the infection often leads to rapid progression. Abnormal soft tissue calcifications may also occur as a result of the infectious process.

Roentgen Findings

The roentgen findings in radiation necrosis of the jawbone and the wall of the maxillary sinus vary considerably, depending upon the stage of the lesion.

1. In the early stage, no radiographic changes may be visible. The diagnosis depends upon the knowledge that external irradiation has been given and upon the clinical symptom of severe pain.

2. In well established cases, the presence of radiation osteitis is manifested by multiloculated areas of sclerosis interspersed with areas of radiolucency (Figs. 19.3 and 19.4).

3. Superimposed osteomyelitis may be present, with irregular areas of bone destruction and sequestration.

4. When teeth are present, radiation necrosis is usually first apparent in the alveolar ridge; the first roentgen evidence of necrosis may be destruction of the dental alveoli.

Differential Diagnosis

The differential diagnosis primarily involves osteomyelitis of pyogenic origin and recurrence of the malignant lesion with secondary involvement of bone. The differentiation may be difficult and is at times impossible.

Postirradiation Sarcoma of Bone

General Considerations

First described by Beck in 1922, postirradiation sarcoma of bone is a well established entity. The lesions can develop after treatment of benign bone lesions such as giant cell tumor, fibrous dysplasia, and bone cyst. They may also develop after treatment of a malignant tumor or in a previously normal bone included in the treatment field of an unrelated lesion.

The tissue dose with which sarcomatous degeneration of the bone may be expected is about 3,000 R. However, a dose as small as 1,550 R given by fractional method can induce sarcomatous change (Cahan et al., 1948). The complica-

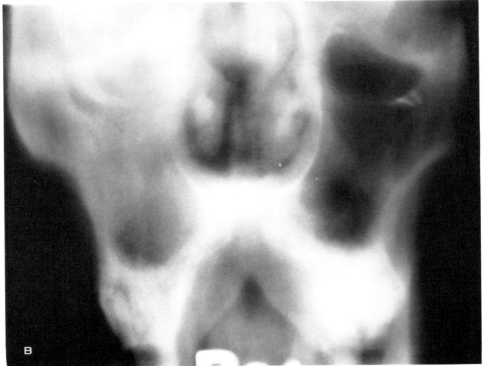

Figure 19.4. Postirradiation necrosis of the maxilla and mandible. *A*, Panorex view of the jaws 2¹/₂ years after external irradiation: there is early postirradiation necrosis of the body of the left hemimandible (*1*) and possible necrosis of the right hemimandible (*2*). Early postirradiation necrosis of both alveolar ridges of the maxillae (*3*) is also present. Radiation damage of the teeth is noted. *B*, Anteroposterior tomogram 4 years after external irradiation: there has been considerable progression of the postirradiation necrosis of the alveolar ridges and the maxillae. There is cloudiness of the right maxillary sinus.

This 48-year-old female was treated with external irradiation for a squamous cell carcinoma of the right retromolar trigone. A tumor dose of 6000 rads was given in 6 weeks with ^{60}Co. Postirradiation necrosis of the bone developed 2¹/₂ years after radiation therapy.

tion is more likely to develop in patients after orthovoltage therapy; it is less common with supervoltage treatment. The incidence of postirradiation sarcoma has been estimated at 0.1% of the 5-year survivors (Phillips and Sheline, 1963), with a variable latent period of from 3 to 40 years.

The postirradiation sarcomas have been classified as osteosarcoma, fibrosarcoma, and, occasionally, chondrosarcoma.

At the University of Texas M. D. Anderson Hospital and Tumor Institute at Houston, eight patients have developed postirradiation sarcoma in an irradiated area after treatment for tumors of the head and neck. Three of these sarcomas were in the maxillary sinus.

Roentgen Findings

The roentgen findings of postirradiation sarcoma of the maxilla or the wall of the maxillary sinus are nonspecific and are as follows (Figs. 19.5 and 19.6): (1) opacity of the maxillary sinus; (2) destruction of the walls of the maxillary sinus; (3) soft tissue mass within and beyond the confines of the sinus; (4) osteoblastic tumor new bone extending beyond the confines

of the sinus; (5) sunburst type of periosteal reaction.

Differential Diagnosis

1. In the presence of bone destruction, postirradiation sarcoma should be differentiated from carcinoma, rhabdomyosarcoma, and advanced granulomatous lesions of the maxillary sinus.
2. When the "sunburst" form of periosteal reaction is present, Ewing's sarcoma must be excluded.
3. Differentiation between radiation necrosis and postirradiation sarcoma may be difficult; however, certain roentgen features may be helpful in the differential diagnosis:

Parameter	Radiation Necrosis	Postirradiation Sarcoma
Latent Period	Several months to 5 years	3 to 40 years
Bone Destruction	No roentgen evidence of bone destruction after 5 years in absence of infection or trauma	Roentgen evidence of bone destruction after 5 years
Tumor new bone	Absent	Present
Soft tissue lesion with calcification	Absent if there is no infection	Present
Pain	Absent if there is no infection or trauma	Present

POSTOPERATIVE CHANGES IN THE MAXILLARY SINUSES

From the radiological standpoint the various operative procedures for treatment of tumors of the maxillary sinus can be grouped as follows: Caldwell-Luc procedure, partial maxillectomy, and total maxillectomy.

It is important to remember that a knowledge of the radiologic changes associated with the various types of surgery is imperative before evaluating posttreatment roentgenograms. It is often impossible to distinguish between postoperative defects and recurrence of the tumor without the benefit of a postoperative base line study for comparative purposes.

Caldwell-Luc Procedure

The Caldwell-Luc procedure consists of two separate surgical openings, one of which is made in the canine fossa to gain access to the maxillary sinus and the other in the nasoantral wall for drainage. This operative procedure is used for removal of teeth and fragments of root in the sinus, for treatment of chronic sinusitis with polyps and cystic degeneration of the mucosa, and for excision of cysts and benign tumors of the maxillary antra.

After the sinus has been surgically drained, thickening of the lining mucosa may persist and a return to the normal radiolucency of the chamber may never occur. Opacification of the maxillary sinus may be demonstrable at any

point in the postoperative period. This may be attributed to recurrent disease, unless the differences between the roentgen appearance of surgically treated and diseased sinuses are understood. Unfortunately, differentiation is not always possible and tissue biopsy may be necessary.

After surgery the sinus appears smaller than average because the lines of demarcation between the sinus and the bony wall have been obliterated by the surgical procedure and scar tissue. Radiologically, the scar tissue appears denser than thickened mucosa, being almost as opaque as the adjacent maxillary bone. If residual air is present in the sinus, it is small in amount and distributed somewhat differently from that in a diseased sinus.

Radiographically the surgical defects from a Caldwell-Luc procedure can be observed on the lateral view as a loss of the bony boundary of the anterior wall of the sinus and, on the anteroposterior view, as a defect in the inferior portion of the nasoantral wall (Fig. 19.7).

Maxillectomy

Maxillectomy is an operation in which the maxillary bone is surgically resected in portion or in toto. Total removal of the sinus is technically impossible without total removal of the maxillary bone. Generally, the entire sinus is

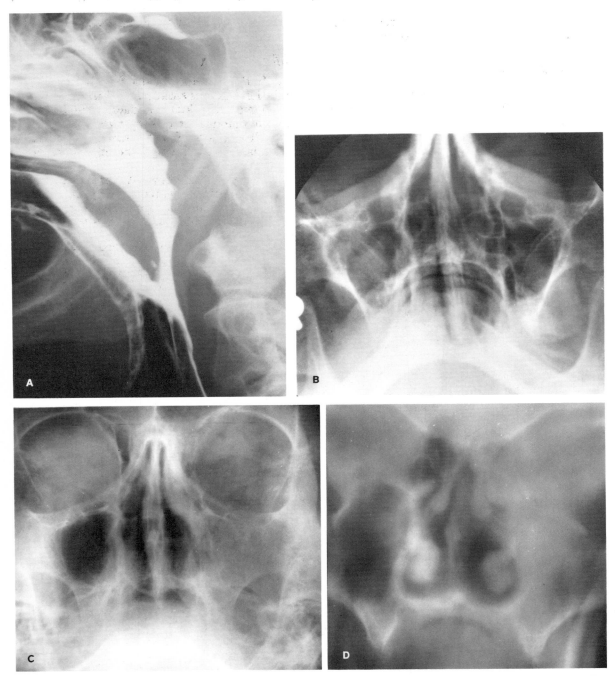

Figure 19.5. Postirradiation osteosarcoma of the left maxillary sinus. *A*, Lateral nasopharyngogram of the nasophar-ynx in 1963: an infiltrative lesion consistent with squamous cell carcinoma involves the superoposterior aspect of the nasopharynx. *B*, Waters view of the paranasal sinuses in 1963: normal-appearing maxillary sinuses. *C*, Waters view in 1968 (58 months after external irradiation): opacification and absence of bone detail of the left maxillary sinus. There is thickening of the lining mucosa of the right maxillary sinus. *D*, Anteroposterior tomogram in 1968: opacification of the left maxillary sinus with destruction of the walls. There is cloudiness of the left ethmoid sinus with destruction of the left lamina papyracea.

This 65-year-old male had squamous cell carcinoma of the nasopharynx treated with external irradiation in 1963. A tumor dose of 6000 rads was given in 6 weeks with ^{60}Co. The patient developed symptoms referable to the left maxillary sinus $4^{1}/_{2}$ years later. Left maxillary resection with orbital exenteration was performed after postirradiation roentgen examination. The pathological diagnosis was osteosarcoma.

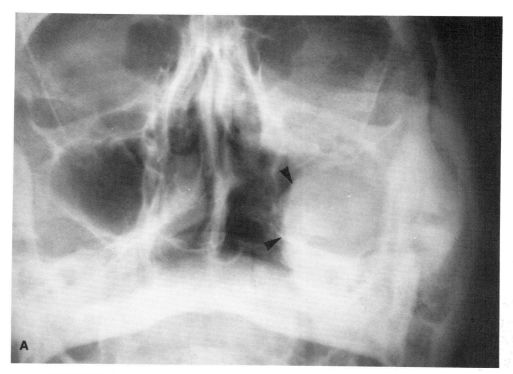

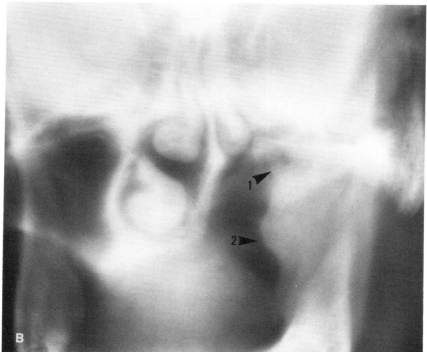

Figure 19.6. Postirradiation osteosarcoma of the left maxillary sinus. *A*, Waters view in 1971 (40 months after external irradiation): Postoperative status of the left maxillary sinus with a large soft tissue mass in the lateral aspect of the remaining sinus (*arrows*). *B*, Anteroposterior tomogram: bone destruction of remaining roof of the left maxillary sinus (*1*). The soft tissue tumor mass is seen on the lateral aspect of the sinus (*2*).

This 75-year-old female had squamous cell carcinoma of the left upper gum, treated with external irradiation to an unknown dosage. Irradiation was followed by partial maxillary resection for residual disease in 1967. She did well until January, 1971. At that time she developed a tumor in the left upper cheek. Tissue biopsy revealed a malignant tumor and a left maxillectomy was performed after roentgen examination. The final diagnosis was osteosarcoma.

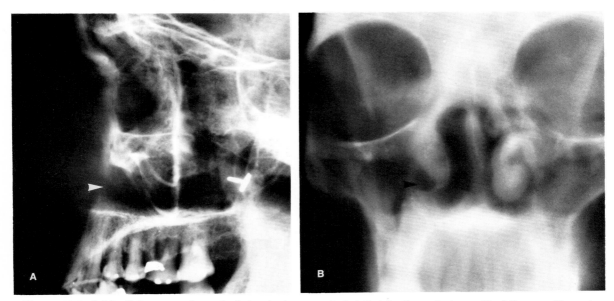

Figure 19.7. Caldwell-Luc procedure. *A*, Lateral view: surgical defect in the anterior wall of the maxillary sinus (*arrow*); *B*, Anteroposterior tomogram: a surgical defect in the inferior part of the right nasoantral wall (*arrow*).

resected en masse without entering its cavity. When the disease has extended beyond the anatomical confines of the sinus, the surgery is often very extensive, with removal of the entire cancerous area. The most frequently used procedure is maxillectomy with exenteration of the eyeball. For early carcinoma or sarcoma of the maxillary sinus, partial maxillectomy may be the procedure of choice. One of the commonly used techniques is the Weber-Ferguson procedure.

In the postoperative state, after partial or total maxillectomy, the roentgen examination provides the following information:

Extent of Surgical Procedure (Figs. 19.8 and 19.9)

The postoperative status of the maxillary sinus can be well evaluated by roentgen examination. In most instances the roentgen examination is the only technique which permits an accurate determination of the extent of the surgical procedure.

Presence of Residual Disease (Fig. 19.10)

The detection of residual disease in the postoperative cavity may be difficult. The presence of an abnormal soft tissue density within the sinus and/or a poorly defined bony margin may be the only roentgen indications of residual disease.

Presence of Recurrent Tumor (Fig. 19.11)

Normally, the postoperative sinus cavity is clearly defined, without evidence of abnormal soft tissue densities. With recurrence, there is usually a soft tissue mass in the cavity and/or destruction of the remaining bone. At times, in an advanced case, there is an extension of the lesion into the surrounding structures.

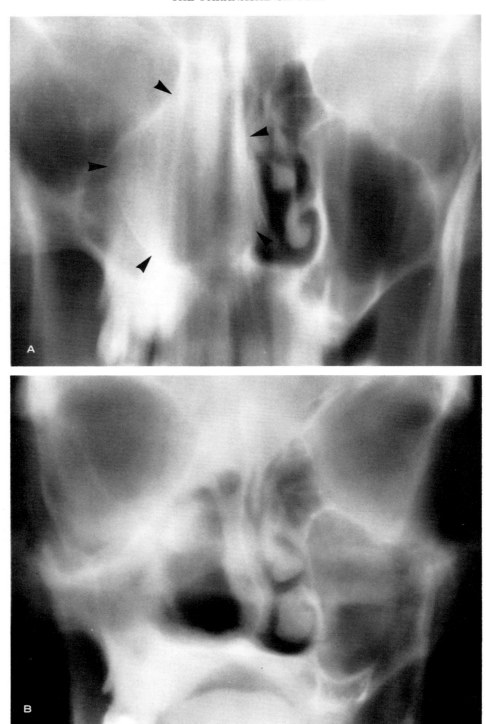

Figure 19.8. Partial maxillectomy for neurofibrosarcoma of the right maxillary sinus without evidence of recurrence. *A*, Preoperative – anteroposterior tomogram in 1969: there is a large expanding soft tissue mass in the medial aspect of the right maxillary sinus with extension medially to the right nasal fossa and superiorly to the right ethmoid sinus (*arrows*). *B*, Postoperative – posteroanterior tomogram in 1971: satisfactory postoperative status of the right maxillary sinus. There is no evidence of recurrence.

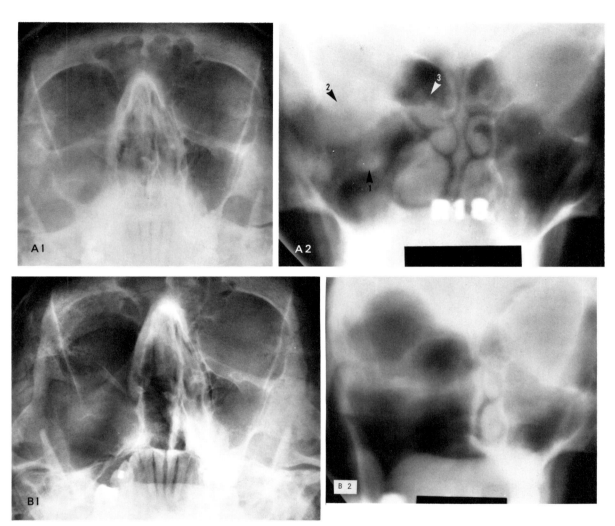

Figure 19.9. Radical maxillectomy for squamous cell carcinoma of the right maxillary sinus without evidence of recurrence. *A*, Preoperative: *1*, Waters view: complete opacification of the right maxillary sinus; *2*, Anteroposterior tomogram: extensive tumor mass arising from suprastructure of the right maxillary sinus (*1*) with extension to the right orbit (*2*) and ethmoid sinus (*3*). *B*, Postoperative (34 months after surgery): *1*, Waters view: a large surgical defect with exenteration of the right eye — no evidence of recurrence; *2*, Anteroposterior tomogram: postoperative defect is shown to good advantage. There is no evidence of recurrence; the remaining bone margins are sharply defined and no soft tissue mass is seen.

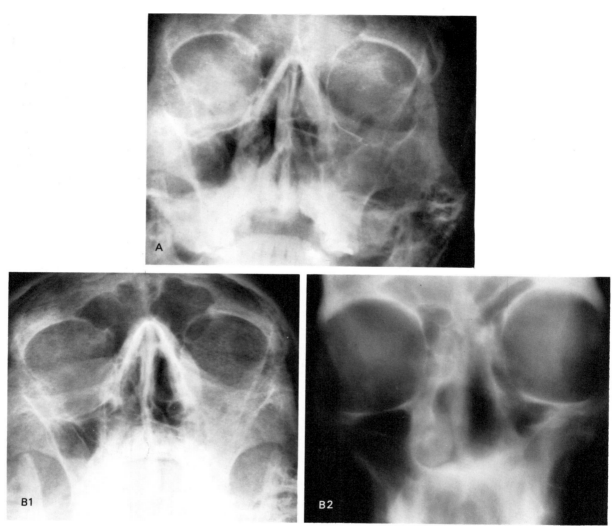

Figure 19.10. Chondrosarcoma of the left maxillary sinus; Caldwell-Luc procedure with residual disease. *A*, Preopera-tive: Waters view: opacification of the left maxillary sinus with possible destruction of the medial wall. *B*, Postoperative (9 months after Caldwell-Luc procedure): *1*, Waters view: opacification of the left maxillary sinus with absence of the left middle and inferior nasal turbinates; *2*, anteroposterior tomogram (anterior cut): postoperative status of the left maxillary sinus and left nasal fossa. The left maxillary sinus is opaque and contains irregular calcific densities.

This 71-year-old female had an excision of a chondrosarcoma of the left nasal fossa and maxillary sinus. A Caldwell-Luc procedure was used. The patient developed swelling of the left side of the nose and left cheek 7 months later. A postoperative paranasal sinus tumor survey revealed the presence of disease in the left maxillary sinus and a partial maxillectomy was performed. The pathological diagnosis was well differentiated chondrosarcoma, arising from the anterior wall of the left maxillary sinus with extension to the soft tissues adjacent to the left ala of the nose.

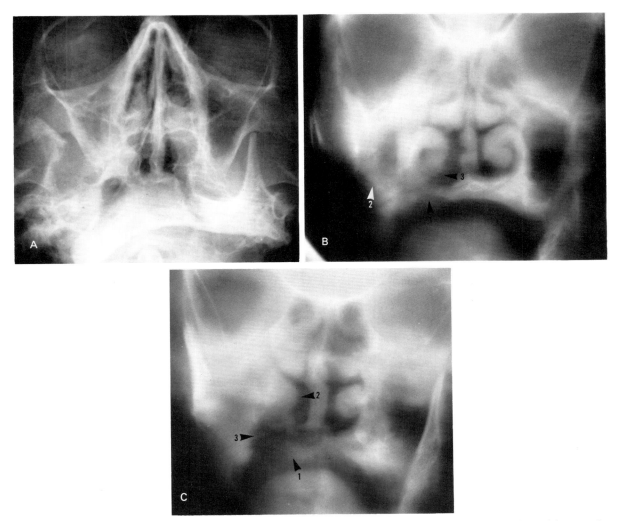

Figure 19.11. Recurrent squamous cell carcinoma of the soft palate (postirradiation and postoperative) with extension to the right maxillary sinus. A, Waters view in 1971: there is thickening of the lining mucosa and clouding of the right maxillary sinus. The right mandible has been partially resected. B, Anteroposterior tomogram (middle cut) in 1971: opacification of the right maxillary sinus. There is partial absence of the right hard palate (1), floor of the right maxillary sinus (2), and inferior part of the right inferior turbinate (3). C, Anteroposterior tomogram (posterior cut) in 1971: a large soft tissue mass involves the hard palate (1), the right inferior turbinate (2), and the lower part of the right maxillary sinus (3).

This 50-year-old-male had squamous cell carcinoma of the soft palate which was treated with external irradiation in 1961. An excision of the soft and hard palate was performed in 1970 for recurrence of the tumor. In 1971 partial resection of the right maxilla was carried out after the radiological studies shown above. Histological confirmation of recurrent squamous cell carcinoma was obtained.

REFERENCES

Beck, A.: Zur frage des roentgensarkoms, Zugleich ein beitrag, zur pathogeneses des sarkoma. Münch. Med. Wochenschr., *69:* 623, 1922.

Bragg, D. G., Shidnia, H., Chu, F. C. H. and Higinbotham, N. L.: The clinical and radiographic aspects of radiation osteitis. Radiology, *97:* 103, 1970.

Cahan, W. B., Woodard, H. Q., Higinbotham, N. L., Stewart, F. W. and Coley, B. L.: Sarcomas arising in irradiated bone: Report of eleven cases. Cancer, *1:* 3, 1948.

Cook, H. P.: Tooth extraction and radiation therapy of the mouth. Br. Dent. J., *120:* 372, 1966.

Cruz, M., Coley, B. L., and Stewart, F. W.: Post-radiation bone sarcoma: report of eleven cases. Cancer, *10:* 72, 1957.

Daland, E. M.: Radiation necrosis of the jaw. Radiology, *52:* 205, 1949

Ewing, J.: Radiation osteitis. Acta Radiol., *6:* 399, 1926.

Gates, O.: Effect of radiation on normal tissues: effects on bone cartilage and teeth. Arch. Pathol., *35:* 323, 1943.

Gorlin, R. J. and Goldman, H. M.: *Thomas' Oral Pathology*, 6th ed., C. V. Mosby Company, St. Louis, 1970.

MacComb, W. S. and Fletcher, G. H.: *Cancer of the Head and Neck.* Williams & Wilkins, Baltimore, 1967.

Niebel, H. and Neenan, E. W.: Dental aspects of osteoradionecrosis. Oral Surg. Oral Med. Oral Pathol., *10:* 1011, 1957.

Phillips, T. L. and Sheline, G. D.: Bone sarcomas following radiation therapy. Radiology, *81:* 992, 1963.

Sabanas, A. O., Dahlin, D. C., Childs, D. S., Jr. and Ivins, J. C.: Postradiation sarcoma of bone. Cancer *9:* 538, 1956.

Sears, W. P., Refft, M., and Cohen, J.: Postirradiation mesenchymal chondrosarcoma: a case report. Pediatrics, *40:* 254, 1967.

Soloway, H. B.: Radiation induced neoplasms following curative therapy for retinoblastoma. Cancer, *19:* 1984, 1966.

Stafne, E. C.: *Oral Roentgenographic Diagnosis* 3rd ed., W. B. Saunders Company, Philadelphia, 1969.

Stafne, E. C. and Bowing, H. D.: The teeth and supporting structures in patients treated by irradiation. Am. J. Orthodont. Oral Surg. (Oral Surg. Sec.), *33:* 567, 1947.

Steiner, G. C.: Postradiation sarcoma of bone. Cancer *18:* 603, 1965.

Thomas, K. H.: *Oral Surgery*, 5th ed., C. V. Mosby Company, St. Louis, 1969.

Watson, W. L. and Scarborough, J. E.: Osteoradionecrosis in Intra-oral cancer. Am. J. Roentgenol., *40:* 524, 1938.

White, G., Sieniewicz, J. and Christensen, W. R.: Improved control of advanced oral cancer with massive roentgen therapy. Radiology. *63:* 37, 1954.

Wildermuth, C. and Cantril, S. T.: Radiation necrosis of the mandible. Radiology, *61:* 771, 1953.

Woodard, H. Q. and Coley, B. L.: The correlation of tissue dose and clinical response in irradiation of bone tumor and of normal bone. Am. J. Roentgenol., *57:* 464, 1947.

Zizmor, J. and Noyek, A. M.: The radiologic diagnosis of postsurgical disease of the sinus and mastoids. Otolaryngol. Clin. North Am., *7:* 251, 1974.

PART III

The Nasopharynx

Chapter 20

Anatomy of the Nasopharynx

GROSS ANATOMY

The nasopharynx is located behind the nasal cavity below the base of the skull and above the level of the soft palate. It is about 4 cm in width, 4 cm in height, and 2 to 3 cm in the anteroposterior diameter. The nasopharynx has five walls and communicates with the oropharynx through the pharyngeal isthmus. The anterior wall of the chamber is formed superiorly by the choanae and the posterior surface of the nasal septum; inferiorly it is delimited by the soft palate. The nasopharynx communicates with the nasal cavity through the choanae and is in close relationship to the posterior aspect of the middle and inferior turbinates. The roof and posterior wall form a continuous slope which sheaths a portion of the sphenoid, the basisphenoid, the basiocciput, and the upper two cervical vertebrae. In the roof, the mucous membrane is thrown into numerous and variable folds by the underlying lymphoid tissue of the pharyngeal tonsil. In the superoposterior midline, near the lower end of the pharyngeal tonsil, there is an inconstant small blind sac, the pharyngeal bursa. More anteriorly, a pharyngeal hypophysis, derived from the pharyngeal end of Rathke's pouch, is usually present. Laterally, the pharyngeal wall contains the pharyngeal orifice of the eustachian tube. This orifice is somewhat triangular in shape and is bound superiorly and posteriorly by a prominence, the torus tubarius, which is produced by the salience of the medial end of the cartilage of the tube. A depression behind the torus tubarius is the pharyngeal recess or fossa of Rosenmüller. The roof of the nasopharynx is firmly attached to the base of the skull by the pharyngeal aponeurosis (Fig. 20.1).

LYMPHATIC DRAINAGE

The nasopharynx possesses a rich lymphatic capillary network. The collecting trunks usually drain into the ipsilateral nodes, but may cross the midline to the opposite side. Most of the collection trunks pass through the pharyngeal wall and drain into the lateral retropharyngeal nodes. These nodes, usually one to three in number, lie in the retropharyngeal space near the base of the skull, anterior to the lateral mass of the atlas. They are related laterally to the internal carotid artery, the last four cranial nerves as they exit from the skull, and the internal jugular vein. The medial retropharyngeal nodes are intercalated on the course of the lymphatics which empty into the lateral group, and are small and inconstant. The efferent lymphatic channels from the lateral retropharyngeal nodes drain into the upper internal jugular chain, especially into the jugulodigastric group. Some of the efferent lymphatic channels from the nasopharynx empty directly into the nodes of the spinal accessory chain (Fig. 20.2).

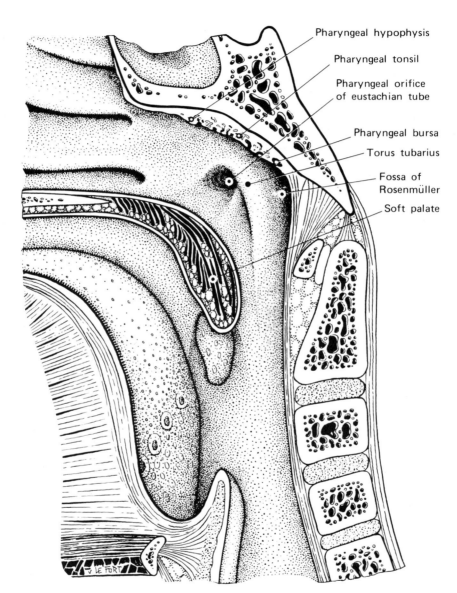

Figure 20.1. Normal anatomy of the nasopharynx.

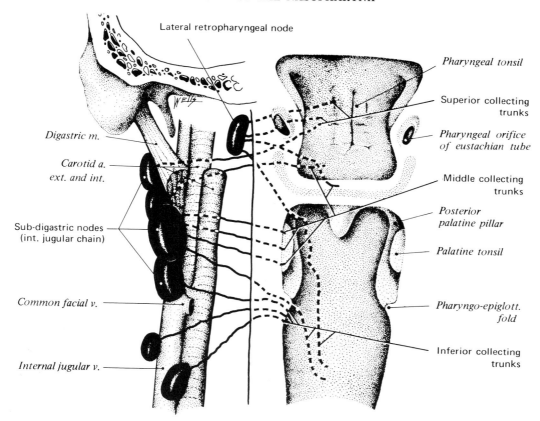

Figure 20.2. Lymphatic drainage of the pharynx.

REFERENCES

Rouviere, H.: *Anatomy of the Human Lymphatic System.* Translated by M. T. Tobiar, J. W. Edwards, Ann Arbor, Michigan, 1938.

Schaeffer, J. P.: *Morris Human Anatomy,* 11th ed., p. 1425–1447. McGraw-Hill Book Company, New York, 1953.

Chapter 21

Roentgen Technique

Radiological examination is essential in any patient suspected of having a lesion of the nasopharynx. The purposes of the examination are 3-fold: (1) to provide a definite diagnosis when clinical symptoms are suggestive; (2) to determine the exact location of the lesion; (3) to reveal the extent of involvement of adjacent structures.

The roentgen examination consists of the following.

LATERAL SOFT TISSUE ROENTGENOGRAM OF NASOPHARYNX

Lateral soft tissue roentgenograms of the nasopharynx are exposed with the patient in the sitting position and with his thigh parallel to the cassette holder. The shoulder is placed against the cassette holder with the chin elevated. The x-ray beam is centered on the temporomandibular joint. A small cone is used. The x-ray exposure is made while the patient inhales through the nose with the mouth closed (Fig. 21.1).

Lateral soft tissue roentgenograms are useful in outlining masses in the nasopharynx and in detecting sclerotic reactions of the basiocciput and basisphenoid. These last changes, however, are more consistently shown in lateral tomograms.

SUBMENTOVERTICAL, OR BASE, VIEW OF SKULL

This examination is carried out with the patient in the sitting position. The head is placed in hyperextension with the canthomeatal line parallel to the cassette. Because of the limitation of flexibility of the neck in some patients, this ideal positioning may not be possible. For practical purposes, the cassette holder is angled 30 to 35° toward the patient and the head is extended backward until the canthomeatal line is parallel to the cassette. The central beam is midway between the ascending rami of the mandible at a right angle to the center of the cassette. Stereoscopic exposures are made during inhalation through the nose. This view is often taken with a modified Valsalva maneuver to distend the nasopharynx and improve visualization of the wall and base structures (Fig. 21.2). This projection is of paramount importance in the demonstration of destructive processes involving the floor of the middle cranial fossa, the foramina, the petrous bone, the walls of the sphenoid sinus, the ethmoid cells, and the pterygoid plates. It is also useful in the detection of soft tissue masses, particularly those limited to or infiltrating the lateral walls of the nasopharynx.

LATERAL TOMOGRAMS OF NASOPHARYNX

Lateral tomograms of the nasopharynx are made with the patient in the lateral recumbent position with the chin raised. The distance from the midline of the nasopharynx to the table top is measured and the proper tomographic layer is calculated. The x-ray beam and Bucky diaphragm are centered on the temporomandibular joint or ³/₄ inch anterior to the external auditory

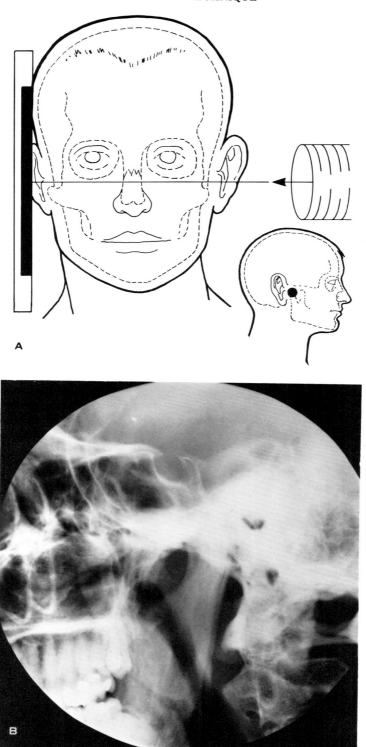

Figure 21.1. Lateral soft tissue roentgenogram of the nasopharynx: *A*, Line drawing to illustrate the radiographic position for the lateral soft tissue roentgenogram; *B*, Radiograph of the nasopharynx in the lateral projection.

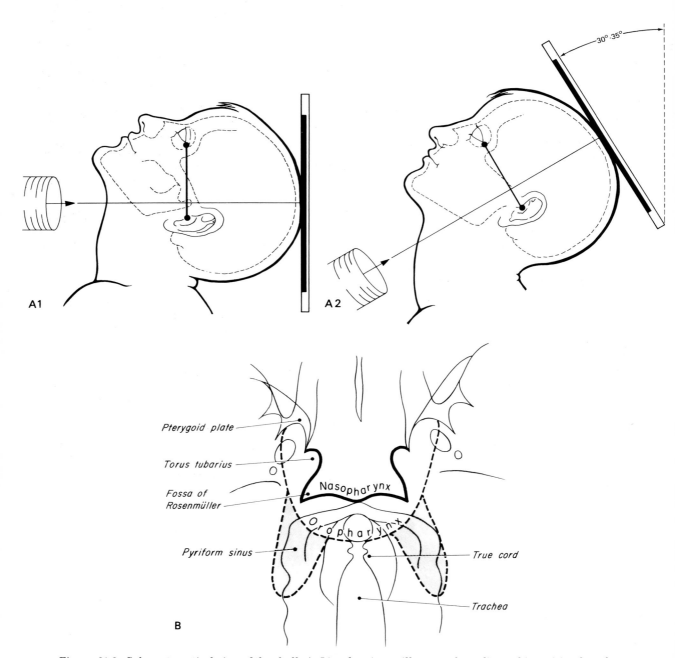

Figure 21.2. Submentovertical view of the skull. *A*, Line drawing to illustrate the radiographic position for submento-vertical view: *1*, standard position; *2*, modified position. *B*, Composite schematic drawing to show the nasopharynx in the submentovertical view. *C*, Radiographs of the nasopharynx in the submentovertical view: *1*, submentovertical view with patient inhaling through the nose—the nasopharyngeal air space and component structures are visualized; *2*, sub-mentovertical view with modified Valsalva maneuver showing distention of the oropharynx—the structure of the nasopharynx is shown to good advantage.

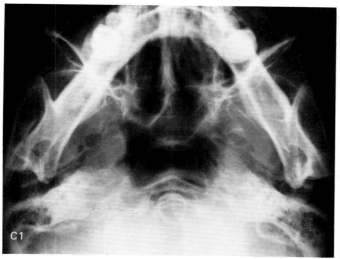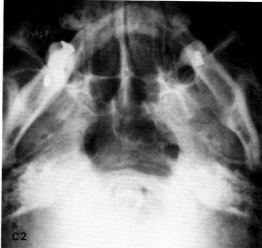

Fig. 21.2 C1–C2.

meatus. In this manner, five 8 × 10 tomograms are obtained. The first exposure is made through the midline of the nasopharynx, the second and third exposures 0.5 cm to either side of the midline plane, and the fourth and fifth exposures 1.5 cm on either side of the midline. All tomographic exposures are made while the patient inhales through the nose with the mouth closed. Occasionally, one will need additional tomograms.

Often lateral tomograms will show soft tissue masses which are not apparent on conventional lateral soft tissue roentgenograms. Changes in the basiocciput, basisphenoid, sphenoid sinus, and dorsum sellae can also be demonstrated.

The destructive changes caused by invasive tumors are well known, but the sclerotic processes attributable to osteoblastic reactions have not been fully recognized. The floor of the sphenoid sinus forms the roof of the nasopharynx and can therefore be involved very early. Bone destruction can be demonstrated in the lateral tomograms by a break in the floor of the sinus. A sclerotic reaction in the basisphenoid or the floor of the sphenoid sinus, seen either in the lateral roentgenogram or the lateral tomogram, is not necessarily caused by tumor involvement but is strongly suggestive of it.

ANTEROPOSTERIOR TOMOGRAMS OF NASOPHARYNX

The positioning of the patient and the technical factors for tomographic sections in this plane are the same as for anteroposterior tomograms of the paranasal sinuses. The first section is made at the level of the pterygoid plates. Two or three sections are made posteriorly at 0.5-cm intervals to a depth of 2.0 cm.

The anteroposterior tomogram is valuable in revealing the location and extent of soft tissue masses not well demonstrated in other projections. Bony changes in the pterygoid plates, the floor of the sphenoid sinus, or the floor of the middle cranial fossa can also be delineated.

OTHER TECHNIQUES

In selected cases, the following additional techniques are often indicated.

Contrast Nasopharynogogram

No premedication is necessary. The patient is placed on the x-ray table in the supine position

with the shoulders and trunk elevated to permit maximal extension of the neck, thus assuring a satisfactory submentovertical view of the base of the skull. After the nostrils are sprayed with 4% lidocaine hydrochloride (Xylocaine), a cannula attached to a syringe is placed at the nostril. The contrast medium, propyliodone (Oily Dionosil), is instilled alternately into each nostril through the cannula. The patient is instructed not to swallow during the instillation. For an adult patient, the optimal total amount of contrast medium is 15 to 20 ml.

Submentovertical and lateral views are taken in the same sequence. An anteroposterior view is occasionally made. For the submentovertical projection the x-ray tube is angled 20° cephalad with the central beam bisecting the midpoint of the canthomeatal line. The central beam is centered to an 8 × 10 cassette used in conjunction with a Bucky diaphragm or a stationary grid. In the lateral view the central beam is centered not to the cassette, but to the point where the central beam of the submentovertical projection bisects the canthomeatal line.

The first set of submentovertical and lateral projections is made when one-half of the contrast medium has been instilled and is repeated on completion of the instillation. The submentovertical projection is made to delineate the lateral walls and roof of the nasopharynx. The lateral projection is made to evaluate the roof and posterior wall of the nasopharynx as well as the nasopharyngeal surface of the soft palate. If involvement of the eustachian tube is suspected, additional films are taken during a modified Valsalva maneuver. An extension of the primary tumor through the posterior choanae into the nasal cavity must be carefully evaluated. A contrast nasopharyngogram is of great aid in determining the precise location and extent of disease.

Anteroposterior Tomograms of Paranasal Sinuses

When there is invasion of the posterior nasal fossa, possible involvement of the ethmoid sinuses should be investigated through careful tomographic study of the paranasal sinuses. Occasionally the maxillary sinus will also be involved; this will usually be apparent on tomographic studies of the paranasal sinuses.

Base Tomograms of Skull

In advanced cases, involvement of the floor of the middle cranial fossa is often present and can be clearly demonstrated by tomographic study of the base of the skull.

Carotid Angiogram

Carotid angiograms have little to offer in the diagnosis of primary malignant tumors of the nasopharynx, because most of the malignancies are squamous cell carcinomas. In general, these are poorly vascularized. An important function of the angiogram, however, is the demonstration of the relationship of the tumor to the major blood vessels of the head and neck. The appreciation of vascular invasion or obstruction before treatment is invaluable.

Among the benign tumors of the nasopharynx, juvenile angiofibroma is the most frequent. The vascular nature of this tumor permits a specific diagnosis to be made by carotid angiography. In an untreated case the characteristic vasular appearance is often demonstrated. After therapy, residual or recurrent disease can be delineated by angiography.

REFERENCES

Binet, E. F. and Moro, J. J.: Additional uses of basal polytomography. Radiology, *104:* 211, 1972.

Epstein, B. S.: Laminography in diagnosis of nasopharyngeal tumors. Radiology, *56:* 355, 1951.

Fletcher, G. H. and Matzinger, K. E.: Value of soft tissue technique in the diagnosis and treatment of head and neck tumors. Radiology, *57:* 305, 1951.

Jing, B. S. and McGraw, J. P.: Contrast nasopharyngography in diagnosis of tumors. Arch. Otolaryngol., *81:* 365, 1965.

Johnson, T. H., Green, A. E. and Rise, E. N.: Nasopharyngography—its technique and uses. Radiology. *88:* 1166, 1967.

Medellin, H. and Wallace, S.: Angiography in neoplasms of the head and neck. Radiol. Clin. North Am., *8:* 307, 1970.

Ruedi, L., and Zuppinger, A.: Zur roentgen-kontrastuntersuchung der nasopharynx. Hals Nas Ohrenheik, *35:* 500, 1934.

Wastie, J. L. The value of tomography in carcinoma of the nasopharynx. Br. J. Radiol., *45:* 470, 1972.

Chapter 22

Roentgen Anatomy

The roentgen anatomy of the nasopharynx in different projections is shown in Figure 22.1.

Several studies have been performed in children and adults in an attempt to establish normal measurements for the soft tissue shadow of the roof and the superoposterior pharyngeal wall of the nasopharynx (Weitz, 1946; Oon, 1964; Ho, 1967; Johannesson, 1968; and Capitanio and Kirkpatrick, 1970). Because technical differences exist between the various studies, absolute normal measurements are difficult to agree upon. The method for measuring the nasopharyngeal soft tissue shadows as suggested by Eller et al., 1971, is shown in Figure 22.2A. The normal measurements in males and females are shown in Figure 22.2B.

The soft tissue shadow is quite thick in children. The involution of the nasopharyngeal soft tissue shadow is not complete by age 25, but continues throughout life. The decrease in the thickness of this soft tissue shadow in adults varies greatly with age, but only slightly with sex. Furthermore, the thickness of the soft tissues is influenced by the configuration of the base of the skull. Because these variations exist, measurement as a means of determining the presence or absence of nasopharyngeal disease is useful only when the increase in the thickness of the soft tissue shadow is substantial. A local thickening should arouse suspicion even if the measurement is normal.

There are two roentgen features which are rather important. One is the relative thickness of the superoposterior pharyngeal wall soft tissues as compared to those of the roof. The second is the concavity of the over-all soft tissue contours of the nasopharynx as seen in the lateral projection. Any changes in these roentgen features is of clinical significance.

On lateral soft tissue roentgenograms of the nasopharynx the lobule of the ear may overlie the nasopharynx and mimic a tumor mass (Fig. 22.3A). This can be clarified by following the perimeter of the shadow outside the nasopharynx. A prominent posterior extremity of the inferior turbinate of the nasal cavity may also protrude into the anterior part of the nasopharynx and simulate a soft tissue tumor mass (Fig. 22.3B).

On the submentovertical view of the skull one can normally see the transverse shadow of the base of the tongue. Occasionally, the shadow is so prominent that it may simulate a soft tissue mass arising from the roof of the nasopharynx. To avoid this, the tongue should be protruded during exposure of the film (Fig. 22.3C).

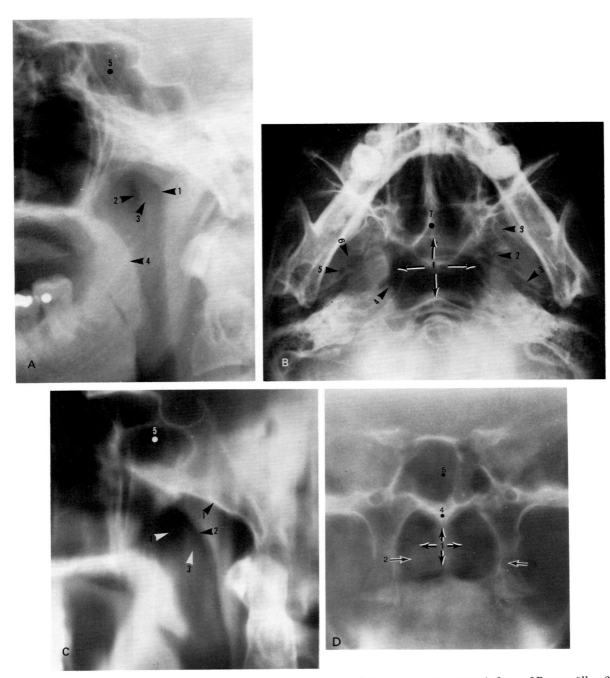

Figure 22.1. Roentgen anatomy of the nasopharynx. *A*, Lateral soft tissue roentgenogram: *1*, fossa of Rosenmüller; *2*, pharyngeal orifice of the eustachian tube; *3*, torus tubarius; *4*, soft palate; *5*, sphenoid sinus. *B*, Submentovertical view: *1*, nasopharynx; *2*, lateral wall of the oropharynx; *3*, eustachian tube; *4*, foramen lacerum; *5*, foramen spinosum; *6*, foramen ovale; *7*, sphenoid sinus; *8*, pterygoid plate. *C*, Lateral tomogram: *1*, clivus; *2*, fossa of Rosenmüller; *3*, torus tubarius; *4*, pharyngeal orifice of eustachian tube; *5*, sphenoid sinus. *D*, Anteroposterior tomogram (0.5 cm posterior to the pterygoid plates): *1*, nasopharynx; *2*, posterior extremity of the inferior turbinate; *3*, pterygoid plate; *4*, sphenoid rostrum; *5*, sphenoid sinus. *E*, Contrast nasopharyngogram in a young child. *1*, lateral projection: (*a*) filling of the lymphoid crypts (crypt sign); (*b*) soft palate. *2*, submentovertical projection: (*a*) pharyngeal orifice of eustachian tube; (*b*) fossa of Rosenmüller; (*c*) air-

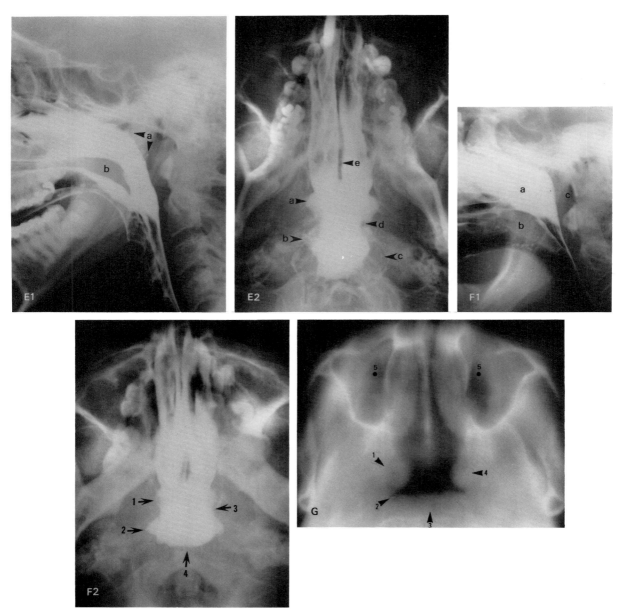

Fig. 22.1 E1, E2, F1, F2, G.

filled oropharynx; (*d*) torus tubarius; (*e*) nasal septum. *F*, Contrast nasopharyngogram in an adult. *1*, lateral projection: (*a*) contrast-filled nasopharynx; (*b*) soft palate; (*c*) posterior pharyngeal soft tissue. *2*, submentovertical projection: (*1*) pharyngeal orifice of the eustachian tube; (*2*) fossa of Rosenmüller; (*3*) torus tubarius; (*4*) posterior pharyngeal wall. *G*, Base tomogram of the skull: *1*, orifice of eustachian tube; *2*, fossa of Rosenmüller; *3*, posterior pharyngeal wall; *4*, torus tubarius; *5*, maxillary sinus.

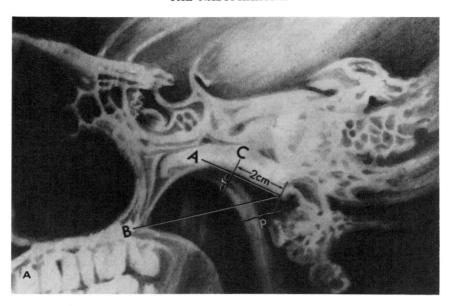

NASOPHARYNGEAL MEASUREMENTS IN 150 MALES

	Age	Superoposterior wall thickness (p)	Roof thickness (r)
Mean	38.8 yr. (16-80)	21.0 mm.	11.5 mm.
Standard Deviation	17.0 yr.	3.8 mm.	4.4 mm.

NASOPHARYNGEAL MEASUREMENTS IN 150 FEMALES

	Age	Superoposterior wall thickness (p)	Roof thickness (r)
Mean	41.6 yr. (16-80)	19.0 mm.	8.9 mm.
Standard Deviation	18.0 yr.	3.6 mm.	3.6 mm.

B

Figure 22.2. *A*, Charcoal drawing to show the method of measuring nasopharyngeal soft tissue shadow (after Eller et al.). *Line A* lies along the base of the skull; *line B* connects the anterior margin of the foramen magnum with the posterior margin of the hard palate; *r* denotes the thickness of the roof and measures 2 cm from the anterior margin of the foramen magnum perpendicular to *line A*; *p* denotes thickness of superoposterior soft tissue shadow and measures from the anterior margin of the foramen magnum along *line B*. *B*, Nasopharyngeal measurements on a lateral roentgenogram of the skull made in the erect position at a 36-inch target film distance.

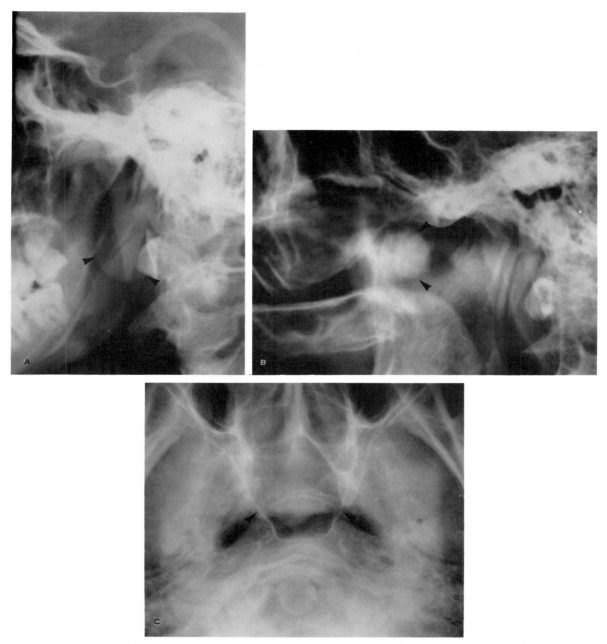

Figure 22.3. *A*, Lateral view of the nasopharynx. The shadow of the ear lobe simulates a nasopharyngeal mass (*arrows*). *B*, Lateral view of the skull. The prominent posterior extremity of the inferior turbinate simulates a nasopharyngeal mass (*arrows*). *C*, Submentovertical view showing the base of the tongue in anterior part of the nasopharynx (*arrows*).

REFERENCES

Capitanio, M. A. and Kirkpatrick, J. A.: Nasopharyngeal lymphoid tissue. Radiology, *96*: 389, 1970.

Eller, J. L, Roberts, J. F., and Ziter,M. H.: Normal nasopharyngeal soft tissue in adults. Am. J. Roentgenol., *112:* 537, 1971.

Epstein, B. S.: Laminography in diagnosis of nasopharyngeal tumors. Radiology, *56:* 355, 1951.

Fletcher, G. H., and Jing, B. S.: *The Head and Neck, An Atlas of Tumor Radiology*. Year Book Medical Publishers, Chicago, 1968.

Jing, B. S.: Tumors of the nasopharynx. Radiol. Clin. North Am., *8:* 324, 1970.

Jing, B. S. and McGraw, J. P.: Contrast nasopharyngogram in diagnosis of tumors. Arch. Otolaryngol., *81:* 365, 1965.

Ho, H. C.: Radiological diagnosis of nasopharyngeal carcinoma with special reference to its spread through the base of the skull. In *Cancer of Nasopharynx; Symposium, International Union Against Cancer*, Vol. 1, edited by C. S. Muir and K. Shanmugaratnam, Medical Examination Publishing Company, Flushing, New York, 1967.

Johannesson, S.: Roentgenological investigation of the nasopharyngeal tonsil in children of different ages. Radiology, *7:* 299, 1968.

Oon, C. L.: Some sagittal measurements of the neck in normal adults, Br. J. Radiol., *37:* 674, 1964.

Rizzuti, R. J. and Whalen, J. P.: The nasopharynx: roentgen anatomy and its alteration in the base view. Radiology, *104:* 537, 1972.

Weitz, H. L.: Roentgenography of adenoids. Radiology, *47:* 66, 1946.

Chapter 23

Inflammatory and Nonneoplastic Diseases of the Nasopharynx

PHARYNGEAL ADENOIDS

General Considerations

The pharyngeal tonsil is located in the midline of the roof of the nasopharynx. In children the pharyngeal tonsil is relatively large and is liable to be hypertrophied because of recurrent upper respiratory infections. Pharyngeal adenoids may occur in adults but are infrequent.

The symptoms related to excessive adenoidal tissue are largely attributable to obstruction of the posterior choanae and to occlusion of the pharyngeal orifices of the eustachian tubes. The mouth is usually open, particularly at night, in an attempt to ensure an adequate air way; this results in snoring sounds. Periodic attacks of deafness are caused by the obstruction of the eustachian tubes. This obstruction may lead to serous or purulent otitis media.

Roentgen Findings

Adenoids usually have a fairly typical appearance which is characterized by a localized soft tissue thickening with a smooth, rounded anterior border located in the roof of the nasopharynx (Fig. 23.1A). However, tumors such as malignant lymphomas may present a similar picture (Fig. 23.1B). Occasionally the adenoids show a soft tissue thickening with an irregular contour in the roof and superoposterior aspect of the nasopharynx (Fig. 23.1C). In such cases the lesion must be differentiated from carcinoma.

Adenoids in children are common and the diagnosis is easy to make because of the rarity of malignant tumors at that age. However, adenoids in the adult cannot be diagnosed with certainty; the final diagnosis often depends upon tissue biopsy.

CYST OF PHARYNGEAL BURSA (THORNWALDT'S CYST)

General Considerations

In the superoposterior aspect of the nasopharynx, at the lower end of the pharyngeal tonsil, a pharyngeal bursa is often located. The orifice of the bursa opens into the pharyngeal mucosa. If the orifice becomes occluded, a cyst often forms. When secondary infection occurs, an abscess may develop.

Roentgen Findings

The findings in cyst of the pharyngeal bursa are relatively characteristic (Fig. 23.2) and are listed below:

1. A sharply defined saccular soft tissue mass is seen projecting from the midline of the superoposterior part of the nasopharynx near the lower end of the pharyngeal tonsil.

2. A small dimple may be seen on the surface of the mass.

3. There is absence of surrounding soft tissue reaction and no osseous involvement.

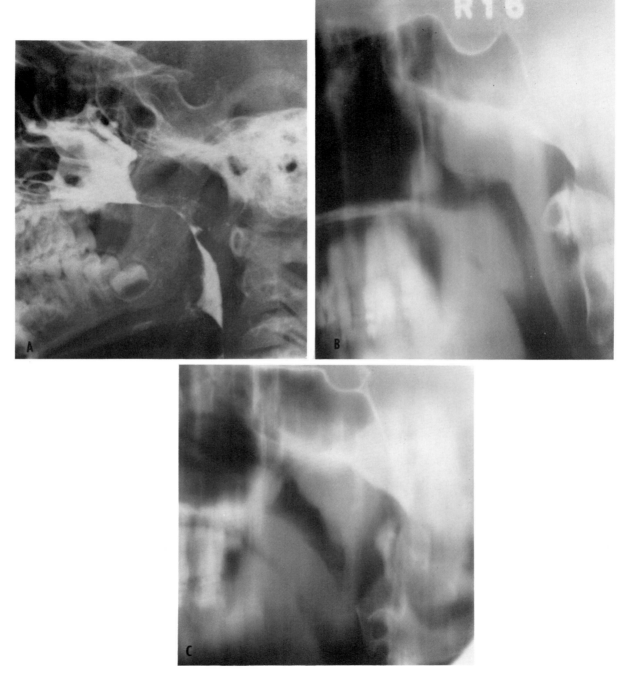

Figure 23.1. Pharyngeal adenoids. *A*, Typical case of pharyngeal adenoids: lateral nasopharyngogram showing a localized soft tissue thickening with a smooth, rounded anterior margin in the roof of the nasopharynx; *B*, Histiocytic lymphoma of the nasopharynx simulating adenoids: lateral tomogram—the roentgen features are similar to those of Figure 8*A*. *C*, Atypical case of pharyngeal adenoids: lateral tomogram showing a more diffuse soft tissue thickening with irregular margins in the roof and the superoposterior aspect of the nasopharynx.

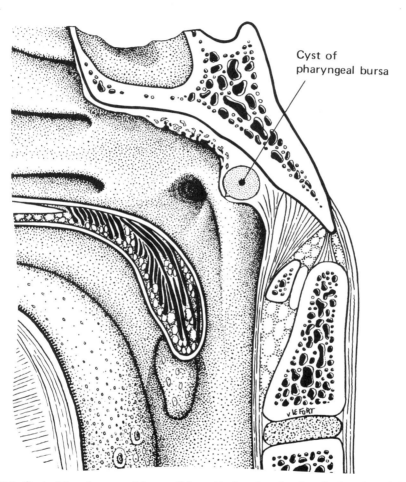

Figure 23.2. Cyst of the pharyngeal bursa. Schematic drawing showing the location of the cyst.

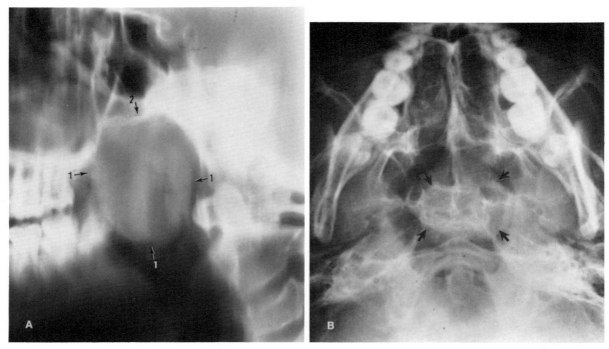

Figure 23.3. Inflammatory polyp of the nasopharynx. *A*, Lateral tomogram: there is a large, rounded, soft tissue mass arising from the roof of the nasopharynx and occupying almost the entire nasopharynx (*1*). The floor of the sphenoid sinus is sclerotic (*2*). *B*, Submentovertical view: the large soft tissue mass lies primarily on the left side of the nasopharynx (*arrows*). Tissue biopsy revealed an inflammatory polyp with extensive thrombosis.

PHARYNGEAL POLYPS

General Considerations

Polyps of the nasopharynx are predominantly inflammatory growths arising from the lining mucosa as a result of hyperplasia. They are probably the sequel of the combined effect of allergy and recurrent upper respiratory infection. The polyps may occur singly as pendunculated or sessile masses in the nasopharynx. The nasopharynx alone may be affected, but an associated sinusitis with polyps and rhinitis is often present.

Roentgen Findings

The common findings are as follows (Fig. 23.3):

1. A well defined rounded soft tissue mass projects into the air space of the nasopharynx.

2. The mass may be pedunculated or sessile and usually arises from the roof of the nasopharynx.

The roentgenographic features are nonspecific and tissue biopsy is often indicated to establish the diagnosis.

RETROPHARYNGEAL ABSCESS

General Considerations

The retropharyngeal space extends from the anterior aspect of the basiocciput downward through the prevertebral space into the posterior mediastinum. The prevertebral fascia and the fascia of the superior constrictor muscle together with the prevertebral muscle become firmly attached to the cervical spine at the level of the prominence of the second cervical vertebra. This anatomical arrangement tends to confine the abscess to the upper portion.

Retropharyngeal abscesses may be located anterior or posterior to the prevertebral fascia.

When a retropharyngeal abscess lies anteriorly it generally results from an injury to the posterior pharyngeal wall by a retained foreign body or from suppuration of the retropharyngeal lymph nodes after an acute respiratory infection. An abscess located posterior to the prevertebral fascia may arise from an infection of the cervical spine.

The patient often complains of painful deglutition, and, if the soft tissue swelling is marked, nasal obstruction, snoring, choking respiration, and dyspnea may occur. In acute cases the body temperature may be elevated, whereas in the chronic type little or no temperature increase may be present.

Roentgen Findings

The findings are very characteristic and are as follows (Fig. 23.4):

1. There is thickening with anterior bulging of the posterior pharyngeal soft tissue shadow and narrowing of the air space of the pharynx.

2. A radiopaque foreign body may be seen within the soft tissues.

3. Air may be present in the posterior pharyngeal soft tissues.

4. A fluid level may appear in the abscess cavity.

5. The cervical spine becomes kyphotic or straight, losing its normal lordosis at the level of greatest swelling.

6. Destruction of one or more cervical vertebral bodies may be present if the abscess is secondary to an infection of the cervical spine.

SARCOIDOSIS OF THE NASOPHARYNX

General Considerations

Sarcoidosis is a systemic disease, mainly affecting the skin, lungs, lymph nodes, spleen, liver, parotid glands, eyes, and bones. The pharyngeal tonsil may be involved; this, however, is rare and not many cases have been reported.

Sarcoidosis is a chronic granulomatous lesion simulating a noncaseating tubercle. In the nasopharynx sarcoid consists of a circumscribed, reddish yellow, lobulated infiltration in the roof at the site of the pharyngeal tonsil or the posterior wall of the nasopharynx. The lesion may be so extensive that it grossly resembles a malignant tumor.

The patient may complain of nasal obstruction. Deafness may be present owing to occlusion of the pharyngeal orifice of the eustachian tube.

Roentgen Findings

The findings of sarcoidosis of the nasopharynx are nonspecific.

1. A small, lobulated soft tissue shadow is seen in the midline of the roof of the nasopharynx at the site of the pharyngeal tonsil. It resembles adenoids.

2. In advanced cases, the lesion may appear as a large, lobulated soft tissue mass projecting from the superoposterior aspect of the nasopharynx. The changes may simulate malignant tumor.

3. There is no demonstrable destruction of the bone at the base of the skull.

4. Evidence of generalized sarcoidosis is often present.

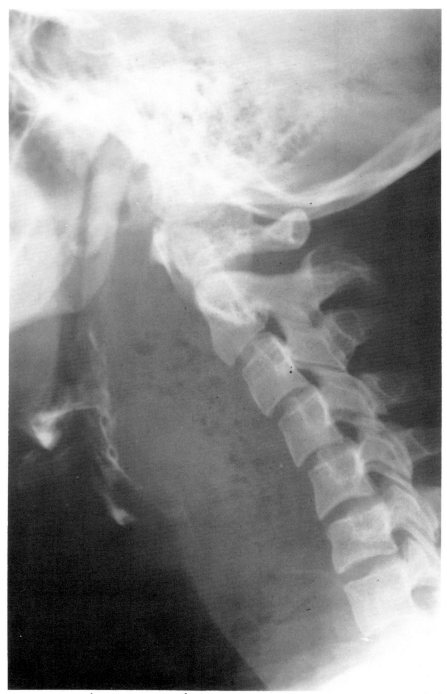

Figure 23.4. Retropharyngeal abscess. Lateral soft tissue roentgenogram of the neck: there is marked thickening of the posterior pharyngeal soft tissue shadows of the pharynx. Irregular radiolucencies are seen within the lesion due to abscess formation. The cervical spine has assumed a kyphotic curvature opposite the inflammatory collection.

TUBERCULOSIS OF THE NASOPHARYNX

General Considerations

Tuberculosis of the nasopharynx is usually secondary in nature. The most frequent site of the primary focus is the cervical lymph nodes. The patient often has chronic tuberculous lymphadenitis with a discharging sinus in the cervical region. The disease usually occurs in early life.

Roentgen Findings

1. It may appear as a well defined soft tissue mass in the region of the pharyngeal tonsil.

2. It may cause a diffuse thickening of the roof and posterior wall of the nasopharynx, without a characteristic appearance (Fig. 23.5A).

3. There is evidence of tuberculous cervical lymphadenitis (Fig. 23.5B).

The changes are nonspecific and cannot be diagnosed radiologically. Biopsy is often required to establish the diagnosis.

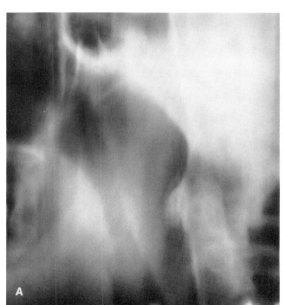
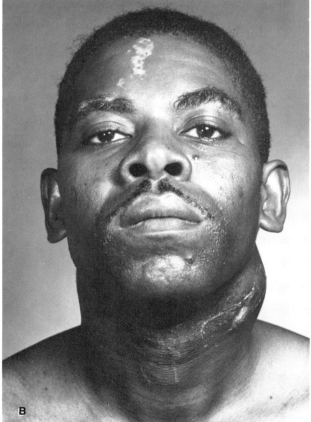

Figure 23.5. Tuberculosis of the nasopharynx: A, Lateral tomogram of the nasopharynx showing diffuse nonspecific thickening of the posterior pharyngeal soft tissue shadows; B, Clinical photograph showing marked swelling of the left neck due to chronic tuberculous lymphadenitis.

REFERENCES

Ballenger, J. J.: *Diseases of the Nose, Throat and Ear.* 11th. ed., p. 231. Lea and Febiger, Philadelphia, 1969.

Chodosh, P. L., and Willis, W.: Tuberculosis of the upper respiratory tract. Laryngoscope, *80:* 679, 1970.

DeLorimier, A. A., Moehring, H. G. and Hannan, J. R.: *Clinical Roentgenology*, Vol. 2, The Head, Neck, and Spinal Column. Charles C Thomas, Springfield, Illinois, 1954.

Guggenheim, P.: Cysts of the nasopharynx. Laryngoscope, *78:* 2147, 1968.

James, A. E., MacMillan, A. S., MacMillan, A. S., Jr., and Momose, K. J.: Thornwaldt's Cyst. Br. J. Radiol., *41:* 902, 1965.

Jonsson, G.: Roentgenographic diagnosis of pathological conditions in the nasopharynx. Acta Radiol., *12:* 651, 1941.

Larsson, L. F.: Nasopharyngeal lesion in sarcoidosis. Acta Radiol., *36:* 361, 1951.

Reynolds, D. F. and Groves, H. J.: A clinical and radiological study of choanal polyps. J. Faculty Radiologists, *7:* 278, 1956.

Chapter 24

Benign Tumors of the Nasopharynx

Benign tumors of the nasopharynx are uncommon. Among them, juvenile angiofibroma is the most frequent. The other lesions are rare. The following is a useful classification:

Epithelial Origin
1. Papillomas
2. Mixed Tumors (Pleomorphic adenoma of accessory gland origin)
3. Adenomas
4. Others

Mesenchymal Origin
1. Myomas
   a. Leiomyoma
   b. Rhabdomyoma
2. Lipomas
3. Fibromas

   a. Myxofibroma
   b. Osteofibroma
   c. Angiofibroma (Juvenile angiofibroma)
4. Angiomas
   a. Lymphangioma
   b. Hemangioma
5. Chondromas
   a. Osteochondroma
   b. Fibrochondroma
6. Neurogenic Tumors
   a. Chordoma
   b. Neurofibroma
   c. Neurilemmoma

Teratoid Origin
1. Teratoma
2. Dermoid

JUVENILE ANGIOFIBROMA

General Considerations

Juvenile angiofibroma is a specific, highly vascular, noninfiltrating, essentially benign but destructive tumor of the nasopharynx. It is usually seen in young males and commonly causes symptoms during the middle teens. Regression may occur spontaneously after adolescence; however, the disease may persist into adulthood. It arises from the basilar fibrocartilage and is situated in the roof of the nasopharynx, usually at a higher level than the malignant tumors.

The current theories of pathogenesis favor a hyperplastic tissue reaction of fibroangiomatous pattern, with hormonal factors playing a dominant role etiologically. Clinically, the incidence is about 0.05% of all neoplasms of the head and neck. Early symptoms such as nasal obstruction and discharge, headache, and deafness are associated with obstruction of the postnasal space. Facial deformity is often present in advanced cases. Severe hemorrhage may result from ulceration and necrosis of the lesion. In the later stage, signs and symptoms are associated with spread of the tumor to surrounding structures. Biopsy may result in catastrophic hemorrhage. It has been stated that this tumor is never malignant; however, there are reports that it may undergo sarcomatous change and that distant metastases to lymph nodes and bone do occur. There is a 35% incidence of local recurrence and an over-all mortality rate of 30%.

Roentgen Findings

Juvenile angiofibroma gives relatively characteristic roentgen manifestations, the features being as follows:

Soft Tissue Mass Arising from Roof of Nasopharynx

The tumor masses usually have well circumscribed and sharply defined margins. The growth may extend anteriorly to the nasal cavity, orbit, and ethmoid and maxillary sinus,

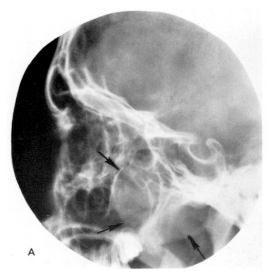
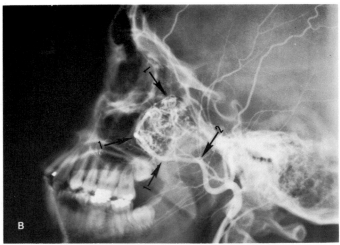

Figure 24.1. Typical nasopharyngeal juvenile angiofibroma. *A*, Left lateral soft tissue roentgenogram of the nasopharynx: a large rounded soft tissue tumor arises from the roof of the nasopharynx and extends anteriorly with anterior bowing of the posterior wall of the left maxillary sinus (*arrows*). *B*, Left carotid angiogram: a tumor on the left side of the nasopharynx is seen to be supplied primarily by the internal maxillary branch of the left external carotid artery. The tumor appears to be very vascular, with a large tumor stain (*1*). The internal maxillary artery (*2*) is slightly dilated, displaced, and bowed (*arrows*). (Reprinted with permission from: G. H. Fletcher and B. S. Jing: The Head and Neck. Chicago: Year Book Medical Publishers, 1968.)

superiorly to the sphenoid sinus, laterally through the pterygopalatine fossa and pterygomaxillary fissure to the infratemporal fossa and the cheek, and inferiorly to the oropharynx (Figs. 17.33, 24.1, and 24.2).

Pressure Erosion of Surrounding Bone

1. Erosion of the floor of the sphenoid sinus may occur with hole formation, and the tumor may extend into the sinus (Fig. 24.2, *A* and *D*).
2. Erosion of the hard palate, clivus, and pterygoid plate may occur.
3. Erosion of the lateral and inferior aspect of the superior orbital fissure, with enlargement of the fissure (Fig. 24.2 *C*), is also seen.

Despite the bone erosion, cranial nerve symptoms seldom occur with this type of tumor.

Bone Displacement

The antral sign may be noted; this is caused by anterior extension of the lesion with forward bowing of the posterior wall of the maxillary sinus (Fig. 24.2*B*). In advanced cases, extension of the lesion into the sinus may occur. Also characteristic is enlargement of the nasal cavity

with thinning of the bony walls and displacement of the septum (Fig. 24.*C*).

Angiographic Appearance

Juvenile angiofibroma has a fairly characteristic appearance.

The blood supply of the juvenile angiofibroma is derived mainly from the internal maxillary and ascending pharyngeal branches of the external carotid artery. The sphenoidal branches of the internal carotid artery frequently make a significant contribution.

The characteristic pattern (Fig. 24.1*B*) of juvenile angiofibroma is as follows:

Early Arterial Phase. There is rapid filling of a rosette of tumor vessels. The tumor vessels are irregular in diameter, but all are of approximately the same size. In cases where the largest contribution is from the internal maxillary artery, it may be dilated, displaced, and bowed laterally.

Capillary Phase. The configuration of the tumor vessels changes to a more homogeneous vascular blush which continues into the venous phase.

Venous Phase. Early filling of the venous system does not occur.

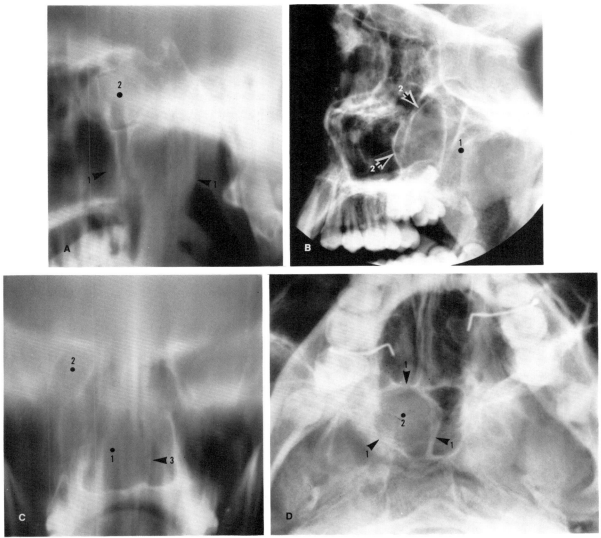

Figure 24.2. Different roentgen manifestations of nasopharyngeal juvenile angiofibroma. *A*, Lateral tomogram: a large, lobulated mass arises from the roof of the nasopharynx (*1*), with extension to the sphenoid sinus (*2*). *B*, Lateral soft tissue roentgenogram: a large tumor mass in the anterior part of the nasopharynx (*1*) with anterior bowing of the posterior wall of the maxillary sinus (*2*). *C*, Frontal tomogram: a tumor mass extends into the nasal cavity (*1*). There is widening of the right superior orbital fissure and erosion of the lateral and inferior aspects of the fissure (*2*). There is also left lateral displacement of the nasal septum (*3*). *D*, Submentovertical view of the skull: sphenoid hole (*1*) with tumor mass extending to the sinus (*2*). (Reprinted with permission from Radiologic Clinics of North America, 8: 340, 1970.)

CHORDOMA

General Considerations

Chordoma of the nasopharynx is a growth of the notochordal remnants and arises in the region of the basisphenoid, the basiocciput, or the upper portion of the cervical spine. It is a rare disease and may occur at any age. Many chordomas which were unequivocally benign when first seen recur and undergo malignant degeneration. Metastases may occur.

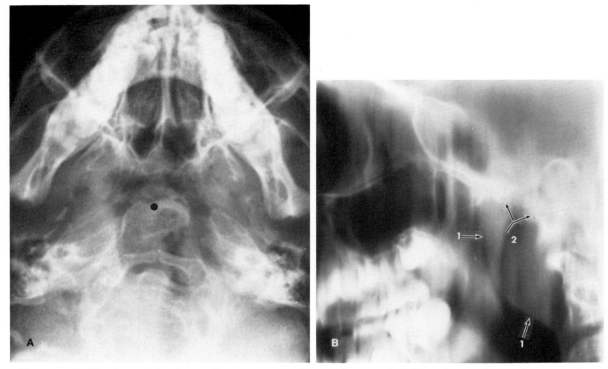

Figure 24.3. Chordoma with a large soft tissue mass in the nasopharynx. *A*, Submentovertical view: destructive lesion of the basisphenoid and basiocciput (*dot*); *B*, Lateral tomogram: a large, lobulated, soft tissue mass arises from the roof and the posterior aspect of the nasopharynx (*1*), and destruction of the basisphenoid and basiocciput is shown to good advantage (*2*).

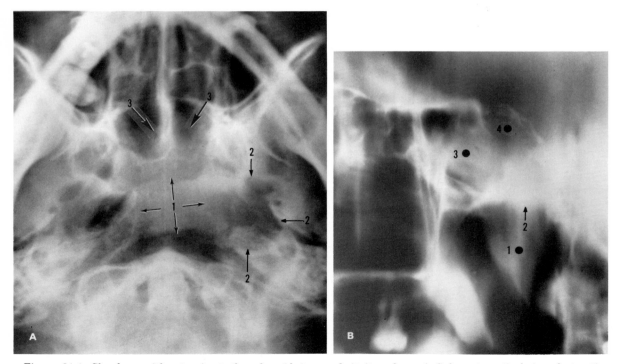

Figure 24.4. Chordoma with extension to the sphenoid sinus and pituitary fossa. *A*, Submentovertical view: destructive lesion of the basisphenoid and basiocciput (*1*) with left lateral extension (*2*). There is destruction of the floor of the sphenoid sinus with mass in the sinus (*3*). *B*, Lateral tomogram: soft tissue mass in the superoposterior aspect of the nasopharynx (*1*). There is destruction of the basisphenoid and the basiocciput (*2*). Extension of the lesion is noted in the sphenoid sinus (*3*) and pituitary fossa, with destruction of the dorsum sellae (*4*).

304

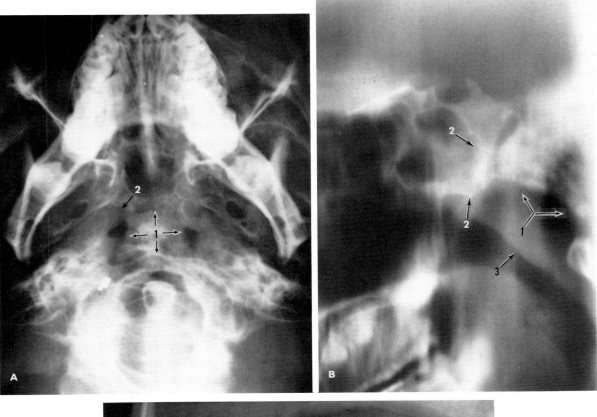

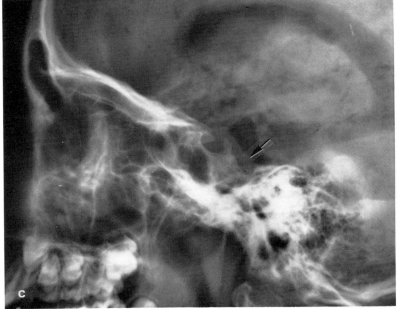

Figure 24.5. Chordoma in basisphenoid region with extension to the basiocciput and base of the brain. *A*, Base view of the skull: destruction of the most part of the basisphenoid and basiocciput (*1*). There is involvement of the anteromedial aspect of the right petrous apex (*2*). *B*, Tomogram of the sella turcica: there is an extensive destruction of the basisphenoid and basiocciput (*1*). There is extension to the sphenoid sinus with partial destruction of its floor and posterior aspect (*2*). There is slight gradual thickening of the soft tissue shadows of the superoposterior aspect of the nasopharynx (*3*). *C*, Pneumoencephalogram: there is a soft tissue density along the posterior aspect of the basisphenoid corresponding to the area of bone destruction (*arrow*). The cisterna pontis shows very little air on the right side, corresponding to the area in which destruction of the right anteromedial portion of the petrous apex is shown on the base view. This is probably due to right-sided extension in the region of petrous pyramid.

Postmortem examination revealed invasive chordoma in the basisphenoidal region with extension into the basiocciput, sphenoid sinuses, and the base of the brain.

Roentgen Findings

The predominant feature of chordoma is bone destruction. In most cases, there is destruction of the basiocciput, the basisphenoid, the sphenoid sinus, and the dorsum sella or the upper cervical spine (Figs. 24.3 to 24.5). The tumor may grow into the nasopharynx with develop- ment of a large soft tissue mass (Fig. 24.3). Calcification in the tumor does occur, but the degree of calcification is generally slight. The roentgen findings of chordoma are nonspecific and closely resemble those of carcinoma. Radiologically the diagnosis can only be suspected on the basis of the location of the tumor.

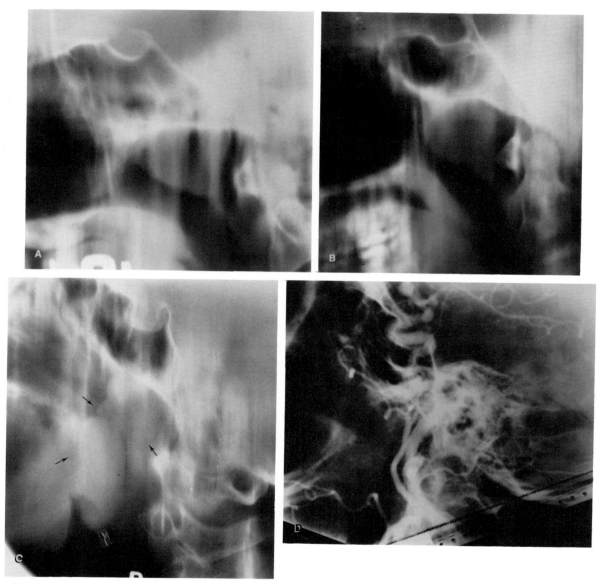

Figure 24.6. Other benign tumors of the nasopharynx. *A*, Chondroma of the nasopharynx. Lateral tomogram of the nasopharynx: The soft tissue shadow of the roof and superoposterior aspect of the nasopharynx is thickened, with a smooth margin. There is no bony involvement. *B*, Mixed tumor of accessory gland origin. Lateral tomogram of the nasopharynx: there is a large soft tissue tumor arising from the superoposterior aspect of the nasopharynx with a smooth rounded margin. Sclerosis of the floor of the sphenoid sinus is noted. The airway of the nasopharynx is narrowed. *C*, Cavernous hemangioma of the nasopharynx and soft palate: *1*, lateral tomogram of the nasopharynx: a large lobulated soft tissue mass involves the posterior aspect and the lateral wall of the nasopharynx and the soft palate (*arrows*); *2*, carotid angiogram: there are no abnormal vessels to suggest the presence of a vascular tumor.

OTHER BENIGN TUMORS

Other benign tumors such as chondroma, neurofibroma, adenoma, angioma, etc., do not produce any characteristic radiographic features and can merely be demonstrated as soft tissue masses in the nasopharynx (Fig. 24.6). Tissue biopsy is the only way to differentiate these lesions.

REFERENCES

Bardwell, J. M., Reynolds, C. T., Ibanez, L. and Luna, M. A.: Report of one hundred tumors of the minor salivary glands. Am. J. Surg., *112:* 493, 1966.

Boies, L. R. and Harris, D.: Nasopharyngeal dermoid of the newborn. Laryngoscope, *75:* 763, 1965.

Bhatia, M. L. and Mishra, S. C.: Intracranial extension of juvenile angiofibroma of the nasopharynx. Laryngoscope, *67:* 1395, 1957.

Bhatia, M. L. and Mishra, S. C. and Prakash, J.: Lateral extension of nasopharyngeal fibroma. J. Laryngol. *81:* 99, 1967.

Cohen, S. R. and Wang, C. I.: Hemangiomas of the head and neck. Ann. Otol. Rhinol. Laryngol. *81:* 584, 1972.

Conley, J., Healey, W. V., Blaugrund, S. M., and Perzin, K. H.; Nasopharyngeal angiofibroma of the juvenile. Surg. Gynecol. Obstet., *126:* 835, 1968.

Dahl, D. J. and Guiss, L. W.: Tumors of Salivary Glands. Am. J. Surg., *118:* 206, 1969.

Dane, W. H.: Juvenile nasopharyngeal fibroma in state of regression. Ann. Otol. Rhinol. Laryngol., *63:* 997, 1954.

Eggston, A. A. and Wolef, D.: *Histopathology of the Ear, Nose and Throat.* Williams & Wilkins, Baltimore, 1947.

Finerman, W. B.: Juvenile nasopharyngeal angiofibroma in female. Arch. Otolaryngol., *54:* 620, 1951.

Foxwell, P. B. and Kelham, B. H.: Teratoid tumors of the nasopharynx. J. Laryngol. Otol., *72:* 647, 1958.

Fitz-Hugh, G. S., Wallenborn, W. M.: Tumors of the nasopharynx: a review of fifty-two cases. Laryngoscope, *71:* 457, 1961.

Halper, H.: Chordomata. Br. J. Radiol., *22:* 88, 1949.

Hanckel, R. W.: Tumors of the nasopharynx: case reports. Laryngoscope, *69:* 415, 1959.

Harma, R.: Nasopharyngeal angiofibroma. Acta Otol. Suppl., *146:* 1, 1958.

Hormia, M., and Koskinen, O. Metastasizing nasopharyngeal angiofibroma. Arch. Otolaryngol., *89:* 107, 1969.

Holman, C. B. and Miller, W. E.: Juvenile nasopharyngeal fibroma, roentgenologic characteristics. Am. J. Roentgenol., *94:* 292, 1965.

House, J. M., Goodman, M. L., Gacek, R. R., and Green, G. L: Chemodectomas of the nasopharynx. Arch. Otolaryngol., *96:* 138, 1972.

Jing, B. S.: Tumors of the nasopharynx. Radiol. Clin. North Am., *8:* 323, 1970.

Massand, G. E. and Awwad, H. K.: Nasopharyngeal fibroma: its malignant potentialities and radiation therapy. Clin. Radiol., *11:* 156, 1960.

McClure, G.: Odontoma of nasopharynx. Arch. Otolaryngol., *44:* 51, 1946.

McGavran, M. H., Sessions, D. G., Derfman, R. F., Davis, D. O. and Ogura, J. H.: Nasopharyngeal angiofibroma. Arch. Otolaryngol., *90:* 90, 1969.

Medellin, H. Wallace, S.: Angiography in neoplasms of the head and neck. Radiol. Clin. North Am., *8:* 307, 1970.

Pendergrass, E. P., Schaeffer, J. P., and Hodge, P. J.: *The Head and Neck in Roentgen Diagnosis*, 2nd. ed., Charles C Thomas, Springfield, Illinois, 1956.

Schindler, M., Hurwitz, S., and Greenhard, H.: Teratoid tumor of the nasopharynx in newborn. Ann. Otol. Rhinol. Laryngol., *63:* 887, 1954.

Ward, G. E., and Hendrick, J. W.: *Diagnosis and Treatment of Tumors of the Head and Neck*. Williams & Wilkins, Baltimore, 1950.

Chapter 25

Malignant Tumors of the Nasopharynx

Malignant tumors of the nasopharynx exhibit a great variety of histological types. In general, squamous cell carcinomas appear to be the most frequent. Other carcinomas and malignant lymphomas are less common. Sarcomas, plasmacytomas, and malignant melanomas are rare. The pathological classification may be simplified as follows:

Carcinomas
1. Squamous Cell Carcinoma
   a. Epidermoid
   b. Lymphoepithelioma
   c. Clear cell variant
   d. Spindle cell variant
   e. Transitional cell carcinoma
2. Adenocarcinoma of the Respiratory Mucosa
3. Carcinoma of Accessory Gland Origin
   a. Adenoid cystic carcinoma
   b. Medullary adenocarcinoma
   c. Mucoepidermoid carcinoma
   d. Malignant mixed tumor
   e. Others
4. Unclassified Carcinoma

Sarcomas
1. Rhabdomyosarcoma
   a. Alveolar type
   b. Embryonal type
2. Fibrosarcoma
3. Chondrosarcoma
4. Unclassified Sarcoma

Malignant Lymphomas
1. Nodular Lymphoma
2. Lymphocytic Lymphoma
3. Histiocytic Lymphoma
4. Burkitt's Lymphoma

Plasmacytomas

Malignant Melanomas, Primary

Others

GENERAL CONSIDERATIONS

The incidence of malignant tumors of the nasopharynx varies both geographically and ethnically. Such tumors have been reported to constitute between 0.25 and 5% of all cancers. At the University of Texas M. D. Anderson Hospital and Tumor Institute, the incidence is about 0.68%. Malignant tumors of the nasopharynx are widespread in Southwest Asia and have a high incidence among the Chinese and Maltese. They are usually found to be 2 to 3 times more common in males; the majority of patients are between the ages of 40 and 70 years. Patients with lymphoepithelioma and malignant lymphomas comprise a younger age group. Rhabdomyosarcoma usually occurs in children and young adults.

The symptomatology of malignant tumors of the nasopharynx varies and often gives rise to different clinical manifestations. In general, the presenting symptoms can be divided into four main groups:

Respiratory Symptoms

These symptoms are related to the primary growth. Nasal obstruction is the common local symptom, especially when a large tumor is present. One or both nares may be obstructed, causing the patient to breathe through his mouth. Epistaxis and nasal discharge may be the first symptom.

Otic Symptoms

These symptoms are also related to the growth at the primary site and are usually attributable to involvement of the nasopharyngeal orifice of the eustachian tube. Tinnitus may occur with an early tumor. Partial filling of the middle ear by a transudate is not infrequent. Impairment of hearing is often observed in large lesions with obstruction of the orifices of the eustachian tubes.

Neurological Symptoms

Extension of the disease to the cranial or orbital cavities may involve the cranial nerves by direct impingement or invasion. Cranial nerve paralysis with eye symptoms often causes the patient to see an ophthalmologist, whereas persistent neuralgic pain affecting the head or face or paralysis of the bulbar cranial nerves may cause the patient to consult a neurologist.

Glandular Symptoms

A swelling in the neck owing to metastasis to the cervical lymph nodes is very common in malignant tumors of the nasopharynx. This symptom may lead the patients to the general surgeon or internist.

ROENTGEN FINDINGS

Site of Origin

Tumors of the nasopharynx arise in the following anatomical sites of the nasopharynx:

Roof and Superoposterior Wall

This is a common site of nasopharyngeal tumors.

Lateral Wall

The most common site of nasopharyngeal tumors is on the lateral wall, in either the fossa of Rosenmüller or the torus tubarius.

Nasopharyngeal Surface of the Soft Palate

The inferior boundary of the nasopharynx is formed by the soft palate. Tumors arising from the nasopharyngeal surface of the soft palate are rare, but they constitute a distinct group and should be separated from those tumors arising from the oropharyngeal aspect of the soft palate.

Tumor of Undetermined Site

When a large exophytic tumor occupies the entire nasopharynx, the actual origin of the lesion often cannot be determined. Occasionally, diffuse infiltrative tumors may involve the entire nasopharynx without evidence of a specific site of origin.

Patterns of Spread

From the site of origin, tumors of the nasopharynx spread by three different routes:

Contiguous Spread

The spread of nasopharyngeal tumors into contiguous structures usually follows anatomical pathways. Therefore, an accurate knowledge of the regional anatomy of the nasopharynx is imperative in order to understand the clinical course of the disease and interpret the roentgen manifestations. Anteriorly, the nasal cavity, including the hard palate, may be invaded. With further extension, the pterygoid plates, the ethmoid cells, and the maxillary sinuses may be involved. Extension through the cribriform plate into the anterior cranial fossa is possible and penetration through the pterygopalatine fossa and the inferior orbital fissure into the orbital cavity may occur. Inferiorly, the tumor may extend to the oropharyngeal wall, the palatine tonsil, and the base of the tongue.

The lateral wall of the nasopharynx is intimately adjacent to the lateral pharyngeal space, which is bounded medially by the buccopharyngeal and prevertebral fasciae and laterally by the mandible, pterygoid muscles, parotid gland, and investing layer of the deep cervical fascia. The space extends from the base of the skull to the level of the hyoid bone at the attachment of the stylohyoid and posterior belly of the digastric muscles. The lateral pharyngeal space is divided into prestyloid and poststyloid compartments by the styloid process and its muscles. The prestyloid compartment contains the internal maxillary artery and the inferior alveolar, lingual, and auriculotemporal branches of the mandibular division of the trigeminal nerve. The superior part of this compartment is directly related to the lateral wall of the nasopharynx and to the fossa of Rosenmüller. The poststyloid compartment contains the internal carotid artery, the internal jugular vein, the 9th, 10th, 11th, and 12th cranial nerves, the cervical sympathetic nerves, and lymph nodes (Fig. 25.1). When a tumor of the nasopharynx extends to the prestyloid compartment, the mandibular division of the trigeminal nerve and the pterygoid and palatine muscles may be involved. Further lateral extension may affect the parotid gland and the masticatory muscles. Involvement of the facial nerve is rare. When a

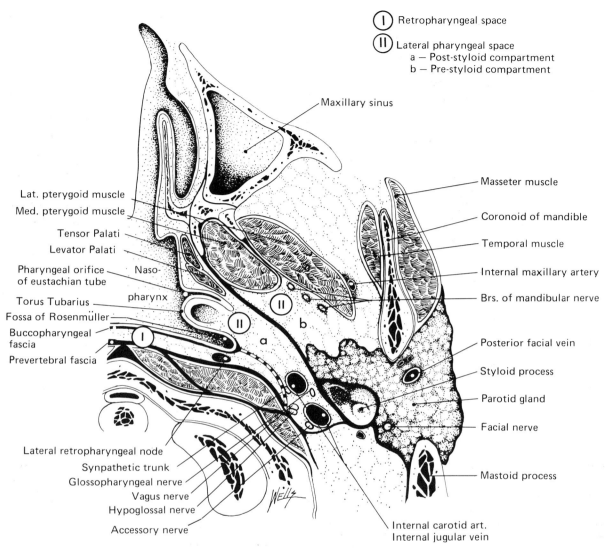

Figure 25.1. Cross-section of the neck through the nasal cavities (viewed from below). *I*, Retropharyngeal space with lateral retropharyngeal node; *II*, lateral pharyngeal space: *a*, poststyloid compartment with its contents; *b*, prestyloid compartment with its contents.

tumor extends to the poststyloid compartment, the last four cranial nerves and the cervical sympathetic nerve may be affected. Lateral extension through the eustachian tube to the middle and external auditory canals may occur.

Superiorly, the extension may involve structures in the base of the skull. Upward spread to the middle cranial cavity is caused by infiltration of the bone at the base of the skull or extension through the basal foramina. The sphenoid sinus is a very vulnerable structure in

relation to the roof of the nasopharynx and is often invaded. Parasellar structures may be invaded, with involvement of the optic nerve. Upward extension from lateral wall lesions may pass through the foramen lacerum. The foramen lacerum is directly related to the lateral aspect of the roof of the nasopharynx and to the fossa of Rosenmüller, and is in direct communication with the cavernous sinus in the middle cranial fossa. Within the cavernous sinus the prominent structures are the internal carotid

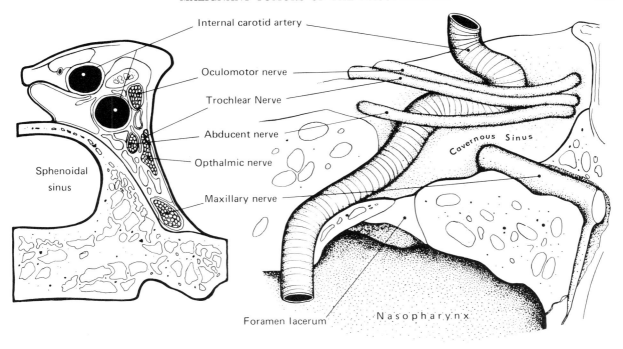

I Coronal Section II Sagittal Section

Figure 25.2. Cavernous sinus with its contents: *I*, coronal section; *II*, sagittal section.

artery and the surrounding sympathetic nerve plexus, the third, fourth, and sixth cranial nerves, and the ophthalmic and maxillary divisions of the trigeminal nerve (Fig. 25.2). Upward spread through the foramen lacerum may cause invasion of the cavernous sinus. The sixth, fourth, and third cranial nerves are, in that order, the most commonly involved, and there may be trigeminal sensory disturbances.

Posteriorly, there is a potential retropharyngeal space between the buccopharyngeal fascia and the prevertebral layer of the deep cervical fascia extending from the base of the skull to the mediastinum. It separates the nasopharynx from the prevertebral muscles and contains lymph nodes, notably the lateral retropharyngeal node which is located anteriorly to the lateral mass of the atlas (Fig. 25.1). Posterior extension to the retropharyngeal space produces few symptoms, but the lateral retropharyngeal node may be affected and secondary involvement and compression of the carotid sheath and its contents may follow. Destruction of the lateral mass of the atlas may occur. Further posterior extension is less readily achieved because of

the dense fascial planes attached to the base of the skull; however, it does occur occasionally. There may be infiltration of the prevertebral fascia and muscles and extension into the cervical spine.

Lymphatic Spread

Malignant tumors of the nasopharynx frequently spread to the cervical lymph nodes; an incidence of 70 to 90% has been reported. The involvement is usually diffuse and often bilateral. The upper internal jugular chain is most commonly involved. There is also a high percentage of involvement of the spinal accessory chain and subjugular, lower jugular, and supraclavicular nodes.

Hematogenous Spread

Blood-borne metastases are frequent and may be found in many tissues and organs, but chiefly in bones, liver, and lungs. The bones commonly involved are those of the spine, pelvis, and skull.

DIAGNOSTIC CRITERIA

Soft Tissue Mass in Nasopharynx

These may be exophytic or ulcerative. In the exophytic type, when the tumor is located in the roof or the posterior wall, the soft tissue mass projects into the air space of the nasopharynx on the lateral soft tissue roentgenogram and lateral and anteroposterior tomograms (Figs. 25.3 and 25.16). Malignant tumors arising from the roof or the posterior wall are nearly always situated eccentrically (Figs. 25.4 and 25.16). If the tumor is asymmetrically situated and extends to the fossa of Rosenmüller, the space of the affected fossa will be obliterated. In lesions of the lateral wall, the profile of the soft tissue mass may or may not be seen on the lateral soft tissue roentgenogram. Lateral and/or anteroposterior tomograms and submentovertical views are often helpful in determining the site and extent of the lesion. Tumors arising from the fossa of Rosenmüller usually project medially into the air space posterior to the torus

tubarius on the submentovertical view (Fig. 25.5). The nasopharyngogram is useful in revealing the location and extent of such lessions (Fig. 25.6). If the tumor is from the torus tubarius, the soft tissue mass will project medially into the air space and lie more anteriorly on the submentovertical view (Fig. 25.7). When the tumor arises from the lateral wall with involvement of the eustachian tube, nasopharyngography is of definite value in demonstrating the patency of the eustachian tube (Fig. 25.8). A huge lateral wall tumor may occupy the entire nasopharyngeal air space and its actual site of origin often cannot be determined on the lateral soft tissue roentgenogram. Under such conditions, the origin of the tumor may be detected by an indirect sign: the soft palate is often displaced anteriorly and the normally convex posterior border becomes concave owing to pressure from the tumor (Fig. 25.9). Lateral and/or anteroposterior tomograms and submentovertical

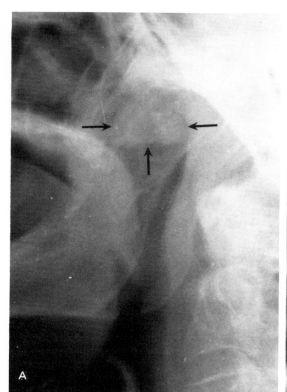
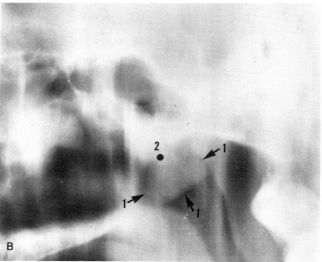

Figure 25.3. Extramedullary plasmacytoma of the nasopharynx. *A*, Lateral soft tissue roentgenogram: there is a well defined polypoid soft tissue mass arising from the roof of the nasopharynx (*arrows*); *B*, Lateral tomogram: the polypoid soft tissue mass is clearly defined (*1*), with a short stalk (*2*).

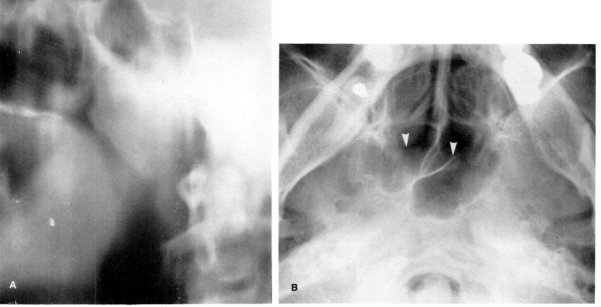

Figure 25.4. Unclassified malignant tumor of the nasopharynx. *A*, Lateral tomogram: there is a soft tissue mass arising from the roof and superoposterior aspect of the nasopharynx; *B*, Base view: the tumor mass lies slightly to the right, but obscures the majority of the nasopharyngeal air space (*arrows*). (Reprinted with permission from Radiologic Clinics of North America, 8: 329, 1970.)

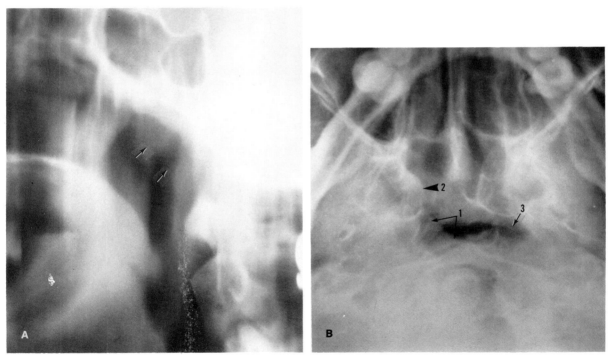

Figure 25.5. Squamous cell carcinoma of the fossa of Rosenmüller. *A*, Lateral tomogram: slight soft tissue thickening of the superoposterior and lateral aspects of the nasopharynx (*arrows*); *B*, Submentovertical view: tumor mass in the region of right fossa of Rosenmüller (*1*) with slight anterior displacement of the torus tubarius (*2*). The left fossa of Rosenmüller is normal in appearance (*3*). (Reprinted with permission from Radiologic Clinics of North America, 8: 330, 1970.)

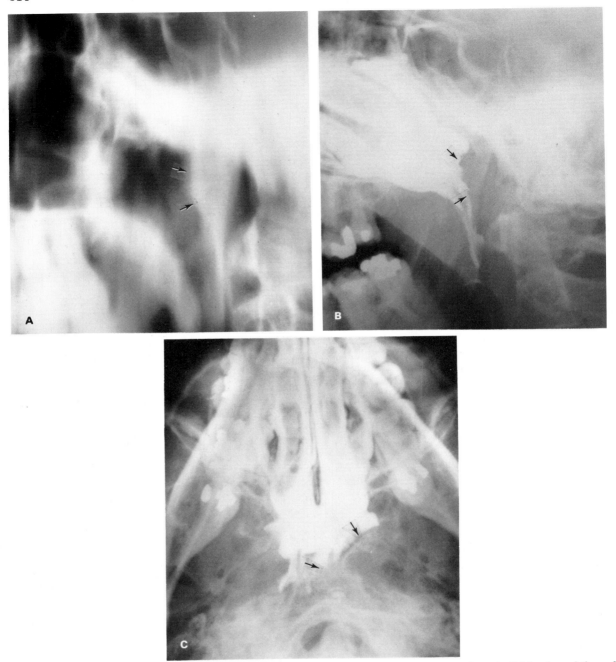

Figure 25.6. Lymphoepithelioma of the fossa of Rosenmüller. *A*, Lateral tomogram: there is thickening of the soft tissue shadows in superoposterior aspect of the nasopharynx (*arrows*); *B*, Nasopharyngogram, lateral view: a filling defect is demonstrable on the superoposterior aspect of the contrast medium (*arrows*); *C*, Nasopharyngogram, submentovertical view: the contrast medium is displaced by a large tumor mass in the left side of nasopharynx. The contours of the torus tubarius and fossa of Rosenmullër are obliterated (*arrows*).

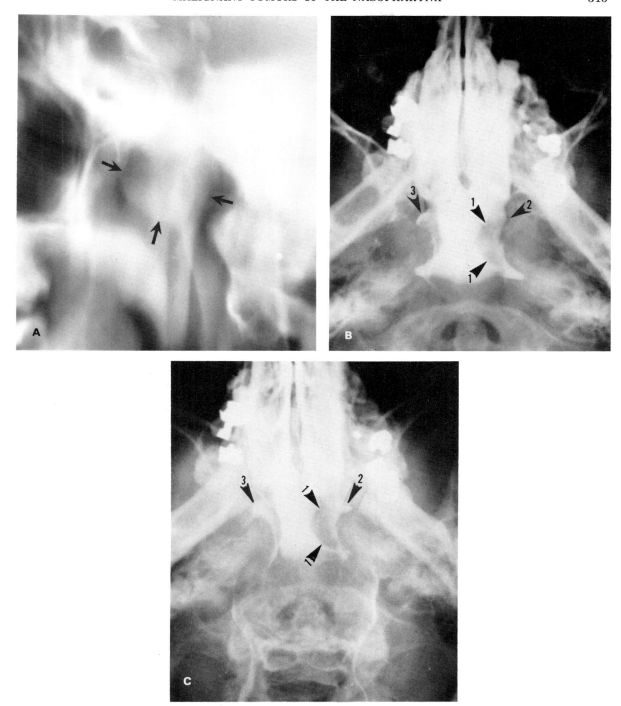

Figure 25.7. Nodular lymphoma of the torus tubarius. *A*, Lateral tomogram: there is a rounded tumor mass in the region of the left torus tubarius (*arrows*). *B*, Submentovertical view of nasopharyngogram during inspiration: there is a rounded filling defect in the region of left torus tubarius (*1*) with obliteration of the orifice of the eustachian tube (*2*). The orifice of the right eustachian tube is normal (*3*). *C*, Submentovertical view of nasopharyngogram during modified Valsalva maneuver: the tumor mass is clearly shown in the region of left torus tubarius (*1*). The left eustachian tube is patent (*2*), and the right eustachian tube remains normal (*3*). (Reprinted with permission from Archives of Otolaryngology, 81: 369, 1965.)

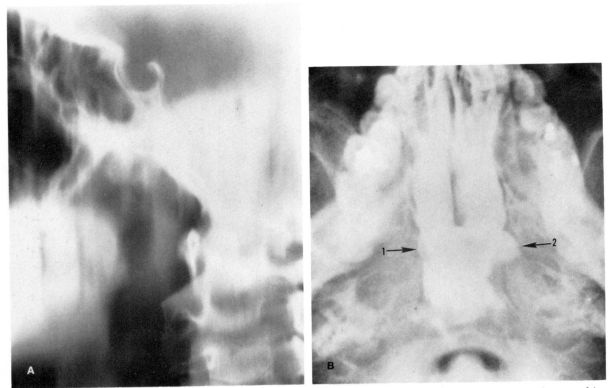

Figure 25.8. Lymphoepithelioma of the right lateral wall of the nasopharynx with involvement of the right eustachian tube. A, Lateral tomogram: there is loss of the soft tissue landmarks on the lateral wall of the nasopharynx. The thickening of the roof of the nasopharynx anterior is probably due to adenoids. B, Submentovertical view of nasopharyngogram: there is an infiltrative tumor mass involving right lateral wall with occlusion of the orifice of the right eustachian tube (1). The left eustachian tube is normal in appearance (2).

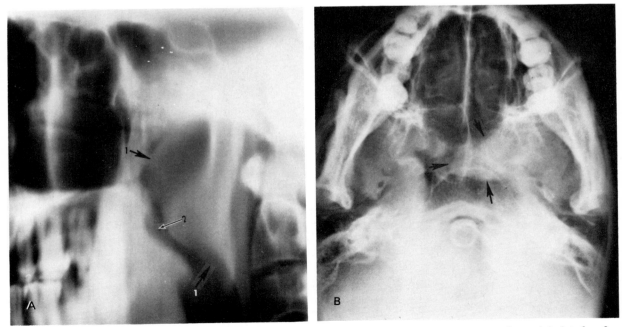

Figure 25.9. Lymphoepithelioma of the lateral wall of the nasopharynx: A, Lateral tomogram: a huge, lobulated, soft tissue mass is seen in the nasopharynx (1) with positive palatal sign (2); B, Submentovertical view: the mass lesion arises from the left lateral wall of the nasopharynx with a polypoid component projecting into the air space (arrows).

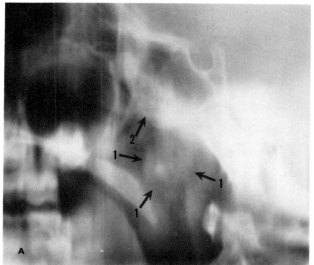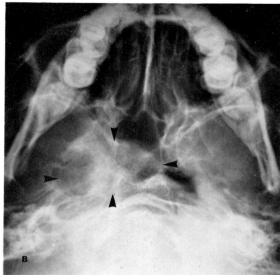

Figure 25.10. Chondrosarcoma of the nasopharynx. *A*, Lateral tomogram: a well defined soft tissue mass with calcification is seen in the superoposterior aspect of the nasopharynx (*1*). There is involvement of the sphenoid sinus (*2*). *B*, Submentovertical view: the calcifying mass lies mainly on the right side of the nasopharynx, with involvement of the posterior and right side of the sphenoid sinus and destruction of the right side of base of the skull (*arrows*).

views usually reveal the site of the tumor (Fig. 25.9; see also Figs. 25.18 and 25.28).

In the ulcerative type of tumors, superficial erosions are usually not demonstrable roentgenographically, but it is often possible to appreciate the deeper craters if they are large enough to alter the contour of the tumor mass. When situated in the roof or posterior wall, the ulcerative tumor may be clearly demonstrated as a saddle-shaped mass on the lateral soft tissue roentgenogram (Figure 25.14). An ulcerative tumor of the torus tubarius may be demonstrated by loss of definition of its shadow and contour on both the lateral soft tissue roentgenograms and the submentovertical view.

Chondrosarcomas of the nasopharynx are extremely rare. When present, they may show lobulated soft tissue masses with calcification. In the advanced case, the sphenoid sinus, clivus, and the floor of the middle cranial fossa may be invaded (Fig. 25.10).

Diffuse Thickening of Posterior Pharyngeal Soft Tissue Shadow

Diffuse thickening of the posterior pharyngeal soft tissue shadow is usually indicative of an infiltrative type of tumor. A tumor arising from the posterior wall is often evident on the lateral soft tissue roentgenogram and lateral tomogram (Fig. 25.11). If the lesion is near the

fossa of Rosenmüller, the air space of the affected fossa will be obliterated on the submentovertical view. A nasopharyngogram is often useful in demonstrating the character of the tumor (Fig. 25.12).

Thickening of Soft Palate

The size of the soft palate has a wide range in the normal subject. On an average, the soft palate varies from 0.5 to 1.0 cm in thickness and from 2.5 to 4.0 cm in length as measured from the posterior margin of the hard palate to the free hanging margin (uvula).

Tumors arising from the nasopharyngeal surface of the soft palate can be exophytic or infiltrative. In the infiltrative type there is thickening of the soft palate with an irregular contour of the nasopharyngeal surface. An exophytic lesion usually appears as a tumor mass arising from the nasopharyngeal surface and protruding into the air space of the nasopharynx (Fig. 25.13).

Sclerosis of Base of Skull

Sclerosis of the base of the skull may result from an osteoblastic response to tumor invasion and/or from secondary inflammatory changes; its significance is not fully understood. Clinically, bony sclerosis does not necessarily mean

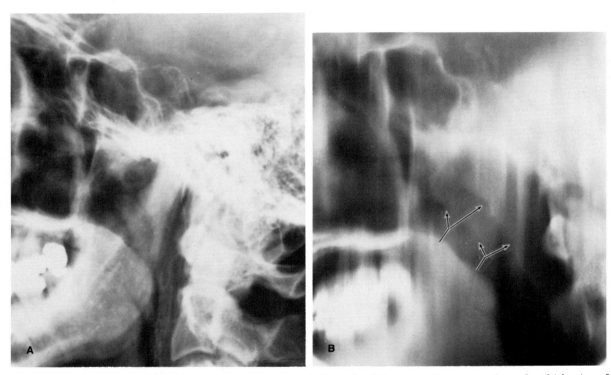

Figure 25.11. Malignant melanoma of the nasopharynx: *A*, Lateral soft tissue roentgenogram: irregular thickening of the soft tissue shadow of the superoposterior aspect of the nasopharynx; *B*, Lateral tomogram: the thickening of the soft tissues is more clearly demonstrated (*arrows*).

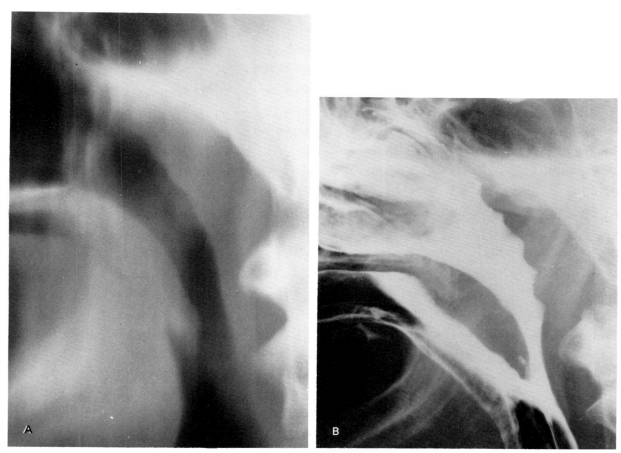

Figure 25.12. Squamous cell carcinoma of the nasopharynx: *A*, Lateral tomogram: there is thickening of the posterior pharyngeal soft tissue shadow with slight irregularity of mucosal surface; *B*, Lateral view of nasopharyngogram: the infiltrating characteristics of the tumor are revealed by the contrast medium. (Reprinted with permission from Radiologic Clinics of North America, 8: 331, 1970.)

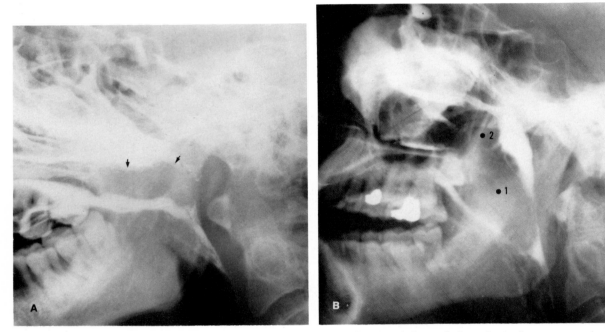

Figure 25.13. Squamous cell carcinoma of the soft palate — two examples. *A*, Lateral view of nasopharyngogram (case 1): considerable thickening of the soft palate is shown, with irregularity of the nasopharyngeal surface (*arrows*). *B*, Lateral view of nasopharyngogram (case 2): thickening of the soft palate (*1*), with a lobulated tumor mass arising from the nasopharyngeal surface and protruding into the anterior part of the nasopharynx and nasal cavity (*2*). (Reprinted with permission from Archives of Otolaryngology, 81: 370, 1965.)

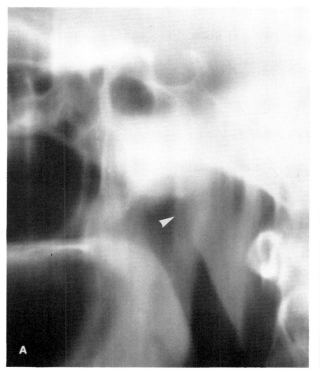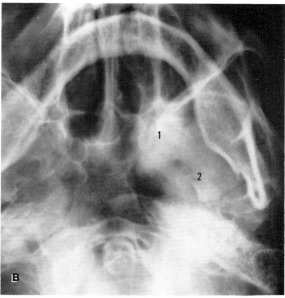

Figure 25.14. Squamous cell carcinoma of the nasopharynx with sclerosis of the pterygoid plate and the floor of middle cranial fossa. *A*, Lateral tomogram: tumor mass with central ulceration (*arrow*) in the roof and superoposterior aspect of the nasopharynx; *B*, Submentovertical view: sclerosis of left pterygoid plate (*1*) and floor of middle cranial fossa (*2*). (Reprinted with permission from: G. H. Fletcher and B. S. Jing: *The Head and Neck, An Atlas of Tumor Radiology*, Ed. by P. Itodes). Chicago: Year Book Medical Publishers, 1968.)

tumor invasion, but is strongly suggestive of it (Figs. 25.14 and 25.15). Even with no demonstrable tumor mass in the nasopharynx, sclerosis of the floor of the sphenoid sinus, clivus, or pterygoid plate, together with clinically positive cervical nodes, cranial nerve involvement, or otic symptoms, is a definite indication for tissue biopsy of the nasopharynx. At The University of Texas M. D. Anderson Hospital and Tumor Institute the majority of cases with sclerosis of the base of the skull have had further invasion of the cranial cavity or local recurrence.

Destruction of Base of Skull

Tumor invasion with destruction of the base of the skull may take place in several regions depending upon the location of the tumor:

Tumors Arising from Roof and Superoposterior Wall of Nasopharynx

These may invade the floor of the sphenoid sinus, the clivus, the pterygoid plate, or the foramen lacerum by superior extension (Fig. 25.16). Posterior extension may result in invasion of the retropharyngeal space and destruction of the lateral mass of the atlas. Lateral extension may cause destruction of the petrosphenoidal region.

Tumors Arising from Lateral Wall

These may involve the following structures:

The Foramen Lacerum. The roof of the fossa of Rosenmüller is directly related both to the foramen lacerum and to the floor of the carotid canal. The osseous floor of the carotid canal is incomplete and the foramen lacerum is partially filled by fibrocartilage which is resistant to tumor invasion. It is, therefore, postulated that the usual mode of invasion of the foramen lacerum is through the carotid canal along the course of the internal carotid artery rather than by direct extension. Radiologically, in the early stage of invasion, the part involved is usually the posterolateral border of the foramen. Eventually the whole petrous apex is destroyed (Fig. 25.17).

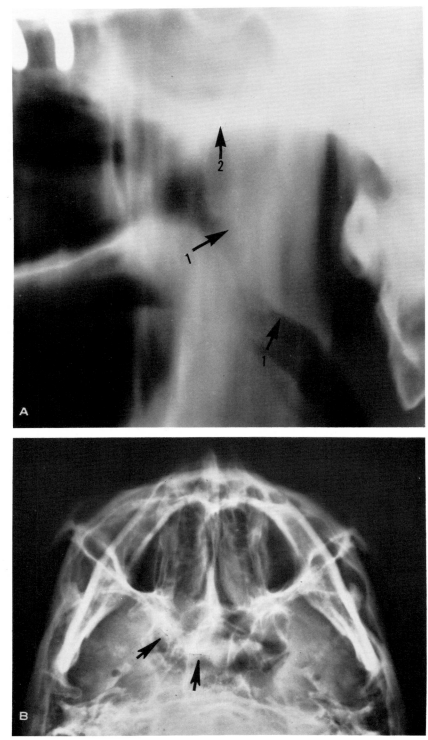

Figure 25.15. Squamous cell carcinoma of the nasopharynx with sclerosis of the floor of the sphenoid. *A*, Lateral tomogram: there is a large soft tissue mass in the nasopharynx (*1*), with sclerosis of the floor of the sphenoid sinus (*2*); *B*, Submentovertical view: sclerosis of the floor and the right lateral wall of the sphenoid sinus (*arrows*).

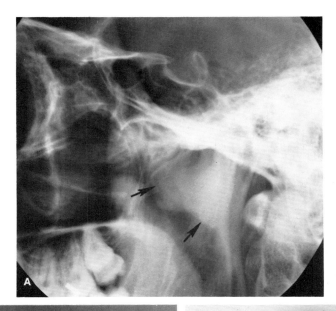

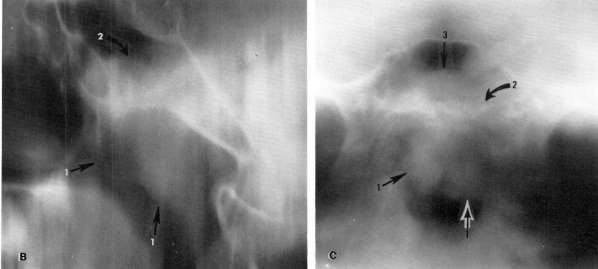

Figure 25.16. Lymphoepithelioma of the nasopharynx with extension to the sphenoid sinus. *A*, Lateral soft tissue roentgenogram: there is a large tumor mass with irregular margins in the roof and superoposterior aspect of the nasopharynx (*arrows*). *B*, Lateral tomogram: the tumor mass is well demonstrated in the roof and superoposterior aspect of the nasopharynx (*1*). There is destruction of the floor of the sphenoid sinus and extension of the lesion into the sinus (*2*). *C*, Anteroposterior tomogram: a large lobulated tumor mass is demonstrated in the roof of the nasopharynx mainly on the left side (*1*). There is destruction of the left side of the floor of the sphenoid sinus (*2*). Extension of the lesion into sphenoid sinus is well shown (*3*).

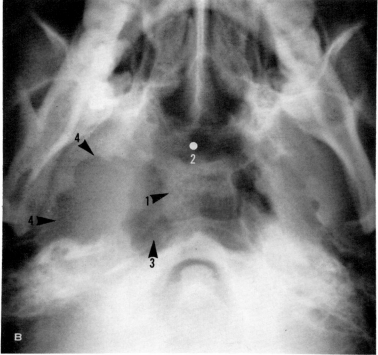

Figure 25.17. Lymphoepithelioma of the nasopharynx with destruction of the foramen lacerum. *A*, Lateral tomogram: tumor involving superoposterior aspect of the nasopharynx (*1*), and destruction of the floor of the sphenoid sinus (*2*); *B*, Submentovertical view: destruction and sclerosis of the right lateral aspect of the clivus (*1*) and of the floor of the sphenoid sinus (*2*), and destruction of the right petrous apex including foramen lacerum (*3*) and of the medial aspect of the greater wing of the sphenoid bone (*4*). (Reprinted with permission from Radiologic Clinics of North America, 8: 333, 1970.)

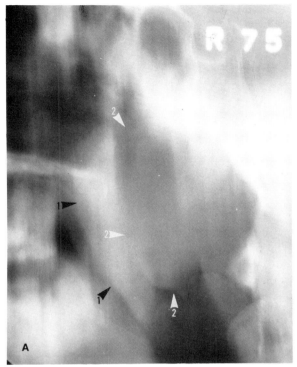
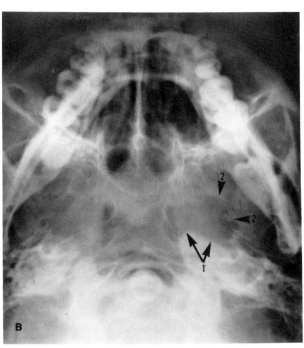

Figure 25.18. Lymphoepithelioma of the lateral wall of the nasopharynx with destruction of the carotid canal. *A*, Lateral tomogram: there is a huge soft tissue mass on the lateral wall of the nasopharynx (*2*) with positive palate sign (*1*); *B*, Submentovertical view: bony destruction of the lateral aspect of the left petrous apex including the carotid canal (*1*) and medial aspect of the left greater wing of the sphenoid bone (*2*). (Reprinted with permission from Radiologic Clinics of North America, 8: 333, 1970.)

The Carotid Canal. When the carotid canal is involved the tumor may not only spread to the cavernous sinus, but may leap the eustachian tube and propagate extradurally toward the greater wing of the sphenoid bone. Involvement of the area adjacent to the foramen ovale may result in trigeminal sensory disturbances, but the third, fourth, and sixth cranial nerves need not be involved. Radiologically, there is destruction of the petrous apex of the temporal bone and of the greater wing of the sphenoid bone near the foramen ovale (Fig. 25.18).

The Foramen Ovale. The lateral pharyngeal space is intimately related to the lateral wall of the nasopharynx and the fossa of Rosenmüller and is limited cranially by the posterior part of the greater wing of the sphenoid containing the foramen ovale. Normally the size and shape of the foramen ovale varies among individuals and may show considerable difference from right to left in the same individual. At times, the bone around the margin of the foramen becomes so thin that it does not cast a shadow on the submentovertical view. Under these circumstances, stereoscopic submentovertical views may demonstrate that the bone around the foramen is thin but intact. Base tomograms of the skull are of value in delineating the foramen and the bony margin. The roentgen features of neoplastic invasion of the foramen ovale are: (1) enlargement with a rounded contour of the foramen; (2) erosion with slight irregularity of the bony margin of the foramen; (3) bony sclerosis around the foramen; and (4) bone destruction in the vicinity of the foramen.

Secondary involvement of the uppermost part of the prestyloid compartment of the lateral pharyngeal space by tumors of the nasopharynx may result in upward spread with invasion of the greater wing of the sphenoid bone around the foramen ovale without obvious extension into the petrous apex of the temporal bone. Sensory disturbances of the mandibular division of the trigeminal nerve may be manifest, but there usually is no involvement of the third, fourth, and sixth cranial nerves. Radiologically, destruction of the greater wing of the sphenoid bone around the foramen ovale is seen, usually without involvement of the petrous apex of the temporal bone (Fig. 25.19).

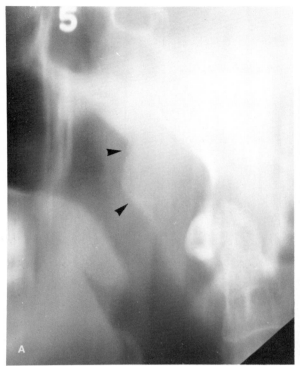

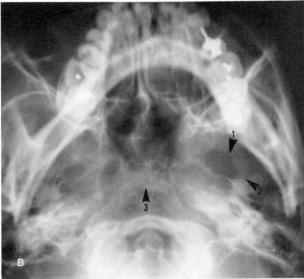

Figure 25.19. Poorly differentiated squamous cell carcinoma of the nasopharynx with destruction of the foramen ovale. *A*, Lateral tomogram: tumor mass with an irregular contour in superoposterior aspect of the nasopharynx (*arrows*). *B*, Submentovertical view: enlargement with irregular margins of the left foramen ovale (*1*). The foramen spinosum is also involved (*2*). There is sclerosis of the clivus (*3*). (Reprinted with permission from Radiologic Clinics of North America, 8: 334, 1970.)

Involvement of Surrounding Structures. In the advanced stage of nasopharyngeal malignancies, involvment of the surrounding structures is often present when the patient is first seen. Anteriorly, the tumors may spread to the nasal cavity, with destruction of the nasal septum and hard palate. Further extension may involve the pterygoid plates, the maxillary and ethmoid sinuses, and occasionally the orbital cavity (Fig. 25.20). Inferiorly, through the pharyngeal isthmus, the tumor can involve the oropharyngeal wall, the tonsilar area, the base of the tongue, and even the hypopharynx. Superiorly the lesion can extend to the sphenoid sinus, the pituitary fossa, and the parasellar structures (Fig. 25.21). Laterally, involvement of the lateral pharyngeal space may occur, with varying clinical manifestations. Through the eustachian tube, lateral wall lesions may spread to the middle and external auditory canals (Fig. 25.22). Posteriorly, destruction of the lateral mass of the atlas may occur and, occasionally, invasion of the prevertebral muscle and the cervical vertebral bodies (Fig. 25.23).

Opacity of Paranasal Sinuses. The changes in the paranasal sinuses may be caused by actual tumor invasion or mechanical obstruction with associated infection secondary to extension of the tumor. In the sphenoid sinus, cloudiness is usually caused by tumor invasion.

Involvement of Mastoid. Clouding of the mastoid cells and destruction of the cell walls are occasionally observed in tumors of the nasopharynx. This change is probably caused by obstruction of the eustachian tube by tumor mass. Its appearance simulates acute or subacute mastoiditis. Rarely, there is a posterior extension with destruction of the temporal bone, including the mastoid process. In our series, actual destruction of the mastoid by tumor invasion has been noted in two cases (Fig. 25.21).

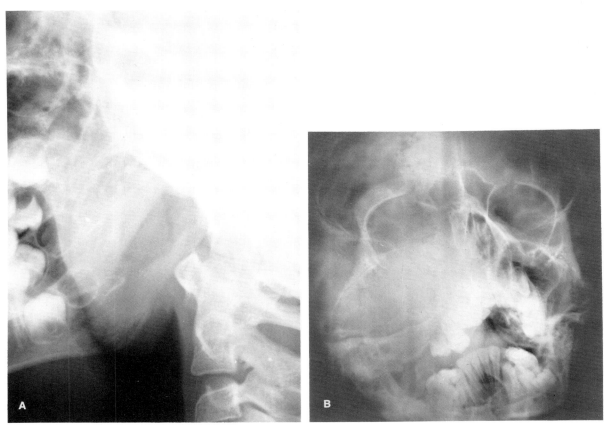

Figure 25.20. Embryonal rhabdomyosarcoma of the nasopharynx with anterior extension. *A*, Lateral soft tissue film: a huge soft tissue mass occupies the entire nasopharynx with extension to the posterior part of the nasal cavity and the oropharynx; *B*, Waters view (4 months later): Extensive anterior extension with destruction of the right orbit, ethmoid and maxillary sinuses, and facial bones. (Reprinted with permission from Radiologic Clinics of North America, 8: 335, 1970.)

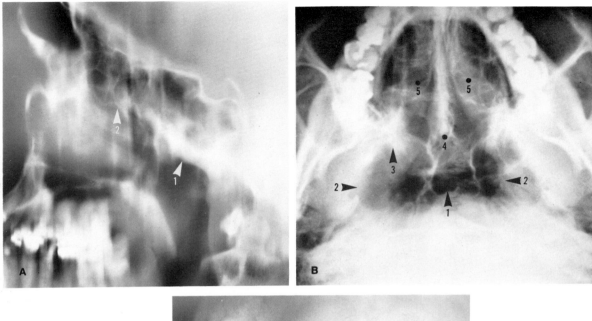

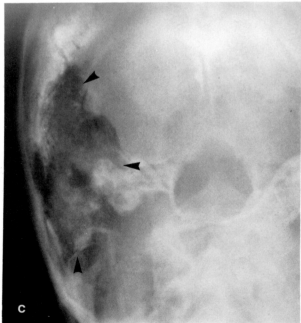

Figure 25.21. Squamous cell carcinoma of the nasopharynx with superior and posterior extension. *A*, Lateral tomogram: an infiltrative tumor is present in the roof and superoposterior aspect of the nasopharynx, with destruction of the floor of sphenoid sinus (*1*) and of the posterior ethmoid cells (*2*). *B*, Submentovertical view: there is destruction of the clivus (*1*), the medial aspect of greater wing of the sphenoid bone on each side (*2*), sphenoid sinus (*4*), and ethmoid sinuses (*5*). Destruction and sclerosis of the right pterygoid plate is also present (*3*). *C*, Towne view (24 months later): destruction of the right temporal bone and mastoid process had occurred (*arrows*). (Reprinted with permission from Radiologic Clinics of North America, 8: 336, 1970.)

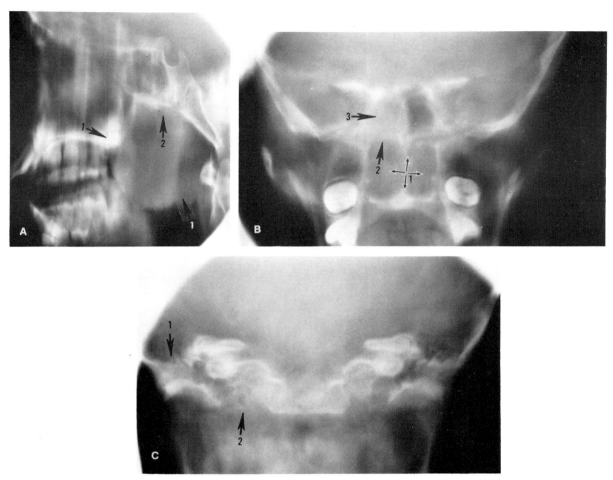

Figure 25.22. Embryonal rhabdomyosarcoma of the nasopharynx with extension to the external auditory canal. *A*, Lateral tomogram: a large rounded soft tissue mass occupies the entire nasopharynx (*1*), with destruction of the floor of the sphenoid sinus and basisphenoid (*2*). *B*, Posterior cut of paranasal sinus tumor survey: tumor mass is noted in the nasopharynx (*1*). There is destruction of the floor of the right sphenoid sinus (*2*) and cloudiness of the right sphenoid sinus (*3*). *C*, Anteroposterior tomogram of the auditory canals: there is extension of the lesion into the right middle and external auditory canals. Bony destruction is noted in the superior aspect of the external canal (*1*). The right jugular fossa (*2*) appears slightly widened, with destruction of the right side of the clivus.

A girl of age 8 complained of nasal obstruction and discharge. There was a whitish soft tissue mass in the right external auditory canal and a large mass lesion in the nasopharynx. Tissue biopsy revealed embryonal carcinoma. (Reprinted with permission from: G. H. Fletcher and B. S. Jing: *The Head and Neck, An Atlas of Tumor Radiology*, Ed. by P. Hodes. Year Book Medical Publisher, Chicago, 1968.

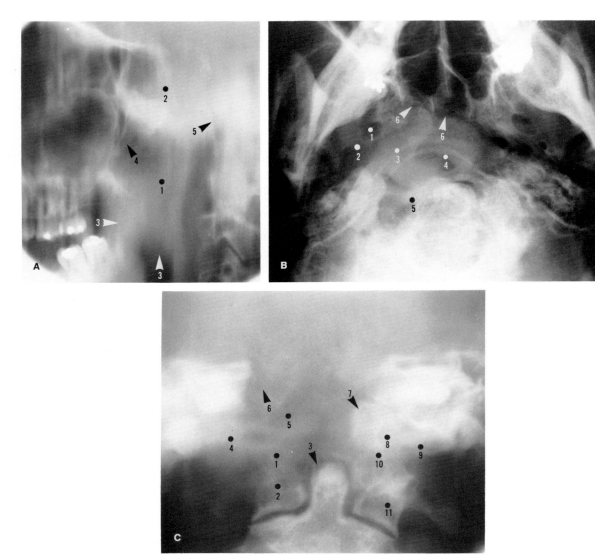

Figure 25.23. Advanced squamous cell carcinoma of the nasopharynx with generalized extension. *A*, Lateral tomogram of the nasopharynx: a large tumor mass occupies the entire nasopharynx (*1*) and extends to the sphenoid sinus (*2*) and oropharynx (*3*). There is destruction of the pterygoid plate (*4*) and the clivus (*5*). *B*, Submentovertical view: a soft tissue mass is apparent in the nasopharynx, mainly on the right side. There is destruction of the middle cranial fossa on the right with involvement of the foramen ovale (*1*), foramen spinosum (*2*), and foramen lacerum (*3*). There is also destruction of the clivus (*4*), the right occipital condyle (*5*), and posterior aspect and right side of the sphenoid sinus (*6*). *C*, Anteroposterior tomogram (posterior cut): destruction of the occipital condyle (*1*), the atlas (*2*), odontoid process (*3*), the jugular fossa (*4*), the tuberculum jugulare (*5*), and the petrous apex (*6*) on the right. On the left, the tuberculum jugulare (*7*), the hypoglossal canal (*8*), the jugular fossa (*9*), the occipital condyle (*10*), and the atlas (*11*) are clearly shown.

DIFFERENTIAL DIAGNOSIS

Pharyngeal Adenoids

This is discussed in Chapter 23.

Granulomatous Disease

This is discussed in Chapter 23.

Juvenile Angiofibroma

This is discussed in Chapter 24.

Chordoma

This is discussed in Chapter 24.

Nasal polyp

A nasal polyp of considerable size may protrude through the choana into the nasopharynx. Roentgenographically, it usually appears as a pedunculated mass lying on the nasopharyngeal surface of the soft palate with a smooth, convex posterior surface. Between the roof of the nasopharynx and the nasopharyngeal aspect of the polyp there is often a radiolucent, curvilinear zone indicating that the polyp does not arise from the roof of the nasopharynx (Fig. 4.5).

Tumors of Sphenoid Sinus

Malignant tumors of the sphenoid sinus are often secondary from extension of tumors of the ethmoid sinus, nasal cavity, or nasopharynx. Primary malignant tumors of the sphenoid sinus are rare and are usually rather advanced when the patient is first seen. The pituitary-palatal line is of considerable value in differentiating sphenoid or sphenoethmoid tumors from nasopharyngeal neoplasms. Using a lateral roentgenogram of the skull, a line is drawn through the center of the pituitary fossa to join the posterior end of the hard palate at its junction with the soft palate. Tumors lying anterior to this line are regarded as of sphenoid or sphenoethmoid origin, whereas tumors below and behind the line are considered to be of nasopharyngeal origin (Fig. 25.24).

Tumors of Parotid Gland

Tumors of the parotid gland, particularly those occurring in the deep lobe, can invade the lateral pharyngeal space and protrude into the nasopharynx. Roentgen examination may reveal a large, well defined soft tissue mass with a sharp convex contour and tapered borders on

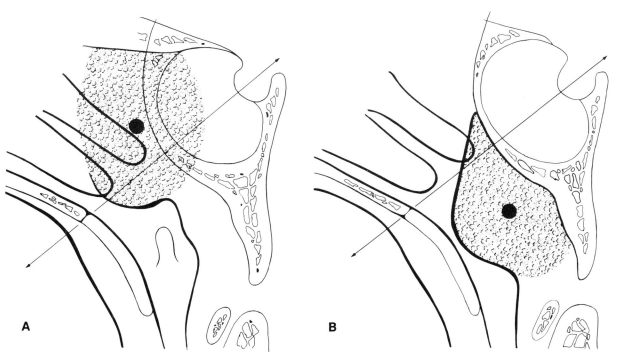

Figure 25.24. The pituitary-palatal line: *A*, Tumors of sphenoid or spheno-ethmoid origin lie anterior to pituitary-palatal line (shaded area); *B*, Tumors of the nasopharynx lie below and behind the pituitary-palatal line (*shaded area*).

the lateral wall of the nasopharynx (extrapharyngeal sign) (Fig. 25.25). The roentgenological differentiation between intrinsic and extrinsic tumor masses of the nasopharynx is sometimes possible. The more intraluminal the mass in proportion to the displacement it causes, the more likely the lesion is to be intrinsic in the nasopharynx. The greater the displacement in proportion to the size of the intraluminal mass, the more likely the lesion is to be extrinsic with secondary extension to the nasopharynx. Occasionally, the clinical findings may suggest the origin of the tumor.

Neurofibromas (Neurinoma, Neurilemmoma) in Neck

Most of the nerve trunks in the neck are located in the lateral pharyngeal space, which is closely related to the lateral wall of the nasopharynx (Fig. 25.1). The neurofibromas can arise from any of the nerve trunks in the lateral pharyngeal space. With an increase in size they may grow medially and project into the nasopharynx. Roentgenographically, the neurofibroma appears as a sharply defined soft tissue mass on the lateral wall of the nasopharynx which may show the extrapharyngeal sign (Fig.

25.26). The findings are nonspecific and may simulate tumor of the parotid gland.

Meningioma of Base of Skull

This lesion is relatively rare. Roentgenographically there is sclerosis of the base of the skull. The floor of the middle fossa may show a mottled destruction. There is partial obliteration with loss of definition of the foramina ovale and spinosum. The petrous apex may be eroded. The changes may simulate carcinoma of the nasopharynx presenting with sclerosis in the base of the skull. Pneumoencephalography and cerebral angiography are of value in differentiating these two lesions (Fig. 25.27).

Tumors of Oropharynx

Tumors arising from the lateral wall of the oropharynx may project into the nasopharynx. The location of the bulk of the tumor may make it possible to determine the origin of the lesion. Roentgenologically, the submentovertical view and anteroposterior tomograms are of paramount importance in delineating the precise origin and extent of the lesion (Fig. 25.28).

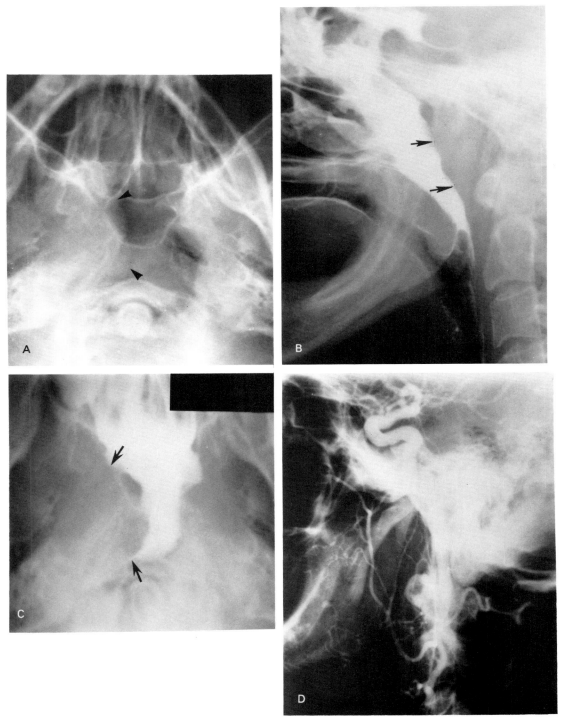

Figure 25.25. Benign mixed tumor of right parotid gland with protrusion into the right side of the nasopharynx. *A*, Base view of skull: There is a large, sharply defined, and gently curved soft tissue mass with tapered borders on the right side of the nasopharynx (*arrows*). *B*, Lateral view of nasopharyngogram: there is a large elongated filling defect in the superoposterior aspect of the nasopharynx (*arrows*). *C*, Submentovertical view of nasopharyngogram: a large filling defect with a convex contour is seen in the right side of the nasopharynx. The tapered borders of the tumor mass are shown to good advantage (*arrows*). *D*, Right common carotid arteriogram: there is an avascular tumor mass in the area of the right parotid gland, with anterior and downward displacement of the internal maxillary branch of the external carotid artery. No abnormal vascularity suggestive of hemangioma or chemodectoma is present in the nasopharynx, the orophaynx, or the soft tissues of the neck.

A 48-year-old female presented with a firm submucosal swelling in the right nasopharyngeal wall. There was some fullness in the right parotid region. At operation, a tumor of the deep lobe of the parotid gland was found. Histopathology revealed benign mixed tumor of the parotid gland.

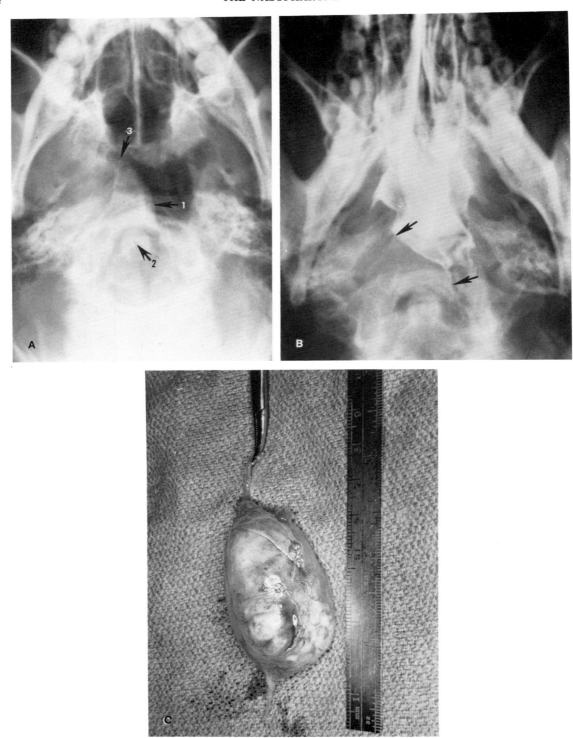

Figure 25.26. Neurofibroma of the cervical sympathetic nerve trunk. *A*, Submentovertical view of the skull: there is a large well defined soft tissue mass on the right side of the nasopharynx (*1*) with downward extension to the oropharynx (*2*). The tapered border of the mass (*3*) suggests an extrapharyngeal origin. *B*, Submentovertical view of nasopharyngogram: a large filling defect with a tapered upper border and a convex contour is present on the right (*arrows*). *C*, Surgical specimen: the tumor mass is ovoid and well encapsulated with a short segment of nerve attached to each end. Histopathology revealed a neurofibroma of the sympathetic nerve trunk.

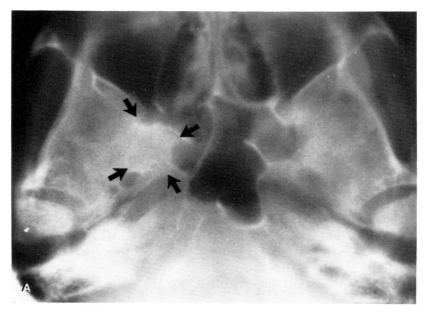

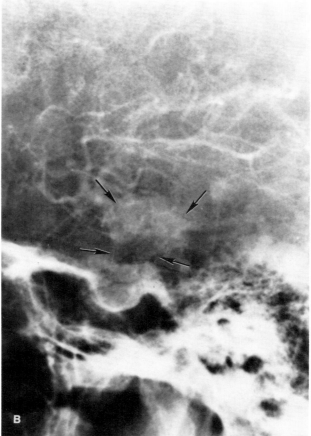

Figure 25.27. Meningioma of the right sphenoid wing. *A*, Base tomogram of the skull: there is sclerosis of the medial aspect of the greater wing of the right sphenoid (*arrows*). The foramen ovale appears normal in size and configuration. *B*, Right carotid angiogram: there is a 3- to 4-cm vascular tumor mass having the characteristics of a meningioma in the right middle fossa along the medial aspect of the right sphenoid ridge (*arrows*). The middle cerebral artery is elevated and the basilar artery is perhaps slightly displaced posteriorly. No tumor encasement of major arteries is identified.

Operative finding: there was a right sphenoid wing meningioma medially placed and involving the right optic canal. A subtotal resection was performed.

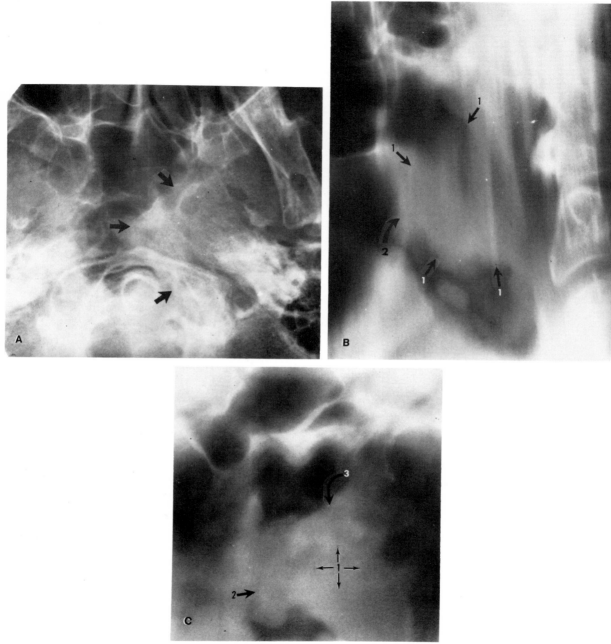

Figure 25.28. Squamous cell carcinoma of left tonsillar area of the oropharynx. *A*, Submentovertical view: there is a large soft tissue mass arising from the left lateral wall of the pharynx and projecting into the air space of the pharynx (*arrows*). *B*, Lateral tomogram of the nasopharynx: a large soft tissue mass with ill defined margins is noted in the pharynx, mainly in the region of the junction of the nasopharynx and oropharynx (*1*). There is anterior displacement of the soft palate (*2*). *C*, Anteroposterior tomogram: the tumor arises from the left lateral wall of the pharynx. The bulk of the lesion is located in the tonsillar area of the oropharynx (*1*) with right lateral displacement of the soft palate (*2*). There is an upward extension of the lesion to the nasopharynx (*3*).

REFERENCES

Abaza, N. A., El-Khashab, M. M., and Fahim, M. S.: Adenoid cystic carcinoma (cylindroma of the palate). Oral Surg., 22: 429, 1966.

Bardwil, J. M., Reynolds, C. T., Ibanez, L., and Luna, M. A.: Report of one hundred tumors of the minor salivary glands. Am. J. Surg., 112: 493, 1966.

Dahl, D. J., and Guiss, L. W.: Tumors of salivary glands. Am. J. Surg., 118: 206, 1969.

David, S. S.: Burkitt's lymphoma of the nasopharynx. J. Laryngol and Otol., 86: 387, 1972.

Dodd, G. D., and Jing, B. S.: Radiographic findings in adenoid cystic carcinoma of the head and neck. Ann. Otol. Rhinol. Laryngol., 81: 591, 1972.

Fitz-Hugh, G. S., and Wallenborn, W. M.: Tumors of the nasopharynx. Laryngoscope, 71: 1, 1961.

Fletcher, G. H., and Million, R. R.: Malignant tumors of the nasopharynx. Am. J. Roentgenol., 93: 44, 1965.

Fu, Y. S., and Perzin, J. H.: Non-epithelial tumors of the nasal cavity, paranasal sinuses and nasopharynx: a clinicopathologic study. III. Cartilaginous tumors (chondroma, chondrosarcoma). Cancer, 34: 453, 1974.

Godtfredsin, E.: Ophthalmologic and neurologic symptoms of malignant nasopharyngeal tumors: clinical study comprising 454 cases with special reference to histopathology and possibility of earlier recognition. Acta Psychiat. Neurol. Suppl., 34: 1, 1944.

Green, R. I.: The radiological appearances of the soft palate with reference to the treatment and cleft palate. J. Faculty Radiologists, 10: 27, 1959.

Ho, H. C.: Radiological diagnosis of nasopharyngeal carcinoma with special reference to its spread through the base of the skull. Cancer of Nasopharynx. UICC Monograph Series, Vol. 1, p. 238. Medical Examination Publishing Company, Flushing, New York, 1967.

Jackson, C.: Primary carcinoma of the nasopharynx. J.A.M.A., 37: 371, 1901.

Jing, B. S.: Tumors of the nasopharynx. Radiol. Clin. North Am., 8: 323, 1970.

Jonsson, G.: Roentgenological findings in malignant tumors of the nasopharynx. Acta Radiol., 15: 1, 1934.

Kasabach, H. H.: Roentgenography of malignant nasopharyngeal tumors. Preliminary report. Laryngoscope, 51: 459, 1941.

Lederman, M.: Cancer of the Nasopharynx: Its Natural History and Treatment. Charles C Thomas, Springfield, Illinois, 1961.

MacComb, W. S., and Fletcher, G. H.: Cancer of the Head and Neck 1st. ed., p. 152., Williams & Wilkins, Baltimore, 1967.

Martin, H. E., and Blady, J. V.: Cancer of the nasopharynx. Arch. Otolaryngol., 32: 692, 1940.

Medellin, H., and Wallace, S.: Angiography in neoplasms of the head and neck. Radiol. Clin. North Am., 8: 307, 1970.

Miller, W. E., Holman, C. B., Dockerty, M. B., and Devine, D. D.: Roentgenologic manifestations of malignant tumors of the nasopharynx. Am. J. Roentgenol., 106: 813, 1969.

Molony, T. J.: Malignant tumors of the nasopharynx. Laryngoscope, 67: 1297, 1957.

Morris, L. F., Hopp, E. S., and Wu, R.: Diagnosis of malignancy of the nasopharynx: cytological studies by the smear technique. Ann. Otol. Rhinol. Laryngol., 58: 18, 1959.

New, G. B.: Highly malignant tumors of the nasopharynx and pharynx. Trans. Am. Acad. Ophthol., 36: 39, 1931.

Owsley, W. C.: Palate and pharynx: roentgenographic evaluation in the management of cleft palate and related deformities. Am. J. Roentgenol., 87: 811, 1962.

Pang, L. Q.: Carcinoma of the nasopharynx. Ann. Otol. Rhinol. Laryngol., 38: 356, 1959.

Pendergrass, E. P., Schaeffer, J. P., and Hodes, P. M.: The Head and Neck in Roentgen Diagnosis, 2nd ed., p. 989. Charles C Thomas, Springfield, Illinois, 1956.

Ranger, D., Thackray, A. C., and Lucas, R. B.: Mucous gland tumors. Br. J. Cancer, 10: 1, 1956.

Raven, R. W.: Cancer of the Pharynx, Larynx and Esophagus, 1st. ed., p. 40. Butterworth and Company Publishers, London, 1956.

Reynolds, D. F., and Groves, H. J.: Clinical and radiological study of choanal polyps. J. Fac. Radiologists, 7: 278, 1956.

Scanlon, P. W., Devine, K. D., and Woolner, L. B.: Malignant lesions of the nasopharynx. Ann. Otol. Rhinol. Laryngol., 67: 1005, 1958.

Smith, L. C., Lane, N., and Rankow, R. M.: Cylindroma (adenoid cystic carcinoma). A report of fifty-eight cases. Am. J. Surg., 110: 519, 1965.

Wang, S. C., Little, J. B., and Schulz, M. D.: Cancer of the nasopharynx. Cancer, 15: 9, 1962.

Ward, G. E., and Hedrick, J. W.: Diagnosis and Treatment of Tumors of the Head and Neck. Williams & Wilkins, Baltimore, 1950.

Soo, Y. S. and Lim, E. C.: Nasopharyngeal carcinoma presenting with sclerosis in the base of the skull. Med. J. Aust., 2: 1126, 1971.

Wong, D. S., Fuller, L. M., Butler, J. J., and Schullenberger, C. C.: Extranodal non-Hodgkins lymphomas of the head and neck. Am. J. Roentgenol., 123: 471, 1975.

Yeh, S.: Histological classification of carcinomas of nasopharynx with critical review as to existence of lymphoepithelioma. Cancer, 15: 895, 1962.

INDEX

Page numbers in italics refer to illustrations